Lecture Notes in Mathematics

Edited by A. Dold and B. Eckmann

828

Probability Theory on Vector Spaces II

Proceedings, Błażejewko, Poland,
September 17 - 23, 1979

Edited by A. Weron

Springer-Verlag
Berlin Heidelberg New York 1980

Editor

A. Weron
Institute of Mathematics
Wrocław Technical University
Wybrzeże Wyspiańskiego 27
50-370 Wrocław
Poland

AMS Subject Classifications (1980): 28CXX, 46B20, 46B30, 46C10, 47B10, 60BXX, 60EXX, 60FXX, 60GXX

ISBN 3-540-10253-1 Springer-Verlag Berlin Heidelberg New York
ISBN 0-387-10253-1 Springer-Verlag New York Heidelberg Berlin

Printed in Germany

Printing and binding: Beltz Offsetdruck, Hemsbach/Bergstr.
2141/3140-543210

F O R E W O R D

The Institute of Mathematics of Wrocław Technical University organized the Second International Conference on Probability Theory on Vector Spaces in Błażejewko from September 17 to September 23, 1979. The first Conference had been organized by the Institute in 1977. At the present Conference there were 74 registered participants from 10 countries, 44 among them from Poland. This Conference was sponsored by the Wrocław Technical University and was organized by the following committee: S.Gładysz, J.Górniak (Secretary), C.Ryll-Nardzewski and A.Weron (Chairman). Mrs. T.Cieślik and Mrs. O.Olak acted as Organizing Secretaries for the Conference.

It was the purpose of this meeting to bring together mathematicians working in Probability Theory on Vector Spaces to discuss the functional analysis aspects of this field. The following (non-disjoint) topics were covered:

Gaussian Processes and Stable Measures	–	*Geometry of Banach Spaces, Special Class of Operators.*
Limit Theorems (CLT, LIL, IP)	–	*Topological Spaces, Geometry of Vector Spaces; C(S), D(0,1)*
Random Fields, Stationary and Vector Valued Processes	–	*Hilbert Space Methods, Dilation Theory and Reproducing Kernels.*
Brownian Motion, Integrability of Random Vectors and Cylindrical Processes	–	*Infinite-Dimensional Calculus and Differential Equations, Semigroups of Measures.*

This volume containes 30 contributions - the written and often extended versions of most lectures gvven at the Conference. A great majority of papers present new results in the field and the rest are expository in nature. The material in this volume complements the material in the earlier volume Probability Theory on Vector Spaces, Proceedings Lecture Notes in Math. vol. 656, 1978, Springer Verlag.

While I take the responsibility for any mistakes in the organization of the Conference and the editing of the Proceedings, I wish to express my gratitude to several persons for their valuable help. On behalf of the Organizing Committee I wish to thank the authorities of the Wrocław Technical University for providing facilities which made it possible to hold the Conference in Błażejewko. I am indepted to Professors S.Gładysz and C.Ryll-Nardzewski for their help in the organization of the program. I wish to thank to Professors: A.Badrikian, S.D.Chatterji, Z.Ciesielski, X.Fernique, S.Gładysz, P.Masani, V.Mandrekar, V.Paulauskas, C.Ryll-Nardzewski and F.H.Szafraniec for presiding over the sessions. I am grateful to my colleagues from the Institute of Mathematics, in particular to Drs. J.Górniak and P.Kajetanowicz, for their help in various administrative matters. Special thanks are due to the contributors to this volume, to those who reviewed the papers and to Springer-Verlag for their excellent cooperation.

Aleksander Weron

CONTENTS

List of Contributors

J. BURBEA,	University of Pittsburgh, Pittsburgh, PA 15260, USA.
T. BYCZKOWSKI,	Wrocław Technical University, 50-370 Wrocław, Poland.
R. CARMONA,	Université de Saint Etienne, 42025 Saint Etienne, France.
S.D. CHATTERJI,	Ecole Polytechnique Fédérale, 1007 Lausanne, Switzerland.
Z. GORGADZE,	Tbilisi State University, 380093 Tbilisi, USSR.
M.D. HAHN,	Tufts University, Medford MA 02155, USA.
A. HULANICKI,	Polish Academy of Sciences, 51-617 Wrocław, Poland.
T. INGLOT,	Wrocław Technical University, 50-370 Wrocław, Poland.
R. JAJTE,	Łódź University, 90-238, Poland.
Z. JUREK,	Wrocław University, 50-384 Wrocław, Poland.
M.J. KLASS,	University of California, Berkeley, CA 94720, USA.
H.H. KUO,	Louisiana State University, Baton Rouge, LA 70803, USA.

V. KVARATSKHELIA,	Academy of Sciences GSSR, 380093 Tbilisi, USSR.
V. LINDE,	Friedrich-Schiller Universität, 69 Jena, DDR.
D. LOUIE,	University of Tennessee, Knoxville, TE 37916, USA.
A. MAKAGON,	Wrocław Technical University, 50-370 Wrocław, Poland.
V. MANDREKAR,	Michigan State University, E. Lansing MI 48824, USA.
P. MASANI,	University of Pittsburgh, Pittsburgh PA 15260, USA.
W. MLAK,	Polish Academy of Sciences, 31-027 Cracow, Poland.
H. NIEMI,	University of Helsinki, 00100 Helsinki 10, Finland.
A. PASZKIEWICZ,	Łódź University, 90-238 Łódź, Poland.
V. PAULAUSKAS,	Vilnius V. Kapsukas University, Vilnius 232042, USSR.
B. S. RAJPUT,	University of Tennessee, Knoxville TE 37916, USA.
E. RYCHLIK,	Warsaw University, 00-901 Warsaw, Poland.
Z. RYCHLIK,	MCS University, 20-031 Lublin, Poland.
F. SCHMIDT,	Technische Universität Dresden, 8027 Dresden, DDR.
R. SHONKWILER,	Georgia Institute of Technology, Atlanta, GA 30332, USA.

Z. SUCHANECKI, Wrocław Technical University, 50-370 Wrocław, Poland.

F.H. SZAFRANIEC, Jagiellonian University, 30-059 Cracow, Poland.

R. SZTENCEL, Warsaw University, 00-901 Warsaw, Poland.

J. SZULGA, Wrocław University, 50-384 Wrocław, Poland.

V. TARIELADZE, Academy of Science of the GSSR, 380093 Tbilisi, USSR.

D. H. THANG, University of Hanoi, Hanoi, Vietnam.

N.V. THU, Institut of Mathematics, 208 D Hanoi, Vietnam.

N. Z. TIEN, University of Hanoi, Hanoi, Vietnam.

A. WERON, Wrocław Technical University, 50-370 Wrocław, Poland.

T. ŻAK, Wrocław Technical University, 50-370 Wrocław, Poland.

R. ŻEBERSKI, Wrocław Technical University, 50-370 Wrocław, Poland.

Titles of non included talks:

The convergence of random variables in topological spaces.

V. BULDYGIN /AN USSR, Kiev/.

Two applications of a lemma on Gaussian covariances.

S. A. CHOBANJAN /AN GSSR, Tbilisi/.

Gaussian covariances in Banach lattices.

S. A. CHOBANJAN and V. I. TARIELADZE /AN GSSR, Tbilisi/.

Weak convergence of sequences of random elements with random indices.

M. CSÖRGO /Carleton University, Ottawa/ and Z. RYCHLIK /UMCS, Lublin/.

Certaines fonctionnelles associées a des fonctions aleatoires Gausiennes.

X. FERNIQUE /Univ. Louis Pasteur, Strasbourg/.

Remarks on CLT in C(S).

E. GINÉ /I.V.I.C., Caracas/.

La loi du logarithme itere dans les espaces de Banach.

B. HEINKEL /Univ. Louis Pasteur, Strasbourg/.

Dilations of Gleason measures.

E. HENSZ /Łódź University/

Aproximation theorems for stochastic operators on L^1.

A. IWANIK /Wrocław Technical University/.

On certain properties of p-uniformly smooth Banach spaces.

A. KORZENIOWSKI /Wrocław University/.

Finite generators for ergodic endomorphisms.

Z. KOWALSKI /Wrocław Technical University/.

Continuity of infinitely divisible distributions.

A. ŁUCZAK /Łódź University/.

Quasi-invariant measures and Markov fields.

V. MANDREKAR /Michigan State University, E. Lansing/.

Remarks on operator stability of probability measures on Banach Spaces

B. MINCER /Wrocław University/.

A characterization of Gaussian process based on differentiable equations.

A. PLUCIŃSKA /Warsaw Technical University/.

On the decomposability semigroups for certain probability measures.

T. RAJBA /Wrocław University/.

On stochastic equations with the reflection boundary condition

M. RUTKOWSKI /Warsaw Technical University/.

Multidimensional dissipative random fields.

M. SŁOCIŃSKI /Jagiellonian University, Cracow/

A short proof of 0-1 Law for stable measures.

W. SMOLEŃSKI /Warsaw Technical University/.

Kolmogorov-Bochner extension theorem for semi-spectral measures.

J. STOCHEL /IM PAN, Cracow/.

On the a.e. convergence of ergodic averages for Bernoulli shifts.

J. WOŚ /Wrocław Technical University/.

Properties of spherically symmetric distributions in R^n.

V. ZOLOTARIEV /AN USSR, Moscow/.

Some new results of V.V. Senatov in multidimensional CLT.

V. ZOLOTARIEV /AN USSR, Moscow/.

HILBERT SPACES OF HILBERT SPACE VALUED FUNCTIONS

by

J. Burbea and P. Masani*

University of Pittsburgh, Pittsburgh, Pa. 15260

1. Introduction

The genesis of this paper lies in two results of Rovnyak [9] asserting the positive definiteness of dually related operator-valued kernels on the Carterian product $D_+ \times D_+$ of the open unit disk D_+ in $\mathbb{C}$[1]. These kernels $L_o(\cdot,\cdot)$, $K_o(\cdot,\cdot)$ involve complex Hilbert spaces W_1, W_2 and a holomorphic function $Y(\cdot)$ on D_+, the values $Y(z)$ of which are linear contractions on W_1 to W_2; their definitions are

(1.1)
$$L_o(z,\zeta) \underset{d}{=} \frac{I_{W_2} - Y(\zeta)\cdot Y(z)^*}{1-\zeta\bar{z}}, \quad z,\zeta \in D_+,$$

$$M_o(z,\zeta) \underset{d}{=} \frac{I_{W_1} - Y(\zeta)^*\cdot Y(z)}{1-\bar{\zeta}z}, \quad z,\zeta \in D_+.$$

---

*Alexander von Humboldt Senior Visiting Scientist at the University of Erlangen, Winter, 1979-1980.

1) In this paper $\mathbb{C}$ is the complex number field, $\mathbb{R}$ the real number field, and $\mathbb{F}$ refers to anyone of these. $\mathbb{N}$ is the set of integers.

The Szegö kernel

(1.2) $$k(z,\zeta) \underset{d}{=} 1/(1-\zeta\bar{z}), \qquad z,\zeta \in D_+,$$

is of course positive definite (PD). But it is easily seen that for non-constant $Y(\cdot)$, the kernel $D(\cdot,\cdot)$ such that $D(z,\zeta) = I_{W_2} - Y(\zeta)\cdot Y(z)^*$ is not PD. Hence the positive definiteness of $L_o(\cdot,\cdot)$ is not trivially deducible from the known result that the product of PD kernels is PD. The same remark applies also to $M_o(\cdot,\cdot)$.

The proofs that $L_o(\cdot,\cdot)$ and $M_o(\cdot,\cdot)$ are PD given by Rovnyak [9], and subsequently by Nagy [6] and Nagy & Foais [7, pp. 231-233] lean heavily on the analytic implications of the kernels, and exploit the theory of holomorphic functions of the Hardy class H_2. In this paper we shall show that analyticity plays only a marginal role: it entails the fulfillment of certain general premises. Formulable in terms of these premises are more abstract and far reaching results on positive-definiteness, which subsume the original.

A convenient way to see this is to ask a general question. Let Λ be any set, let $K_1(\cdot,\cdot)$, $K_2(\cdot,\cdot)$ be PD kernels on $\Lambda\times\Lambda$ to $CL(W_1, W_1)$, $CL(W_2, W_2)$,[2] respectively, and let $Y(\cdot)$ be a function on Λ to $CL(W_1, W_2)$. Under what conditions will the kernel $L(\cdot,\cdot)$ defined $\forall\, \lambda,\lambda' \in \Lambda$ by

(1.3) $$L(\lambda,\lambda') \underset{d}{=} K_2(\lambda,\lambda') - Y(\lambda')\cdot K_1(\lambda,\lambda')\cdot Y(\lambda)^*$$

be PD on $\Lambda\times\Lambda$ to $CL(W_2, W_2)$?

Our answer involves the <u>function Hilbert spaces</u> $\mathcal{F}_1, \mathcal{F}_2$ for which the <u>reproducing kernels</u> are $K_1(\cdot,\cdot)$, $K_2(\cdot,\cdot)$, and also the multiplication operator M_Y from $\mathcal{F}_1$ to $\mathcal{F}_2$ induced by the given function $Y(\cdot)$. Briefly, we find (Thm. 3.4) that

(1.4) $L(\cdot,\cdot)$ is PD iff M_Y is a contraction.

Rovnyak's assertion that $L_o(\cdot,\cdot)$ is PD follows at once from this general result on taking

---

2) $CL(\mathcal{X}\ \mathcal{Y})$ stands for the space of continuous linear operators on the Banach space $\mathcal{X}$ to the Banach space $\mathcal{Y}$.

(1.5) $\Lambda = D_+$, $K_j(z,\zeta) = k(z,\zeta)\cdot I_{W_j}$, $j = 1,2$,

where $k(\cdot,\cdot)$ is the Szegö kernel of (1.2). We have only to check the validity of the premises of our general theorem (3.4) in this special case. This verification naturally calls for some lemmas which lean (infact heavily) on the analytic side of the problem (§7).

We should point out that the spaces F_1, F_2 consist of functions on Λ taking values in the Hilbert spaces W_1, W_2, respectively, and that a preliminary question we have to consider is the extension of the Aronszajn concepts of "function Hilbert space" and "reproducing kernel" to this vectorial setting. We must also demonstrate that every operator-valued PD kernel is the reproducing kernel of some Hilbert space valued function Hilbert space (Thm. 2.7). In essence we have to adapt the theory given in Pedrick's unpublished report [8].

The abstract treatment of the dual kernel $M_o(\cdot,\cdot)$ is slightly more complex. We have to start with a set Λ endowed with an involution $*$ and let $Y^*(\lambda) \underset{d}{=} \{Y(\lambda^*)\}^*$ and $K_j^*(\lambda,\lambda') \underset{d}{=} K_j(\lambda^*,\lambda'^*)$, and to consider the kernel $M(\cdot,\cdot)$ defined $\forall \lambda,\lambda' \in \Lambda$ by

(1.6) $$M(\lambda,\lambda') = K_1^*(\lambda,\lambda') - Y(\lambda')^{*}\cdot K_2^*(\lambda,\lambda')\cdot Y(\lambda).$$

We then have, cf. Thm. 4.4, the exact analogue of (1.4), viz.

(1.7) $M(\cdot,\cdot)$ is PD iff M_{Y^*} is a contraction.

The assertion that $M_o(\cdot,\cdot)$ is PD follows at once from (1.7) on taking Λ, K_1, K_2 as in (1.5). We do not as yet know whether our recourse to an involutory Λ reflects an intrinsic aspect of the problem or merely paucity of understanding.

The idea to consider the multiplication operators M_Y and M_{Y^*} was suggested by the proof of Nagy & Foias [7, pp. 231-233], and indeed as (1.4) and (1.7) show, their role is intrinsic. It is, however, a drawback of these results that they place restraints on M_Y and M_{Y^*} rather than on $Y(\cdot)$ and $Y^*(\cdot)$ themselves. To remove this blemish it is reasonable, following Rovnyak, to impose on $Y(\cdot)$ the condition that its values $Y(\lambda)$ be contractions on W_1 to W_2. But this is not enough, and we faced with the problem of filling the blanks in the implication

$$\sup_{\lambda\in\Lambda} |Y(\lambda)| \leq 1 \quad \& \quad \ldots\ldots \quad \Longrightarrow \quad |M_Y| \leq 1,$$

and in the corresponding one for M_{Y*}. We find, cf. Prop. 3.6(iv), that one insertion for the blank is "F_1 and F_2 are norm-related" in the sense that $\forall f_1 \in F_1$ & $\forall f_2 \in F_2$,

$$|f_1(\lambda)|_{W_1} \leq |f_2(\lambda)|_{W_2}, \forall\lambda\in\Lambda \quad \Longrightarrow \quad |f|_{F_1} \leq |f|_{F_2},$$

and correspondingly with the inequality $\geq$, cf. Def. 3.5. But this again is not enough: also needed is a condition ensuring that the domain of M_Y is everywhere dense in F_1, cf. 3.6(iii).

The simple relationship between operator-valued and scalar-valued kernels exemplified in (1.5) is of some general interest. When such a relationship prevails we call the vector-valued function Hilbert space an inflation of the scalar-valued function Hilbert space (Def. 6.1). There is a natural nexus between such inflations and tensor products (Thm. 6.6).

In this paper several lemmas are not enunciated, and proofs of results are either only scetched or altogether omitted. A more complete version will appear elsewhere. In §2 we study the concepts of Hilbert-space valued functions and its reproducing kernel. In §3 we turn to the multiplication operator from one function Hilbert space to another, and enunciate the Main Thm. 3.4, which is central to the entire paper. §4 deals with the dual theory stemming from the presence of an involution. §§5,6 are devoted to scalar function Hilbert spaces and their inflations. Finally in §7 we study the Hardy spaces from the standpoint of the earlier sections and deduce Rovnyak's results.

Our work suggests that the Rovnyak results have general analogues for all smooth domains of a suitable type in the space $\mathbb{C}^n$ of n complex variables. Another open problem which emerges is that of the existence of a function Hilbert space which is not norm-related to itself. We hope to examine these questions in another paper.

2. Hilbert spaces of functions with values in a Hilbert space

In this section we shall adhere to the following notation:

(2.1)
- (i) Λ is an arbitrary non-void set
- (ii) W is a Hilbert space over $\mathbb{F}$
- (iii) $W^\Lambda = \{f\colon f \text{ is a function on } \Lambda \text{ to } W\}$
- (iv) $\forall \lambda \in \Lambda \ \& \ \forall f \in W^\Lambda, \quad E_\lambda(f) \underset{d}{=} f(\lambda) \in W.$

We shall call E_λ the evaluation operator at λ. It plays a significant role in the formulation of the following concept:

2.2 Def. We say that F is a function Hilbert space, more fully a Λ,W function Hilbert space, iff (i) $F \subseteq W^\Lambda$, (ii) F is a Hilbert space over $\mathbb{F}$,
(iii)[3] $\quad \forall \lambda \in \Lambda, \quad E^F_\lambda \underset{d}{=} \text{Rstr.}_F\, E_\lambda \in CL(F, W)$.

Let F be a Λ,W function Hilbert space. Then by 2.2(iii) and the fact that W and F are Hilbert spaces, we see that for each λ, $(E^F_\lambda)^* \in CL(W, F)$. The correspondence $\lambda \to (E^F_\lambda)^*$ thus defines a hyper-surface or variety in the vector space $CL(W, F)$, cf. [5, Def. 2.8]. Together with the covariance kernel it plays a central role and merits nomenclature:

2.3 Def. Let F be a Λ,W function Hilbert space. Then (a) the correspondence $\lambda \to (E^F_\lambda)^*$, also written $((E^F_\lambda)^*\colon \lambda \in \Lambda)$, is called the reproducing variety of F; (b) the covariance kernel of this variety, i.e. the kernel $K(\cdot,\cdot)$ defined $\forall \lambda,\lambda' \in \Lambda$ by

$$K(\lambda,\lambda') \underset{d}{=} [(E^F_{\lambda'})^*]^* \cdot (E^F_\lambda)^* = E^F_{\lambda'} \cdot (E^F_\lambda)^*$$

is called the reproducing kernel of F.

The reason for this particular nomenclature lies in the parts (a), (e) of the following proposition, the proof of which is quite straight forward.

---

3) $\text{Rstr}_S F$ means the restriction of the function F to the set S.

2.4 Prop. Let F be a Λ,W function Hilbert space with reproducing kernel $K(\cdot,\cdot)$. Then

(a) $\forall\ (\lambda,w) \in \Lambda\times W\ \&\ \forall f \in F,\quad ((E_\lambda^F)^*(w),\ f)_F = (w,\ f(\lambda))_W$;

(b)[4] $S_{E^*} \underset{d}{=} \mathfrak{S}\{(E_\lambda^F)^*(w) : (\lambda,w) \in \Lambda\times W\} = F$;

(c) $K(\cdot,\cdot)$ is a PD kernel on $\Lambda\times\Lambda$ to $CL(W,W)$;

$\&\ \forall \lambda \in \Lambda,\quad |E_\lambda^F| = \sqrt{|K(\lambda,\lambda)|}$;

(d) $\forall\ (\lambda,w) \in \Lambda\times W,\quad (E_\lambda^F)^*(w) = K(\lambda,\cdot)(w)$;

(e) $\forall\ (\lambda,w) \in \Lambda\times W\ \&\ \forall f \in F,\quad (K(\lambda,\cdot)(w),\ f)_F = (w,\ f(\lambda))_W$;

(f) $\forall f \in F\ \&\ \forall \lambda \in \Lambda,\quad |f(\lambda)|_W \leq |f|_F \sqrt{|K(\lambda,\lambda)|}$.

The frequent occurence of the expression $(E_\lambda^F)^*(w)$ in the preceeding result suggests the utility of an abbreviation:

(2.5) $\forall (\lambda,w) \in \Lambda\times W,\quad E^*_{\lambda,w} \underset{d}{=} (E_\lambda^F)^*(w)$.

Clearly $(E^*_{\lambda,w} : (\lambda,w) \in \Lambda\times W)$ is a variety in F parametrized over $\Lambda\times W$, and the result 2.4 (b) shows that it fundamental in F. We shall therefore call it the fundamental variety of F. Rephrased in terms of the notation (2.5) our results 2.4(a),(d) read:

$$
(2.6)\ \begin{cases} (a) & \forall(\lambda,w) \in \Lambda\times W\ \&\ \forall f \in F,\quad (E^*_{\lambda,w},\ f)_F = (w,\ f(\lambda))_W \\ (b) & \forall (\lambda,w) \in \Lambda\times W,\quad E^*_{\lambda,w}(\cdot) = K(\lambda,\cdot)(w). \end{cases}
$$

We shall now take a different standpoint and consider as given a PD kernel $K(\cdot,\cdot)$ on $\Lambda\times\Lambda$ to $CL(W, W)$. We ask if there exists a Λ,W function Hilbert space of which the reproducing kernel is $K(\cdot,\cdot)$. The answer is affirmative as shown by Aronszajn [1] (for $W = \mathbb{F}$) and Pedrick [8] (unpublished), and as may also be seen by a deduction from the Kernel Thm. [5, Thm. 2.10 & App. C]. More precisely we have the following theorem:

---

4) For $A \subseteq F$, $\langle A \rangle$ and $\mathfrak{S}(A)$ denote the linear manifold, respectively closed linear manifold, spanned by A in F.

2.7 Kernel Thm. (Functional form). Let Λ, W be as in (2.1) (i), (ii) and let $K(\cdot,\cdot)$ be a PD kernel on $\Lambda\times\Lambda$ to $CL(W, W)$. Then $\exists |$ a Λ, W function Hilbert space F such that the reproducing kernel of F is $K(\cdot,\cdot)$.

3. The multiplication operator over function Hilbert spaces. Norm-relatedness

We shall now be concerned with two function Hilbert spaces F_1, F_2 with the same domain space Λ but differing range spaces W_1, W_2, and a function $Y(\cdot)$ on Λ to $CL(W_1, W_2)$. Our goal will be to investigate the kernel $L(\cdot,\cdot)$ described in (1.3). The following will be our notation:

(3.1)
- (i) Λ is an arbitrary non-void set
- (ii) W_1 W_2 are Hilbert spaces over $\mathbb{F}$
- (iii) $Y(\cdot)$ is a function on Λ to $CL(W_1, W_2)$
- (iv) for $i = 1,2$, F_i is a Λ, W_i function Hilbert space and $K_i(\cdot,\cdot)$ is its reproducing kernel
- (v) $\mathcal{D}_1 \underset{d}{=} \{f: f \in F_1 \ \& \ Y(\cdot)\{f(\cdot)\} \in F_2\}$
- (vi) $\forall \lambda,\lambda' \in \Lambda, \quad L(\lambda,\lambda') \underset{d}{=} K_2(\lambda,\lambda') - Y(\lambda')\cdot K_1(\lambda,\lambda')\cdot Y(\lambda)^*$.

3.2 Def. Let $\forall f \in \mathcal{D}_1$, $M_Y(f) \underset{d}{=} Y(\cdot)\{f(\cdot)\}$. Then M_Y is called the multiplication operator due to $Y(\cdot)$.

A straight forward argument shows that

(3.3)
- (a) M_Y is a single-valued closed linear operator with the domain $\mathcal{D}_1$ in F_1 and a range R in F_2
- (b) $\mathcal{D}_1 = F_1 \Rightarrow M_Y \in CL(F_1, F_2)$.

We are now ready to enunciate our main theorem:

3.4 Main Thm. Let (i) Λ, W_1, W_2, F_1, F_2, K_1, K_2 Y, $\mathcal{D}_1$, L be as in (3.1), and (ii) $\mathcal{D}_1 = F_1$. Then the following conditions are equivalent:

(α) $|M_Y| \leq 1$;

(β) the self-adjoint operator

$$D_2 \underset{d}{=} I_{F_2} - M_Y \cdot (M_Y)^*$$

is non-negative;

(γ) the kernel $L(\cdot,\cdot)$ is PD on $\Lambda \times \Lambda$ to $CL(W_2, W_2)$.

<u>Proof remarks.</u> The premise (ii) entails that $M_Y \in CL(F_1, F_2)$ and $(M_Y)^* \in CL(F_2, F_1)$. A crucial step is that $\forall \lambda \in \Lambda$,

$$(M_Y)^* \cdot (E_\lambda^{F_2})^* = K_1(\lambda, \cdot) \cdot Y(\lambda)^* \quad \text{on } W_2.$$

This entails that $\forall \lambda, \lambda' \in \Lambda$,

$$Y(\lambda') \cdot K_1(\lambda, \lambda') \cdot Y(\lambda)^* = E_{\lambda'}^{F_2} \cdot M_Y (M_Y)^* \cdot (E_\lambda^{F_2})^*$$

and

(1) $$L(\lambda, \lambda') = E_{\lambda'}^{F_2} \cdot D_2 \cdot (E_\lambda^{F_2})^*.$$

The equation (1) shows that

$$L(\cdot, \cdot) \text{ is PD iff } D_2 \succcurlyeq 0.$$

Also obviously, $D_2 \succcurlyeq 0$ iff $|M_Y| \leq 1$. □

As indicated §1, this result is very useful but has the blemish of prescribing conditions on M_Y rather than on $Y(\cdot)$ itself. Towards eradicating this blemish we shall assume that each $|Y(\lambda)| \leq 1$, and seek additional conditions in order to ensure that $|M_Y| \leq 1$. Our search leads us to the relationship between F_1 and F_2 mentioned in the next definition:

3.5 <u>Def.</u> Let Λ, W_1, W_2, F_1, F_2 be as in (3.1). We say that F_1, F_2 are <u>norm-related</u>, iff $\forall f_1 \in F_1$ & $\forall f_2 \in F_2$,

$$|f_1(\cdot)|_{W_1} \leq |f_2(\cdot)|_{W_2} \text{ on } \Lambda \implies |f_1|_{F_1} \leq |f_2|_{F_2}$$

&

$$|f_1(\cdot)|_{W_1} \geq |f_2(\cdot)|_{W_2} \text{ on } \Lambda \implies |f_1|_{F_1} \geq |f_2|_{F_2}.$$

Unfortunately, as the next proposition shows, norm-relatedness together with the condition that each $|Y(\lambda)| \leq 1$ is not sufficient to ensure that $|M_Y| \leq 1$; we again need a condition on $\mathcal{D}_1$:

3.6 Prop. Let (i) Λ, W_1, W_2, F_1, F_2 and Y be as in (3.1);

(ii) $Y(\cdot)$ be bounded on Λ:

$$\beta \underset{d}{=} \sup_{\lambda \in \Lambda} |Y(\lambda)| < \infty;$$

(iii) $\{E^*_{\lambda,w_1} : (\lambda, w_1) \in \Lambda \times W_1\} \subseteq \mathcal{D}_1$;

(iv) F_1, F_2 be norm-related.

Then $M_Y \in CL(F_1, F_2)$ and $|M_Y| \leq \beta$.

Proof remarks. Let $f \in \mathcal{D}_1$. Then by (ii) we have

$$\forall \lambda \in \Lambda, \qquad |\{M_Y(f)\}(\lambda)|_{W_2} = |Y(\lambda)\{f(\lambda)\}|_{W_2} \leq \beta|f(\lambda)|_{W_1},$$

whence by the norm-relatedness (iv),

$$|M_Y(f)|_{F_2} \leq \beta|f|_{F_1}.$$

Thus M_Y is continuous on its domain $\mathcal{D}_1$. Since by (iii) and 2.4(b), $\mathcal{D}_1$ is everywhere dense in F_1, a standard argument yields the desired result. □

An obvious corollary of this proposition and Thm. 3.4 is the following:

3.7 Cor. Let (i) - (iv) be as in Prop. 3.6 with $\beta \leq 1$. Then the kernel $L(\cdot,\cdot)$ of 3.1 (v) is PD on $\Lambda \times \Lambda$ to $CL(W_2, W_2)$.

4. Involutory parameter sets and the dual theory

We now turn to the dual of the main Thm. 3.4 which emerges when Λ is endowed with an involution, i.e. with a one-one function * on Λ onto Λ such that $\lambda^{**} = \lambda$ for all $\lambda \in \Lambda$.

Let * be an involution on the set Λ, and W, W_1, W_2 be Hilbert spaces over $\mathbb{F}$, Y(·) be a function on Λ to CL(W_1, W_2) and K(·,·) a kernel on $\Lambda\times\Lambda$ to CL(W, W). Then we define their duals Y*(·) and K*(·,·) by:

$$(4.1)\quad \begin{cases} Y^*(\lambda) \underset{d}{=} \{Y(\lambda^*)\}^*, & \lambda \in \Lambda, \\ K^*(\lambda,\lambda') \underset{d}{=} K(\lambda^*,\lambda'^*), & \lambda,\lambda' \in \Lambda. \end{cases}$$

It is a triviality that

$$(4.2)\quad \begin{cases} \text{(a) } Y(\cdot) \text{ is on } \Lambda \text{ to } CL(W_1, W_2) \\ \qquad\qquad \Longrightarrow Y^*(\cdot) \text{ is on } \Lambda \text{ to } CL(W_2, W_1) \\ \text{(b) } K(\cdot\ \cdot) \text{ is PD on } \Lambda\times\Lambda \text{ to } CL(W, W) \\ \qquad\qquad \Longrightarrow K^*(\cdot,\cdot) \text{ is PD on } \Lambda\times\Lambda \text{ to } CL(W, W). \end{cases}$$

It is now necessary to lay down our notation:

(4.3)

(i) Λ is an arbitrary non-void set and * is an involution on Λ

(ii) W_1, W_2, F_1, F_2, K_1, K_2, Y, are as in (3.1)

(iii) $\mathcal{D}_2 \underset{d}{=} \{f: f \in F_2 \ \ \& \ \ Y(\cdot)^*\{f(\cdot)\} \in F_1\}$

(iv) $\forall \lambda,\lambda' \in \Lambda, \quad M(\lambda,\lambda') \underset{d}{=} K_1^*(\lambda,\lambda') - Y(\lambda')^*\cdot K_2^*(\lambda,\lambda')\cdot Y(\lambda)$.

We can then state our second main theorem, which is the dual of 3.4, the first:

4.4 <u>Main Thm.</u> Let (i) Λ, W_1, W_2 F_1, F_2, K_1, K_2 Y, $\mathcal{D}_2$ be as in (4.3), (ii) $\mathcal{D}_2 = F_2$. Then the following conditions are equivalent:

(α) $\quad |M_{Y*}| \le 1$;

(β) the self-adjoint operator

$$D_1 \underset{d}{=} I_{F_1} - M_{Y*}\cdot(M_{Y*})^*$$

is non-negative;

(γ) the kernel $M(\cdot,\cdot)$ is PD on $\Lambda\times\Lambda$ to $CL(W_1, W_1)$.

Proof remarks. In essence, we apply Thm. 3.4 with F_1, F_2 interchanged and with $Y^*(\cdot)$ instead of $Y(\cdot)$. We then get three equivalent conditions (α'), (β'), (γ'), the last of which involves the kernel $N(\cdot,\cdot)$ defined by

$$\forall\, \lambda,\lambda' \in \Lambda, \qquad N(\lambda,\lambda') \underset{d}{=} K_1(\lambda,\lambda') - Y^*(\lambda')K_2(\lambda,\lambda')\{Y^*(\lambda)\}^*.$$

A simple computation shows that $N^*(\cdot,\cdot) = M(\cdot,\cdot)$. In view of (4.2) (b), the conditions (α'), (β'), (γ') reduce to (α), (β), (γ). □

Every set Λ possesses the trivial involution $\lambda^* = \lambda$. But in hardly any situation will the conditions 4.4(ii) and (α) be valid with the trivial involution. The utility of the last theorem stems from the fact that there are applications in which there exist non-trivial involutions which validate the conditions 4.4(ii) and (α).

It is again a drawback of Thm. 4.4 that it lays conditions on M_{Y^*} instead of $Y^*(\cdot)$ itself. It is clear that we can again eradicate this blemish by assuming that each $|Y(\lambda)| \leq 1$, F_1, F_2 are norm-related and the fundamental variety of F_2 lies in $\mathcal{D}_2$. We leave it to the reader to formulate the resulting proposition and corollary, which are dual to Prop. 3.6 and Cor. 3.7.

5. Scalar function Hilbert spaces

The concepts and results of §2 remain valid of course in the special case in which W is the number field $\mathbb{F}$. But some simplify and it is worth recording these simplifications before proceeding further.

For a Λ, $\mathbb{F}$ function Hilbert space F, the reproducing variety $((E_\lambda^F)^* : \lambda \in \Lambda)$ is a hypersurface in $CL(\mathbb{F},F)$ and the reproducing kernel takes values in $CL(\mathbb{F},\mathbb{F})$, cf. Def. 2.3. If we identify $CL(\mathbb{F},F)$ with F, and $CL(\mathbb{F},\mathbb{F})$ with $\mathbb{F}$ in the obvious way, we can redefine the reproducing variety more simply as a hypersurface in F, and the reproducing kernel more simply as one with values in $\mathbb{F}$. Rigorously speaking, these reformulations rest on the following triviality:

5.1 Triv. Let F be a Λ, $\mathbb{F}$ function Hilbert space. Then (a) $\exists|$ kernel $k(\cdot,\cdot)$ on $\Lambda\times\Lambda$ to $\mathbb{F}$ such that $\forall\lambda \in \Lambda$, the operator $(E^F_\lambda)^*$ is just multiplication by $k(\lambda,\cdot)$:

$$\forall c \in \mathbb{F}, \qquad (E^F_\lambda)^*(c) = k(\lambda,\cdot)\cdot c \quad \in \quad F;$$

(b) the reproducing kernel $K(\cdot,\cdot)$ of F is just multiplication by this $k(\cdot,\cdot)$:

$$\forall \lambda,\lambda' \in \Lambda, \qquad K(\lambda,\lambda)(c) = k(\lambda,\lambda)\cdot c, \quad c \in \mathbb{F}.$$

This triviality leads us to the following definition:

5.2 Def. Let F be a Λ, $\mathbb{F}$ function Hilbert space and let $k(\cdot,\cdot)$ be the unique kernel of 5.1(a). Then (a) $(k(\lambda,\cdot)\colon \lambda \in \Lambda)$ will be called the ordinary reproducing variety of F; (b) $k(\cdot,\cdot)$ will be called the ordinary reproducing kernel of F.

The qualification "ordinary" is merited by the fact that in the literature on scalar-function Hilbert spaces, it is $k(\cdot,\cdot)$ which is called the "reproducing kernel", cf. e.g. Halmos [2, §30] or Hille [3, p.343]. Let us reformulate some of the results in §2 in terms of the ordinary variety and kernel:

5.3 Prop. Let F be a Λ, $\mathbb{F}$ function Hilbert space and $k(\cdot,\cdot)$ be its ordinary reproducing kernel. Then

(a) $\quad \forall\lambda \in \Lambda \;\&\; \forall f \in F, \qquad (k(\lambda,\cdot), f)_F = \overline{f(\lambda)}$

(b) $\quad \mathfrak{S}\{k(\lambda,\cdot)\colon \lambda \in \Lambda\} = F$

(c) $\quad \forall\lambda \in \Lambda, \qquad |k(\lambda,\cdot)|_F = \sqrt{k(\lambda,\lambda)}.$

These results are immediate consequences of 5.1 and the results 2.4(a), (b), (d).

6. Inflations of scalar function Hilbert spaces

The vector-valued, Λ, W, function Hilbert spaces often encountered in analytic applications have reproducing kernels of the simple form $k(\cdot,\cdot)I_W$, where $k(\cdot,\cdot)$ is a $\mathbb{F}$-valued PD kernel on $\Lambda\times\Lambda$. In this section we shall study the simplifications of our previous results that accrue in this special case. By Thm. 2.7 this $k(\cdot,\cdot)$ is the reproducing kernel of some Λ, $\mathbb{F}$ function Hilbert space, and this fact suggests the following definition.

6.1 Def. Let (i) F_0 be a Λ, $\mathbb{F}$ function Hilbert space with ordinary reproducing kernel $k(\cdot,\cdot)$, (ii) W be a Hilbert space over $\mathbb{F}$, and (iii) F be a Λ,W function Hilbert space with reproducing kernel $K(\cdot,\cdot)$. We say that F is a Λ,W inflation of F_0, iff $K(\cdot,\cdot) = k(\cdot,\cdot)\cdot I_W$ on $\Lambda\times\Lambda$.

The next result provides some equivalent renderings of the concept of inflation. Its proof is quite straightforward.

6.2 Lemma. Let (i) F_0 be a Λ, $\mathbb{F}$ function Hilbert space with ordinary reproducing kernel $k(\cdot,\cdot)$ and (ii) F be a Λ,W function Hilbert space with reproducing kernel $K(\cdot,\cdot)$. Then the following conditions are equivalent:

(α) F is a Λ,W inflation of F_0;

(β) the reproducing variety $((E_\lambda^F)^*: \lambda \in \Lambda)$ consists of multiplications by $k(\lambda,\cdot)$, i.e.

$$\forall(\lambda,w) \in \Lambda\times W, \quad (E_\lambda^F)^*(w) = k(\lambda,\cdot)\cdot w;$$

(γ) $\forall(\lambda,w) \in \Lambda\times W$, $k(\lambda,\cdot)\cdot w \in F$, and

$$\forall f \in F, \quad (k(\lambda,\cdot)\cdot w, f)_F = (w, f(\lambda))_W.$$

We must once again fix our notation:

$$(6.3)\begin{cases} \text{(i)} & \Lambda, W_1, W_2, Y(\cdot) \text{ are as in (3.1)} \\ \text{(ii)} & F_0 \text{ is a } \Lambda, \mathbb{F} \text{ function Hilbert space and } k_o(\cdot,\cdot) \text{ is its ordinary reproducing kernel,} \\ \text{(iii)} & F_1 \text{ and } F_2 \text{ are the } \Lambda,W_1 \text{ and the } \Lambda,W_2 \text{ inflations of } F_0 \\ \text{(iv)} & \mathcal{D}_1 \underset{d}{=} \{f: f \in F_1 \ \& \ Y(\cdot)\{f(\cdot)\} \in F_2\} \\ \text{(v)} & \forall \lambda,\lambda' \in \Lambda, \quad L(\lambda,\lambda') \underset{d}{=} k_o(\lambda,\lambda')[I_{W_2} - Y(\lambda')\cdot Y(\lambda)^*] \end{cases}$$

We then obviously have for inflated function Hilbert spaces and the kernel L of (6.3)(v) the following analogue of the Main Thm. 3.4:

6.4 Thm. Let (i) Λ, W_1, W_2, Y, F_0, k_o, F_1, F_2, $\mathcal{D}_1$, L be as in (6.3), and (ii) $\mathcal{D}_1 = F_1$. Then the following conditions are equivalent:

(α) $\quad |M_Y| \leq 1$;

(β) the self-adjoint operator

$$\mathcal{D}_2 = I_{F_2} - M_Y \cdot (M_Y)^*$$

is non-negative;

(γ) the kernel $L(\cdot,\cdot)$ is PD on $\Lambda \times \Lambda$ to $CL(W_2, W_2)$.

Likewise for inflated function Hilbert spaces and involutory Λ, we obviously have the following analogue of the dual Thm. 4.4:

6.5 Thm. Let (i) Λ, W_1, W_2, Y, F_0, k_o, F_1, F_2 be as in (6.3), (ii) * be an involution on Λ, and

$$\lambda, \lambda' \in \Lambda, \quad M(\lambda,\lambda') = k_o^*(\lambda,\lambda')\{I_{W_1} - Y(\lambda')^* \cdot Y(\lambda)\},$$

(iii) $\mathcal{D}_2 \underset{d}{=} \{f\colon f \in F_2 \ \& \ Y^*(\cdot)\{f(\cdot)\} \in F_1\} = F_2$.

Then the following conditions are equivalent:

(α) $\quad |M_{Y^*}| \leq 1$;

(β) the self-adjoint operator

$$\mathcal{D}_1 = I_{F_1} - M_{Y^*} \cdot (M_{Y^*})^*$$

is non-negative;

(γ) the kernel $M(\cdot,\cdot)$ is PD on $\Lambda \times \Lambda$ to $CL(W_1, W_1)$.

The last two theorems again have the drawback of stipulating conditions on M_Y and M_{Y^*} instead of conditions on $Y(\cdot)$ and $Y^*(\cdot)$ themselves. As in §§3,4 this situation can be mitigated by requiring that each $|Y(\lambda)| \leq 1$, that F_1, F_2 be norm-related and that the fundamental varieties of F_1 and F_2 be within the domains $\mathcal{D}_1$ and $\mathcal{D}_2$. We leave the formulation of these results to the reader.

Finally we state without proof the following result which establishes a nexus between inflations and tensor products:

6.6 Thm. Let (i) Λ be a non-void set, (ii) W be a Hilbert space over $\mathbb{F}$, (iii) $F_{\mathbb{F}}$ be a Λ, $\mathbb{F}$ function Hilbert space, and (iv) F_W be the Λ,W inflation of $F_{\mathbb{F}}$. Then

(a) $F_{\mathbb{F}}\cdot W \underset{d}{=} \{\Phi(\cdot)\, w : \Phi \in F_{\mathbb{F}}\ \&\ w \in W\}$ is fundamental in F_W;

(b) $\forall\ \Phi(\cdot)w\ \&\ \Phi'(\cdot)w' \in F_{\mathbb{F}}\cdot W$,

$$(\Phi(\cdot)\cdot w,\ \Phi'(\cdot)\cdot w')_{F_W} = (\Phi,\Phi')_{F_{\mathbb{F}}}\cdot(w,\ w')_W,$$

in particular, $|\Phi(\cdot)\cdot w|_{F_W} = |\Phi|_{F_{\mathbb{F}}}\cdot|w|_W$

(c) if $T: \Phi(\cdot)\cdot w \to w\otimes\Phi$, then $\overline{T} \underset{d}{=} \mathfrak{S}(T)$ is a linear isometry on F_W onto the tensor product $W\otimes F_{\mathbb{F}}$, i.e. onto the space of Hilbert-Schmidt operators on W to $F_{\mathbb{F}}$.

7. The Hardy classes $H_2(D_+, \mathbb{C})$ and $H_2(D_+, W)$

In this section

$$(7.1)\quad \begin{cases} D_+ \underset{d}{=} \{z:\ z \in \mathbb{C}\ \&\ |z| < 1\} \\ C \underset{d}{=} \{z:\ z \in \mathbb{C}\ \&\ |z| = 1\} \\ W \text{ is a Hilbert space over } \mathbb{C}. \end{cases}$$

We emphasise that $\mathbb{F}$ is now $\mathbb{C}$, Λ is D_+, the open unit disk in $\mathbb{C}$, and that W need not be separable. The definition of the Hardy classes involves the Lebesgue Bochner class $L_2(C, W)$. This is defined to be the set of all separably-ranged, Borel measurable functions on C to W for which

$$(7.2)\quad |f|_2^2 \underset{d}{=} \frac{1}{2\pi}\int_0^{2\pi} |f(e^{i\vartheta})|_W^2 d\vartheta < \infty.$$

A standard result, cf. e.g. [7, p.183][5], is that

$$(7.3)\quad \begin{cases} L_2(C,\ W) \text{ is a Hilbert space over } \mathbb{C} \text{ under the inner product} \\ (f,g)_2 \underset{d}{=} \frac{1}{2\pi}\int_0^{2\pi}(f(e^{i\vartheta}),\ g(e^{i\vartheta}))_W\ d\vartheta. \end{cases}$$

---

5) The separability assumption made in [7, p.183] is removable as far as (7.3) is concerned.

Let f be in $L_2(C, W)$. Then so is the function $\zeta \to \zeta^{-n}f(\zeta)$, $\zeta \in C$, for all $n \in \mathbb{N}$. The nth Fourier coefficient of f is defined by

$$(7.4) \qquad w_n(f) \underset{d}{=} \frac{1}{2\pi} \int_0^{2\pi} f(e^{i\vartheta})e^{-ni\vartheta} d\vartheta \in W.$$

We define

$$(7.5) \qquad L_2^{o+}(D, W) \underset{d}{=} \{f: f \in L_2(C, W) \ \& \ \forall n<0, w_n(f) = 0\}.$$

It is easily seen that $L_2^{o+}(C, W)$ is a closed subspace of $L_2(C, W)$.

Next let Hol.(D_+, W) be the set of all holomorphic functions on D_+ to W, cf. [4, pp.92,93], and let

$$(7.6) \qquad \forall f \in \text{Hol.}(D_+, W), \quad \|f\|^2 \underset{d}{=} \sup_{0<r<1} \frac{1}{2\pi} \int_0^{2\pi} |f(re^{i\vartheta})|_W^2 \, d\vartheta.$$

We define the Hardy class $H_2(D_+, W)$ by

$$(7.7) \qquad H_2(D_+, W) \underset{d}{=} \{f: f \in \text{Hol.}(D_+, W) \ \& \ \|f\| < \infty\}.$$

We shall take for granted the following well-known and fundamental theorem on $H_2(D_+, W)$. (For parts (a) and (b), cf. Nagy & Foias [7, pp.183-186].)

7.8. Thm. (a) $\forall f_+ \in H_2(D_+, W)$, $\exists | \, f \in L_2^{o+}(C, W)$ such that

$$\lim_{r \to 1-} \int_0^{2\pi} |f_+(re^{i\vartheta}) - f(e^{i\vartheta})|_W^2 \, d\vartheta = 0.$$

(b) $H_2(D_+, W)$ is a Hilbert space over $\mathbb{C}$ under the inner product and norm:

$$(f_+, g_+) \underset{d}{=} \frac{1}{2\pi} \int_0^{2\pi} (f(e^{i\vartheta}), g(e^{i\vartheta}))_W \, d\vartheta$$

$$|f_+|^2 \underset{d}{=} \frac{1}{2\pi} \int_0^{2\pi} |f(e^{i\vartheta})|_W^2 \, d\vartheta,$$

where $f_+, g_+ \in H_2(D_+, W)$, and f,g are related to f_+, g_+ as in (a);

(c) $\forall f_+ \in H_2(D_+, W)$ and f as in (a), $\|f_+\| = |f_+|$; more fully

$$\sup_{0<r<1} \frac{1}{2\pi} \int_0^{2\pi} |f_+(re^{i\vartheta})|_W^2 \, d\vartheta = \frac{1}{2\pi} \int_0^{2\pi} |f(e^{i\vartheta})|_W^2 \, d\vartheta.$$

Given f_+ in $H_2(D_+, W)$, we shall speak of any function f on C to W satisfying the equation in 7.8(a) as its <u>radial limit</u>. Thus the radial limit is unique only up to sets of zero Lebesgue measure; more precisely, f_1, f_2 are radial limits of $f_+ \Rightarrow f_1 = f_2$ a.e. (Leb.) on C.

Our next theorem shows that the Hardy classes fall under the purview of the theory developed in §§2,6.

7.9 <u>Thm.</u> (a) $H_2(D_+, W)$ is a D_+,W function Hilbert space.
(b) $H_2(D_+, W)$ is the D_+,W inflation of the $D_+,\mathbb{C}$ function Hilbert space $H_2(D_+, \mathbb{C})$, and the latter has the Szegö reproducing kernel $k(\cdot,\cdot)$ of (1.2).

<u>Proof remarks.</u> (a) Write $F = H_2(D_+, W)$ and $F_0 = H_2(D_+,\mathbb{C})$. In view of 7.8(b) we have only to show that $\forall z \in D_+$, $E_z^F \in CL(F, W)$. But this is clear from the inequality:

$$(2) \qquad |f_+(z)|_W \le \frac{|f_+|_F}{\sqrt{(1-|z|^2)}}, \quad f \in F, \quad z \in D_+,$$

which easily follows from the paraphrased Cauchy formula:

$$(1) \qquad f_+(z) = \frac{1}{2\pi} \int_0^{2\pi} f_+(e^{i\theta}) \overline{k(z,e^{i\theta})} d\theta.$$

(b) That $H_2(D_+, \mathbb{C})$ is a function Hilbert space with the reproducing kernel $k(\cdot,\cdot)$ is well known, cf [2, p32]. (This also follows on taking $W=\mathbb{C}$ in (1)). Next from (1) we can show that

$$(k(z,\cdot)w, f)_F = (w, f(z))_W, \quad f \in F, \quad z \in D_+, \ w \in W.$$

This, cf. 6.2, shows that F is the inflation of F_0. □

To get to Rovnyak's result we must recall the definition of a holomorphic operator-valued function:

7.10 <u>Def.</u> Let W_1, W_2 be Hilbert spaces over $\mathbb{C}$, and $Y(\cdot)$ be a function on D_+ to $CL(W_1, W_2)$. We say that $Y(\cdot)$ is <u>holomorphic on D_+</u>, iff $\exists$ a sequence $(A_n)_1^\infty$ in $CL(W_1, W_2)$ such that

$$\forall z \in D_+, \quad Y(z) = \sum_{k=0}^{\infty} A_k z^k,$$

the convergence being in the strong operator topology.

Deducible from the theory of holomorphy given in Hille & Phillips [4, pp.92-98] is the standard result that

$$(7.11)\begin{cases} f(\cdot) \in \text{Hol.}(D_+, W_1) \quad \& \quad Y(\cdot) \in \text{Hol.}(D_+, CL(W_1, W_2)) \\ \qquad\qquad\qquad \Longrightarrow \quad Y(\cdot)\{f(\cdot)\} \in \text{Hol.}(D_+, W_2). \end{cases}$$

Since complex conjugation is an involution on D_+, each function $Y(\cdot)$ on D_+ to $CL(W_1, W_2)$ has a dual $Y^*(\cdot)$ on D_+ to $CL(W_2, W_1)$ given by $Y^*(z) \underset{d}{=} \{Y(\bar{z})\}^*$, $z \in D_+$. We now ask what kind of multiplication operators $Y(\cdot)$ and $Y^*(\cdot)$ induce over the Hardy classes. The following lemma provides the answer for bounded holomorphic $Y(\cdot)$.

7.12 Lemma. Let $Y(\cdot) \in \text{Hol.}(D_+, CL(W_1, W_2))$ and $Y(\cdot)$ be bounded on D_+:

$$\beta \underset{d}{=} \sup_{z \in D_+} |Y(z)| < \infty.$$

Then

(a) $M_Y \in CL(H_2(D_+, W_1), H_2(D_+, W_2)) \quad \& \quad |M_Y| \leq \beta$;

(b) $Y^*(\cdot) \in \text{Hol.}(D_+, CL(W_2, W_1))$ and $Y^*(\cdot)$ is bounded on D_+:

$$\beta^* \underset{d}{=} \sup_{z \in D_+} |Y^*(z)| = \beta < \infty;$$

(c) $M_{Y^*} \in CL(H_2(D_+, W_2), H_2(D_+, W_1)) \quad \& \quad |M_{Y^*}| \leq \beta$.

Proof remarks. (a) Let $f \in H_2(D_+, W_1)$. Then for $0<r<1$,

$$\int_0^{2\pi} |\{M_Y(f)\}(re^{i\theta})|^2_{W_2}\, d\theta \;\leq\; \beta^2 \int_0^{2\pi} |f(re^{i\theta})|^2_{W_1}\, d\theta.$$

Taking the supremum over r in $[0, 1)$, we see, cf. (7.6), that $\|M_Y(f)\| \leq \beta\, \|f\| < \infty$. This and (7.11) show that $M_Y(f) \in H_2(D_+, W_2)$. Also by 7.8(c), $|M_Y(f)| \leq \beta |f|$; thus $|M_Y| \leq \beta$.

(b) Let $Y(z) = \Sigma_0^\infty A_k z^k$, $z \in D_+$. Then we have $Y^*(z) \supseteq \Sigma_0^\infty A_k^* z^k$. But since $|A_k^*| = |A_k|$, and the radius of convergence of a power series in any Banach space depends solely on the sequence of absolute values of its Taylor coefficients [4, p.96, Thm. 3.11.4], we see that the last series has a radius $r \geq 1$. It follows that $Y^*(z) = \Sigma_0^\infty A_k^* z^k$, $z \in D_+$. Also $\beta^* = \beta$ trivially.

(c) follows from (b) in the same way that (a) follows from the hypothesis. □

This lemma shows that for <u>contractive</u>, <u>holomorphic</u> $Y(\cdot)$ on D_+ to $CL(W_1, W_2)$, the premises (ii) and (α) of Thms. 6.4 and 6.5 are fulfilled (since now $\beta^* = \beta = 1$). Hence the conclusions (γ) of these theorems hold. Since for the Szegö kernel $k^*(z,\zeta) = k(\zeta,z)$, cf. (1.2), we have proved the following theorem:

7.13 <u>Thm.</u> (Rovnyak). Let $Y(\cdot)$ be a holomorphic function on D_+ to $CL(W_1, W_2)$ such that $\sup_{z \in D_+} |Y(z)| \leq 1$. Then the kernels $L_o(\cdot,\cdot)$, $M_o(\cdot,\cdot)$ of (1.1) are PD on $D_+ \times D_+$ to $CL(W_2, W_2)$ and $CL(W_1, W_1)$, respectively.

Finally, we should point out that it is an easy consequence of Thm. 7.8(c) that

(7.14) $\begin{cases} \text{the function Hilbert spaces } H_2(D_+, \mathbb{C}) \text{ and} \\ H_2(D_+, W) \text{ are norm-related.} \end{cases}$

<u>References</u>

1. N. Aronszajn, Theory of reproducing kernels, Trans. Amer. Math. Soc. 68(1950), 337-404.

2. P. Halmos, A Hilbert space problem book, van Norstrand, New York, 1967.

3. E. Hille, Introduction to the general theory of reproducing kernels, Rocky Mountain J. of Math. 2(1971), 321-368.

4. E. Hille and R.S. Phillips, Functional analysis and semi-groups, Coll. Pub. Vol. 31, Amer. Math. Soc. Providence R.I, 1957.

5. P. Masani, Dilations as propagators of Hilbertian varieties, SIAM J of Math. Anal. 9(1978), 414-456.

6. B. Sz.-Nagy, Positive definite kernels generated by operator-valued analytic functions, Acta Sci. Math. 26(1965), 191-192.

7. B. Sz.-Nagy and C. Foias, Harmonic analysis of operators on Hilbert space, North Holland, New York, 1970.

8. G. B. Pedrick, Theory of reproducing kernels in Hilbert spaces of vector-valued functions, Univ. of Kansas Tech. Rep. 19, Lawrence, 1957.

9. J. Rovnyak, Some Hilbert spaces of analytic functions, Dissertation Yale Univ., 1963.

ON THE INTEGRABILITY OF GAUSSIAN RANDOM VECTORS

T.Byczkowski and T.Żak

Let μ be a Gaussian measure on a separable Banach space. The famous result of Fernique - Landau & Shepp states that

$$\int \exp(\alpha\, q(x)^2)\, d\mu < +\infty$$

for sufficiently small $\alpha > 0$, where q is a homogeneous measurable seminorm.

Since there are linear spaces, being natural spaces of sample paths of stochastic processes (such as $D[0,1]$ or $L_\phi[0,1]$), which are no longer Banach nor locally convex, one needs some generalization of this result for non-homogeneous seminorms. First result in this direction has been obtained by Inglot and Weron [10] : namely, if μ is a (symmetric) Gaussian measure on a separable metric linear space E then

$$\int \exp(\alpha\, q(x))\, d\mu < +\infty$$

for sufficiently small $\alpha > 0$, where q is a seminorm (nonnecessarily homogeneous) generating the topology of E. If q is p-homogeneous then

$$\int \exp(\alpha\, q(x)^{2/2-p})\, d\mu < +\infty$$

for $\alpha > 0$ small enough (see [9], [11]).
The above results have been obtained by a slight modification of Fernique's method [7].

The purpose of this note is to prove a slightly stronger result; namely, we prove that for every Gaussian measure μ and every ε with $0 < \varepsilon < 2$

$$\int \exp(\, q(x)^{2-\varepsilon})\, d\mu < +\infty \;;$$

moreover, if q is p-homogeneous then

$$\int \exp(\alpha\, q(x)^{2/p})\, d\mu < +\infty$$

for $\alpha > 0$ sufficiently small. The proof consists in adaptation of de Acosta estimates (given in [1]) to Fernique's method.

Let us start with introducing some terminology. We will deal with a measurable vector space (E,B) i.e. E is a real vector space and B is a σ-field of subsets of E such that:

(i) The mapping $(x,y) \longrightarrow x+y$ from $(E \times E, B \otimes B)$ into (E,B) is measurable.

(ii) The mapping $(\lambda,x) \longrightarrow \lambda\cdot x$ from $(R \times E, \mathcal{R} \otimes B)$ into (E,B) is measurable, where $(R, \mathcal{R})$ is the real line with the Borel σ-field.

A function $q : E \longrightarrow R^+$, $q(0) = 0$, will be called a seminorm if it is subadditive, that is $q(x+y) \leqslant q(x) + q(y)$ for every $x,y \in E$, and nondecreasing, that is $q(\alpha x) \leqslant q(\beta x)$ if $|\alpha| \leqslant |\beta|$, for every $\alpha, \beta \in R$ and every $x \in E$.
A seminorm q is called p-homogeneous, $0 < p \leqslant 1$, if for every $x \in E$ and every $\alpha \in R$, $q(\alpha x) = |\alpha|^p q(x)$.

Definition. A probability measure μ is said to be Gaussian (in the sense of Fernique) if it is stable of index 2 and if for every independent random variables X_1 and X_2 with the distribution μ, $X_1 + X_2$ and $X_1 - X_2$ are independent.

The following lemmas will be basic for our considerations. The first one is a modification of the main inequality in Fernique's paper [7].

Lemma 1. Let X be a symmetric strictly stable Gaussian random element with values in (E,B) and q a measurable seminorm. Then for every $s > 0$ and for every $\varepsilon > 0$

$$P\{q((1/2^{1/2})X) > s\cdot(1+\varepsilon)\} \cdot P\{q((1/2^{1/2})X) \leqslant s\cdot\varepsilon\} \leqslant [P\{q(X) > s\}]^2 .$$

Lemma 2 and Lemma 3 are taken from de Acosta [1] (Lemma 3.1 (a) and Lemma 3.3).

Lemma 2. Let X be a symmetric strictly stable of index 2 random element with values in (E,B) and q a measurable seminorm. Then for every $s > 0$ and every $\varepsilon > 0$

$$P\{q(X) > s\} \geqslant 2\cdot P\{q((1/2^{1/2})X) > s\cdot(1+\varepsilon)\} \cdot P\{q((1/2^{1/2})X) \leqslant s\varepsilon\}.$$

Lemma 3. Let X be a symmetric strictly stable random element with values in (E,B) and q a measurable seminorm. Then for every $a > 0$ and every $b > 1$

$$\sum_{n=1}^{\infty} P\{q((1/2^{1/2})^n X) > ab^n\} < +\infty .$$

Now we are able to prove our theorem.

<u>Theorem 1</u>. Let μ be a Gaussian measure in (E,B) and q a measurable seminorm in (E,B). Then for every ε, $0<\varepsilon<2$ there exists $C_\varepsilon > 0$ such that for every $t>0$

$$\mu\{x : q(x) > t\} \leqslant C_\varepsilon \cdot \exp(-t^{2-\varepsilon}).$$

If q is p-homogeneous, $0<p\leqslant 1$, then there exist $C>0$ and $\alpha > 0$ such that for every $t>0$

$$\mu\{x : q(x) > t\} \leqslant C \cdot \exp(-\alpha t^{2/p}).$$

Proof. We prove the theorem only for μ symmetric and strictly stable. The general case can be easily obtained by symmetrization (see [1]).
Let X be a random element with the distribution μ. Using Lemma 1 we obtain by induction :

$$(1) \quad P\{q((1/2^{1/2})^n X) > s\prod_{i=1}^{n}(1+\varepsilon_i)\} \cdot \prod_{i=1}^{n}\delta_i^{2^{n-i}} \leqslant [P\{q(X) > s\}]^{2^n}$$

for every $s>0$, $\varepsilon_i > 0$, $i = 1,2,\dots,n$; where

$$\delta_i = P\{q((1/2^{1/2})^i X) \leqslant s\cdot\varepsilon_i \prod_{j=1}^{i-1}(1+\varepsilon_j)\}.$$

Take $\delta > 0$, $\varepsilon_i = 2^\delta - 1$, $i = 1,2,\dots,n$.

We show that $\prod_{i=1}^{n}(\delta_i)^{1/2^i} \xrightarrow[n\to\infty]{} \rho > 0$.

We have $\prod_{i=1}^{n}(\delta_i)^{1/2^i} = \prod_{i=1}^{n}[P\{q((1/2^{1/2})^i X) \leqslant s(2^\delta - 1)\cdot(2^\delta)^{i-1}\}]^{1/2^i}$.

It is easy to see that if we take $a = \frac{s(2^\delta - 1)}{2^\delta}$ and $b = 2^\delta$ in Lemma 3 then

$$\sum_{i=1}^{\infty} P\{q((1/2^{1/2})^i X) > \frac{s(2^\delta - 1)}{2^\delta}\cdot(2^\delta)^i\} < +\infty.$$

Hence,

$$\prod_{i=1}^{n}\left[1 - P\{q((1/2^{1/2})^i X) > \frac{s(2^\delta - 1)}{2^\delta}\cdot(2^\delta)^i\}\right] = \prod_{i=1}^{n}\delta_i \xrightarrow[n\to\infty]{} \rho_1 > 0,$$

which implies that $\prod_{i=1}^{n}\delta_i^{1/2^i} \xrightarrow[n\to\infty]{} \rho > 0$.

Therefore, from (1) we obtain

$$P\{q(X) > s\} \geqslant \rho \cdot \left[P\{q((1/2^{1/2})^n X) > s \cdot 2^{\delta \cdot n}\}\right]^{1/2^n}$$

for n = 1,2,...

Let us take the subsequence (n_k), where $n_k = 2k$.

In virtue of the subadditivity of q we obtain

$$P\{q(X) > s \cdot 2^k \cdot 2^{2\delta \cdot k}\} \leqslant P\{q((1/2^k)X) > s \cdot 2^{2\delta \cdot k}\} \leqslant \left[\frac{1}{\rho} P\{q(X) > s\}\right]^{2^{2k}}$$

for every k = 1,2,...

Next, observe that $1/\rho$ is a nonincreasing function of s. Hence, we can take s so large that

$$0 < 1/\rho \cdot P\{q(X) > s\} < 1 .$$

Using the standard method of interpolation and choosing C'_ε large enough we obtain

$$P\{q(X) > t\} \leqslant C'_\varepsilon \cdot \exp(-\alpha t^{2-\varepsilon})$$

for $\alpha > 0$ sufficiently small. Since ε is an arbitrary number in (0,2), this implies the conclusion.

If q is p-homogeneous, $0 < p \leqslant 1$, we can take a δ in the inequality (1), such that $1/2^{p/2} < \delta < 1$. Put $\varepsilon_i = \delta^i$,

$\rho_2 = \prod_{i=1}^{\infty}(1+\varepsilon_i) < +\infty$. Using the p-homogeneity of q we can easily check, arguing as above that

$$\prod_{i=1}^{\infty}(\delta_i)^{1/2^i} = \rho > 0 .$$

Then inequality (1) reduces to :

$$P\{q(X) > s \cdot \rho_2 \cdot 2^{pn/2}\} \leqslant \left[\frac{1}{\rho} P\{q(X) > s\}\right]^{2^n} .$$

Repeating the arguments we have just used we obtain the following conclusion :

$$P\{q(X) > t\} \leqslant C \cdot \exp(-\alpha t^{2/p}) ,$$

where $\alpha > 0$ is sufficiently small.

From our theorem we obtain two corollaries concerning the integrability of seminorms.

Corollary 1. If μ is a Gaussian measure on a measurable vector space and q is a measurable seminorm then for every ε, $0 < \varepsilon < 2$, and every $\alpha \in R$

$$\int \exp(\alpha\, q(x)^{2-\varepsilon})\, d\mu(x) < +\infty .$$

Corollary 2. Let μ be as above. If q is a measurable p-homogeneous seminorm then there exists $\alpha > 0$ such that

$$\int \exp(\alpha\, q(x)^{2/p})\, d\mu(x) < +\infty .$$

It seems to be interesting whether in Corollary 1 one can put $\varepsilon = 0$ for α small enough; more precisely, we have the following

Conjecture 1. Let μ and q be as in Corollary 1. Then there exists $\alpha > 0$ such that

$$\int \exp(\alpha\, q(x)^2)\, d\mu(x) < +\infty .$$

In the following example we will show that our conjecture is valid for certain Orlicz spaces.

Example. Let $(T, \mathcal{F}, m)$ be a separable, σ-finite measure space and Φ be a subadditive, nondecreasing continuous function defined on $[0,\infty)$ such that $\Phi \geq 0$ and $\Phi(t) = 0$ if and only if $t = 0$. Let L_Φ be the space of all $\mathcal{F}$-measurable real-valued functions f with the property

$$[f]_\Phi = \int \Phi(|f|)\, dm < +\infty .$$

$(L_\Phi, [\ ]_\Phi)$ is a complete separable linear metric space (it is a particular case of Orlicz space). Let ξ be a symmetric measurable Gaussian stochastic process. Write $\sigma^2(t) = E\,\xi^2(t)$. Assume that $\sigma \in L_\Phi$. Then almost all sample paths of ξ belong to L_Φ ([4],[8]). Let X be the L_Φ- valued Gaussian random element induced by ξ. We show that for every $\alpha < 1/2[\sigma]_\Phi^2$

$$E \exp(\alpha [X]_\Phi^2) < +\infty .$$

Without loss of generality we can assume that $\xi(t) = 0$ whenever $\sigma(t) = 0$. Let $\lambda = \Phi(\sigma(t))/[\sigma]_\Phi$. Then $\int \lambda\, dm = 1$. By Jensen's Inequality we obtain

$$E \exp(\alpha [X]_\Phi^2) = E \exp\left(\alpha\left(\int \Phi(|\xi(t)|)\, m(dt)\right)^2\right) =$$

$$E \exp\left(\alpha\left(\int \Psi(\xi(t))\lambda(t)\, m(dt)\right)^2\right) \leq E\left(\int \exp(\alpha\, \Psi(\xi(t))^2)\lambda(t)\, m(dt)\right) =$$

$$\int \left(E \exp \alpha\, \Psi(\xi(t))^2\right) \lambda(t)\, m(dt) ,$$

where $\Psi(\xi(t)) = \Phi(|\xi(t)|)/\lambda(t)$ when $\lambda(t) > 0$ and $= 0$ when $\lambda(t) = 0$.

Let now n denote the standard normal distribution. By the subadditivity of Φ we have $\Phi(xc) \leqslant (|x|+1)\Phi(c)$ whence for t such that $\lambda(t) > 0$

$$E\exp\left(\alpha\frac{\Phi(|\xi(t)|)^2}{\lambda(t)^2}\right) = \int \exp\left(\alpha\frac{\Phi(\sigma(t)x)^2}{\lambda(t)^2}\right)dn(x) \leqslant$$

$$\frac{2}{\sqrt{2\pi}}\int_0^\infty \exp\left(\alpha\frac{(x+1)^2\Phi(\sigma(t))^2}{\lambda(t)^2} - \frac{1}{2}x^2\right)dx = \frac{2}{\sqrt{2\pi}}\int_0^\infty \exp\left(\alpha[\sigma]_\Phi^2(x+1)^2 - \frac{1}{2}x^2\right)dx$$

If $\alpha < 1/2[\sigma]_\Phi^2$ then $\alpha\cdot[\sigma]_\Phi^2 = (1-\varepsilon)/2$ for a certain ε, $0 < \varepsilon < 1$, which implies that the last integral is finite and ends the proof of our statement.

Remark 1. Let $(T, \mathcal{F}, m)$ be as in the above example. Assume that Φ satisfies (Δ_2) condition (which is weaker than subadditivity). Let $\rho_\Phi(f) = \inf\{u > 0 : \int\Phi(|f/u|)\,dm \leqslant 1\}$. Suppose that there exists $p \in (0,1]$ such that $\rho_\Phi(f+g) \leqslant (\rho_\Phi(f)^p + \rho_\Phi(g)^p)^{\frac{1}{p}}$. Then $\|f\|_\Phi = \rho_\Phi(f)^p$ is a p-homogeneous seminorm. In this case it is possible to prove that if $\alpha < 1/2\|\sigma\|_\Phi^{2/p}$ then $E\exp\alpha\|X\|_\Phi^{2/p} < +\infty$ and this estimate for α is the best possible. This fact can be obtained by a modification of Marcus and Shepp method [13] (see [12]).

Remark 2. If q is p-homogeneous or if it is generated by a sequence of p_i-homogeneous seminorms then the above results provide useful information about asymptotic behavior of Gaussian measures. In particular, in these situations the following holds:

$$(*) \qquad \lim_{t\to 0+}(1/t)\,\mu\{x : q(t^{1/2}x) > \varepsilon\} = 0$$

for every $\varepsilon > 0$.

If E is complete separable metric linear space and q is a seminorm generating the topology of E then this condition guarantees the validity of the so-called Invariance Principle (see [3]). This implies that there exists the E-valued Wiener process $W(t)$, $t \in [0,1]$ such that $W(1)$ has the distribution μ. However, if q is no longer p-homogeneous the property $(*)$ cannot be deduced from the integrability results. The recent result of the authors ([5], Th.3.1.) shows that the property $(*)$ is still valid in this more general situation.

For application of this result to several functional random limit theorems the reader is referred to [6].

In connection with the property (*) one can state the following

Conjecture 2. Let μ be a Gaussian measure and q a measurable seminorm on a measurable vector space (E,B). Assume that the mapping $\alpha \longrightarrow q(\alpha x)$ is continuous at 0, for every $x \in E$. Then there exists $C > 0$ such that for every $\varepsilon > 0$ we can find a positive constant $\alpha(\varepsilon)$ for which the following holds

$$(**) \qquad \mu\{x : q(x/t) > \varepsilon\} \leqslant C \exp(-\alpha(\varepsilon)t^2)$$

for every $t > 0$.

Let us mention that the positive answer to this conjecture would imply that Conjecture 1 is valid.

Observe that if $E = L_0[0,1]$ (the space of all measurable functions on the unit interval) and

$$q(f) = \int \frac{f(t)}{1 + f(t)}\, dt$$

then (**) holds, because every Gaussian measure on $L_0[0,1]$ is induced from $L_2[0,1]$ by a continuous linear mapping (Example in [3]). If the covariance of μ is bounded by K then $\alpha(\varepsilon)$ can be taken $\varepsilon^2/(1+K)$.

Remark 3. If X is a symmetric Gaussian (more generally: strictly stable of index r, $0 < r \leqslant 2$) with distribution μ then there exists the natural convolution semigroup μ_t connected with μ; namely take μ_t to be the distribution of $t^{1/2}X$ ($t^{1/r}X$, respectively). In this notation the property (*) takes the following form:

$$\lim_{t \to 0+} (1/t)\, \mu_t\{x : q(x) > \varepsilon\} = 0$$

for every $\varepsilon > 0$.

In the papers [1], [2] de Acosta showed that if μ is strictly stable of index $\neq 2$ and q is homogeneous and satisfies some non-degeneracy condition then $\lim_{t \to 0+} (1/t)\, \mu_t\{x : q(x) > \varepsilon\}$ exists and is strictly positive.

Before stating a generalization of this result we need the notion of the q-continuity of a semigroup.

Let $(\mu_t)_{t>0}$ be a convolution semigroup on a measurable space (E,B) and let q be a measurable seminorm. $(\mu_t)_{t>0}$ will be called q-continuous if for every $\varepsilon > 0$

$$\lim_{t\to 0+} \mu_t\{x : q(x) > \varepsilon\} = 0 .$$

It can be observed that all interesting semigroups of probability measures are q-continuous under suitable q; in particular, if q is p-homogeneous and $(\mu_t)_{t>0}$ is the natural semigroup connected with a strictly stable measure then $(\mu_t)_{t>0}$ is automatically q-continuous. Hence the following theorem can be regarded as a generalization of de Acosta result :

Theorem 2. Let q be a measurable seminorm and $(\mu_t)_{t>0}$ a q-continuous semigroup of probability measures on a measurable space. There exists a right-continuous, nonincreasing function θ such that

$$\lim_{t\to 0+} (1/t)\, \mu_t\{x : q(x) > s\} = \theta(s)$$

for every $s > 0$ at which θ is continuous.
If μ_t, $t > 0$ are Gaussian, then $\theta \equiv 0$. If $q > |f|$ for a measurable linear functional f such that $f(\cdot)$ is not Gaussian (with respect to μ_1) then $\theta \not\equiv 0$.

This result seems to be new even if $E = R^n$ and q is non-homogeneous. The proof is contained in [5] .

References

[1] de Acosta, A. (1975) Stable measures and seminorms, Ann. Probability 3, 865-875.

[2] de Acosta, A. (1977) Asymptotic behavior of stable measures, Ann. Probability 5, 494-499.

[3] Byczkowski, T. (1976) The invariance principle for group-valued random variables, Studia Math. 56, 187-198.

[4] Byczkowski, T. (1979) Norm convergent expansion for L_ϕ-valued Gaussian random elements, Studia Math. 64, 87-95.

[5] Byczkowski, T. and Żak, T. Asymptotic properties of semigroups of measures on vector spaces, (to appear in Ann. Probability).

[6] Byczkowski, T. and Inglot, T. The invariance principle for vector-valued random variables with application to functional random limit theorems, (to appear).

[7] Fernique, X. (1970) Intégrabilité des vecteurs Gaussiens, C.R. Acad. Sci. Paris Ser. A 270, 1698-1699.

[8] Gorgadze, Z.G. (1976) On measures in Banach spaces of measurable functions, (Russian) Trudy Tbliss. Univ. 166, 43-50.

[9] Helm, W. (1978) On Gaussian measures and the central limit theorem in certain F-spaces, Springer Lecture Notes in Math. 656. 59-65.

[10] Inglot, T and Weron, A. (1974) On Gaussian random elements in some non-Banach spaces, Bull. Acad. Polon. Sci. Ser. Math. Astronom. Phys. 22, 1039-1043.

[11] Inglot, T. (1976) Ph. D. Thesis, Technical University, Wrocław.

[12] Litwin, T. (1980) M. Sc. Thesis, Technical University, Wrocław.

[13] Marcus, M.B. and Shepp, L.A. (1971) Sample behavior of Gaussian processes, Proc. Sixth Berkeley Symp. Math. Statist. Prob. 2, 423-442.

Technical University, Wrocław

INFINITE DIMENSIONAL NEWTONIAN POTENTIALS (†)

by René CARMONA

Département de Mathématiques
Université de Saint Etienne
23 rue Paul Michelon
42023 SAINT ETIENNE Cédex
FRANCE

Summary

We give a survey of various curiosities and problems concerning potential theory of infinite dimensional Brownian motion processes.

I. INTRODUCTION:

It is very surprising to note that very few results have been published (see nevertheless [12], [2] and [6]) on infinite dimensional Newtonian potential theory after the fundamental work of L.Gross (see [7] and [8]). It is all the more bizarre as these papers initiated a wave of interest in the study of partial differential equations and stochastic processes in infinite dimensional Banach spaces. Our feeling is that the reasons have to do with the highly pathological character of the situation. This has been demonstrated by V.Goodman [6], who proved, among other things,

(†) Talk given at the Second International Conference "Probability Theory on Vector Spaces" held in Blazejewko (Poland) in September 1979

the existence of bounded non constant harmonic functions.

Here we present a review of these pathologies. Some of them are new, but the reader familiar with infinite dimensional analysis and measure theory will not be surprised. All the proofs are very elementary. They are based on a simple property of Gaussian measures on Banach spaces, which is proved in section II, and which we believe is essentially known.

II. GAUSSIAN MEASURES IN BANACH SPACES:

We begin by fixing some notations and assumptions which will be used subsequently. We would like to lay emphasis on the fact that they will not be recalled explicitely.

E will be an infinite dimensional Banach space, the Borel σ-field of which will be denoted by $\mathcal{B}_E$ and γ will be a fixed Gaussian measure on E. That is to say, γ is a probability measure on the measure space $(E, \mathcal{B}_E)$ such that each element of the dual space E^* of E is a centered Gaussian random variable on the probability space $(E, \mathcal{B}_E, \gamma)$. Note that for us measure will mean non negative measure.

To make our life easier we will assume that E is separable and that the topological support of γ is the whole space E (in fact these assumptions are not necessary and the following results apply to more general situations).

Then E^* can be considered as a subset of $L^2(E,\gamma)$ and we will call H^* its closure. It is well known that H^* can be viewed as the dual of a Hilbert space H which is a dense subset of E (H is often called the reproducing kernel Hilbert space, R.K.H.S. for short, of γ), the inclusion map into E being continuous.

$<\ ,\ >$ will denote the duality pairing between E^* and E.

Lemma1:

For each $x \in E \setminus H$ there is a complete orthonormal system of H^*, say $\{e_j^*; j \geq 1\}$ contained in E^* such that:

$$\sup_{j \geq 1} |<e_j^*, x>| = \infty \tag{2.1}$$

Proof:

Let U denote the closed unit ball of H^*. Since x is not a continuous linear form on H^* we have:

$$\sup_{x^* \in E^* \cap U} |\langle x^*, x\rangle| = \infty. \tag{2.2}$$

As a preliminary to the very proof of the lemma let us check that for each finite subset of E^*, say $\{e_1^*, \ldots, e_n^*\}$, which is orthonormal in H^*, and for each $\varepsilon > 0$, there exists an element e_{n+1}^* of E^* such that:

(i) $\{e_1^*, \ldots, e_n^*, e_{n+1}^*\}$ is an orthonormal system, and

(ii) $\max_{j=1,\ldots,n+1} |\langle e_j^*, x\rangle| \geqslant \varepsilon.$ (2.3)

This is an easy consequence of the relation:

$$\sup_{x^* \in E^* \cap U,\ x^* \perp \{e_1^*, \ldots, e_n^*\}} |\langle x^*, x\rangle| = \infty,$$

(where $\perp$ denotes orthogonality in H^*) which is obtained from (2.2) and the infinite dimensionality of the spaces.

Now we proceed to the proof of the lemma. Let $\{x_n^*; n \geq 1\}$ be a dense countable subset of the closed unit ball of E^* equipped with the weak star topology (without any loss of generality we assume $x_1^* \neq 0$). We use an induction argument to construct the desired complete orthonormal system.

Let us set $e_1^* = \|x_1^*\|_{H^*}^{-1} x_1^*$, and only if $|\langle e_1^*, x\rangle| < 1$, let us use the above argument (see (2.3)) to choose e_2^* such that $\{e_1^*, e_2^*\}$ is a subset of E^* which is orthonormal in H^* and such that:

$$\max \{ |\langle e_1^*, x\rangle|,\ |\langle e_2^*, x\rangle| \} \geq 1.$$

Now, at each step of the induction we add one or two elements e_j^* to our system in such a way that:

(i) at least one more of the elements of the set $\{x_n^*; n \geq 1\}$ is now contained in the linear span of the e_j^*, and

(ii) the maximum of the $|\langle e_j^*, x\rangle|$ is increased by an amount greater than 1.

Finally we obtain a subset $\{e_j^*; j \geq 1\}$ of E^* which is an orthonormal system of H^*, which satisfies (2.1), and the linear span of which contains all of the x_n^*'s. This last property forces the system to be complete.■

The numerical sequence $\{\langle e_j^*, x\rangle; j \geqslant 1\}$ is unbounded. Moreover we may assume that all of its elements are non negative. So, there exists a sequence $\{\omega_j ; j \geq 1\}$ of positive numbers which satisfies:

$$\sum_{j=1}^{\infty} \omega_j < \infty \qquad \text{and} \qquad \sum_{j=1}^{\infty} \omega_j \langle e_j^*, x \rangle = \infty. \tag{2.4}$$

Let $\{e_j ; j \geq 1\}$ be the complete orthonormal system of H which corresponds to $\{e_j^* ; j \geq 1\}$ in the Riesz identification of H^* and H. The next result is concerned with Borel subsets of E on which the measure γ is concentrated.

Proposition 1:

Let $x \in E \setminus H$ and let the sequences $\{e_j^* ; j \geq 1\}$ and $\{\omega_j ; j \geq 1\}$ be as above and let us set:

$$L_x = \left\{ z \in E ;\ \sum_{j=1}^{\infty} \langle e_j^*, z \rangle e_j = z \text{ and } \sum_{j=1}^{\infty} \omega_j |\langle e_j^*, z \rangle| < +\infty \right\}. \tag{2.5}$$

Then L_x is a Borel subset of E such that $\gamma(L_x)=1$, $H \subset L_x$, and $x \notin L_x$. Furthermore, equipped with the norm:

$$\|z\|_x = \max \left\{ \sum_{j=1}^{\infty} \omega_j |\langle e_j^*, z \rangle| \ ,\ \sup_{n \geq 1} \left\| \sum_{j=1}^{n} \langle e_j^*, z \rangle e_j \right\|_E \right\} \tag{2.6}$$

it becomes a real separable Banach space and the inclusion maps of H into L_x and of L_x into E are continuous and have dense ranges.

Proof:

The e_j^* 's being continuous linear forms, formula (2.5) defines a Borel vector subspace of E. Clearly $x \notin L_x$ by (2.4). Furthermore (2.6) defines a norm because the e_j^*'s separate the points of E. If $z \in H$ we have:

$$\|z\|_x \leq \max \left\{ \left(\sum_{j=1}^{\infty} \omega_j^2\right)^{1/2} \left(\sum_{j=1}^{\infty} \langle e_j^*, z\rangle^2\right)^{1/2},\ \sup_{n \geq 1} a \left\| \sum_{j=1}^{n} \langle e_j^*, z \rangle e_j \right\|_H \right\}$$

(where a denotes the norm of the inclusion map of H into E)

$$\leq \max \left\{ \left(\sum_{j=1}^{\infty} \omega_j^2\right)^{1/2},\ a \right\} \|z\|_H ,$$

which proves that H is contained in L_x with continuous inclusion map. Moreover it is clear that:

$$\|z\|_E \leq \|z\|_{L_x}$$

whenever $z \in L_x$. Now it is a straightforward exercise to show that L_x is complete. It is not more difficult to check that the finite linear combinations of the e_j's, with rational coefficients constitute a dense set in L_x for the norm $\| \|_x$. So L_x is separable and the inclusion maps have dense ranges. To finish the proof it remains to argue $\gamma(L_x)=1$. $\{<e_j^*,\cdot>;j\geq 1\}$ is a sequence of independent identically distributed normal random variables on the probability space $(E,\mathcal{B}_E,\gamma)$. The family $\{\omega_j;j\geq 1\}$ being summable, the series:

$$\sum_{j=1}^{\infty} \omega_j |<e_j^*,\cdot>|$$

converges γ-almost surely. Moreover it is well known that $\left\{\sum_{j=1}^{n} <e_j^*,\cdot>e_j;n\geq 1\right\}$ is a vector valued martingale in E which has integrability properties enough (see [5]) to converge γ-almost surely to the identity. The proof is complete.∎

Remarks:

1. The restriction of γ to L_x is clearly Gaussian. Its R.K.H.S. is again H. So, if i denotes the inclusion map of H into L_x, and if we use the terminology of [7], the triplet (i,H,L_x) is an abstract Wiener space.
2. The biorthogonal system $\{(e_j,e_j^*);j\geq 1\}$ is a monotone Schauder basis of L_x (compare with [10.Remark p.67]).
3. Having in mind the applications of the following sections, we designed the statement and the proof of Proposition 1. We must confess that by that time we were not aware of the fact that, at least part of it, was already known. Indeed, it follows for example from [1] that if $x\notin H$, there exists a Borel vector subspace of E which has full γ-measure and which contains H without containing x. But this is not enough for our purpose because we will need this subspace to be equipped with a Banach structure which will have to be finer than the one of E.

As an aside consequence of proposition 1 we mention the following characterization of the R.K.H.S. of a Gaussian measure in a Banach space (see also [8.Propositions2,3])

Corollary 1:

An element x of E belongs to H if and only if for all Borel subsets A of E the map $\lambda \longrightarrow \gamma(A-\lambda x)$ is continuous at the origin of $\mathbb{R}$.

III. γ-BROWNIAN MOTION TRANSITION KERNELS:

Definition 1:

The kernels p_t *defined by:*

$$p_t(x,A) = \gamma\left(\frac{1}{\sqrt{t}}(A-x)\right) \qquad t>0,\ x\in E,\ A\in\mathcal{B}_E$$
$$p_0(x,A) = 1\!\!1_A(x) \quad (1) \qquad x\in E,\ A\in\mathcal{B}_E \tag{3.1}$$

for $x\in E$ *and* $A\in\mathcal{B}_E$ *constitute a convolution semigroup. They are called the transition kernels of the* γ*-Brownian motion in* E.

If f is a real measurable function on E we set:

$$\left[P_t f\right](x) = \int_E f(y)\ p_t(x,dy) \qquad x\in E,\ t>0 \tag{3.2}$$

whenever the integral makes sense. It is easy to check that the map

$$[0,\infty[\times\mathbb{R}_+ \ni (t,x) \longrightarrow \left[P_t f\right](x)\in\mathbb{R}$$

is jointly continuous if f is bounded and continuous, and is uniformly continuous on any set of the form $[t_0,t_1]\times B$ for $0<t_0\leq t_1<+\infty$ and $B\subset E$ bounded provided f is assumed to be bounded and locally uniformly continuous.

Formula (3.2) defines semigroups of operators on various function spaces. These semigroups are sometimes strongly continuous: this is for example the case on the Banach space of bounded uniformly continuous functions on E ([8.Proposition 6]).

Our semigroup of kernels is of Feller type. Unfortunately Corollary 1 tells us that it is not of strong Feller type. More specifically we have:

Proposition 2:

Whatever element $x\in E$ *is chosen, there is no* σ*-finite measure on* $(E,\mathcal{B}_E)$ *with respect to which all the measures* $p_t(x,.)$ *for* $t>0$ *are absolutely continuous.*

Proof:

Without any loss of generality we may assume x=0 and we may restrict ourselves to finite measures. Let $\{e_j^*;j\geq 1\}$ be any subset of E^* which is a complete orthonormal

---

(1) The notation $1\!\!1_A$ stands for the indicator function of the set A.

system of H^* and for each $t>0$ let us set:

$$A_t = \left\{ y\in E;\ \limsup_{j\to\infty} \frac{\langle e_j^*, y\rangle}{\sqrt{2\operatorname{Log} j}} = t \right\}.$$

$\left\{\frac{1}{\sqrt{t}} e_j^*; j\geq 1\right\}$ is a sequence of independent identically distributed normal random variables on the probability space $(E, \mathcal{B}_E, p_t)$ (*). So, classical probability calculus tells us that $p_t(A_t)=1$ for all $t>0$. Consequently there is no finite measure which can charge all of the uncountably many Borel sets A_t which are disjoint.∎

Remark 4:

Note that the classical construction:

$$\Theta(.) = \sum_{m,n\geq 1} 2^{-(m+n)}\ p_{t_n}(x_m,.)$$

where $\{t_m; m\geq 1\}$ and $\{x_n; n\geq 1\}$ are countable dense sets in $[0,\infty[$ and E respectively, gives rise to a finite measure which satisfies:

$$\Theta(A) = 0 \implies \left(\forall t>0,\ \forall x\in E,\ \ p_t(x,A) = 0\right)$$

for all open sets A in E, but not for all Borel subsets of E.

We end this section with a result of V.Goodman (see [6.Corollaries 2 and 3]).

Definition 2:

A measurable function f on E is said invariant if for each $t>0$ and for each $x\in E$, $[P_t f](x)$ makes sense and equals $f(x)$.

In the finite dimensional case the invariant functions are the harmonic functions and consequently every bounded invariant function is constant (see for example [13]). This is no longer the case in the present situation.

Proposition 3:

There are non constant bounded invariant functions. Moreover there are invariant functions which are unbounded in every neighborhood of each point.

Proof:

Since E is infinite dimensional, $H\neq E$ and we can find $x\in E\setminus H$. Let L_x be the Banach

(*) We use the notation p_t for $p_t(0,.)$. So, $p_t(x,A)=p_t(A-x)$ if $t\geq 0$, $x\in E$ and $A\in\mathcal{B}_E$.

space given by Proposition 1. If we set:

$$f = \sum_{j=1}^{\infty} \gamma_j \mathbb{1}_{n^{-1}x+L_x} \tag{3.3}$$

where $\{\gamma_j;j\geq 1\}$ is any non constant bounded numerical sequence, it is clear that f is a non constant bounded invariant function. Now, if the sequence $\{\gamma_j;j\geq 1\}$ satisfies:

$$\lim_{j\to\infty} \gamma_j = \infty$$

instead of being bounded, the same formula (3.3) proves the second claim of the proposition.■

Remarks:

5. It is very easy to prove that any bounded invariant function is constant on each H-coset (i.e. of the form $y+H$ with $y\in E$).
6. If we set $f=\mathbb{1}_{L_x}$ with $x\in E\setminus H$ and L_x given by Proposition 1, f is invariant, bounded, identically equal to one on H and zero at x. This proves a conjecture of V.Goodman (see [6.p.219]).

IV. γ-BROWNIAN MOTION MARKOV PROCESS:

Let Ω denote the space of continuous functions from $\mathbb{R}_+$ into E and let us set $X_t(\omega)=\omega(t)$ for $t\in\mathbb{R}_+$ and $\omega\in\Omega$. Now let $\mathcal{F}$ be the smallest σ-field with respect to which all the X_t are measurable ($'''$), and let $\mathcal{F}_t$ be the smallest σ-field with respect to which all the X_s with $0\leq s\leq t$ are measurable. There is a unique probability measure, say $\mathbb{P}_0$, on $(\Omega,\mathcal{F})$ such that:

(i) $\mathbb{P}_0\{X_0=0\} = 1$, and

(ii) for each $0=t_0<t_1<\ldots<t_n$ the random variables $(t_j-t_{j-1})^{-1/2}(X_{t_j}-X_{t_{j-1}})$ for $j=1,\ldots,n$ are independent identically distributed with common law γ when considered on the probability space $(\Omega, \mathcal{F}, \mathbb{P}_0)$.

The construction of this measure has been argued in [8]. It can also be done using Chevet's result on the ε-tensor product of abstract Wiener spaces (see [3.Example2]). Now, if $x\in E$ we define the measure $\mathbb{P}_x$ by:

($'''$) Note that $\mathcal{F}$ is the Borel σ-field of Ω equipped with the topology of uniform convergence on the compact subsets of $\mathbb{R}_+$.

$$\mathbb{P}_x = \tau_{-x}(\mathbb{P}_0)$$

where the space translation τ_x is defined by the formula:

$$[\tau_x \omega](t) = x + \omega(t) \qquad t \geq 0, \ \omega \in \Omega, \ x \in E.$$

$(\Omega, \mathcal{F}, \mathcal{F}_t, X_t, \mathbb{P}_x)$ is a strong Markov process in E (†). Its transition kernels are the $p_t(x,A)$ we introduced in section III. As usual, if Θ is a measure on $(E, \mathcal{B}_E)$ we define the measure $\mathbb{P}_\Theta$ by:

$$\mathbb{P}_\Theta(A) = \int_E \mathbb{P}_x(A)\ \Theta(dx) \qquad A \in \mathcal{F}.$$

Now, if A is a subset of E the first hitting time of A is defined by:

$$T_A = \inf\{t>0;\ X_t \in A\}$$

For simplicity we will consider here that T_A is a $\mathcal{F}_t$-stopping time whenever $A \in \mathcal{B}_E$. Nevertheless see footnote (†). Let us recall some more standard definitions.

A _is said to be polar if we have:_ $\forall x \in E,\ \mathbb{P}_x\{T_A < +\infty\} = 0.$

x _is said to be regular for_ A _if we have:_ $\mathbb{P}_x\{T_A = 0\} = 1.$

It is clear that polar sets have no regular point. The converse is true in the finite dimensional case (see for example [13.Theorem 6.4]) but we do not know what is going on in the present situation.

Among the properties that are true in the finite dimensional case and for which we do not know a proof in the present infinite dimensional setting let us quote: is it true that

$$\forall t>0, \quad \forall x \in E, \quad \mathbb{P}_x\{T_A = t\} = 0$$

whenever A is a Borel sub set of E?

Here is an example of an infinite dimensional polar set:

Proposition 4:

H _is polar_.

Proof:

First, let us fix $x \in E \setminus H$ and let the Banach space L_x be given by Proposition 1. It

(†) In order to keep this survey in a readable form we ignore deliberately the problems of adding a cemetery to E, making the filtration $\mathcal{F}_t$ right continuous, and completing the σ-fields with respect to the measures $\mathbb{P}_x$, even though we implicitely assume that they have been taken into account.

is clear that:

$$\mathbb{P}_x\left\{\forall t\geq 0,\ X_t\in x+L_x\right\} = 1,$$

so that:

$$\mathbb{P}_x\left\{T_H=\infty\right\} = 1. \tag{4.1}$$

In order to prove (4.1) when $x\in H$, we use Markov property and the fact that $p_t(y,H)=0$ whenever $t>0$ and $y\in H$. ■

Remark 7:

We used the fact that γ-Brownian motion starting from x never leaves the set $x+L_x$. This idea is taken from [6] where it is utilized in the proof of the disconnectness of E for the fine topology.

V. NEWTONIAN POTENTIALS:

We introduce the various potential kernels associated with the transition kernels of the γ-Brownian motion in E.

Definition 3:

If $\lambda>0$ we set: (¹)

$$g_\lambda(x,A) = \int_0^{+\infty} e^{-\lambda t}\ p_t(x,A)\ dt \qquad x\in E,\ A\in\mathcal{B}_E.$$

If $\lambda=0$, g_λ is simply denoted g *and it is called the γ-Newtonian potential kernel in* E.

Here are some nice properties of the potential kernels (see also [8], [12] and [2]). If $\lambda>0$ and if f is a real measurable function on E we set:

$$[G_\lambda f](x) = \int_E f(y)\ g_\lambda(x,dy) \qquad x\in E,$$

whenever the integral makes sense. Again when $\lambda=0$, G_λ is simply denoted by G and Gf is called the γ-Newtonian potential of f. The operators G_λ constitute the resolvent of the semigroups $\{P_t;t\geq 0\}$ introduced in section III and the results of general semigroup theory apply. When studying potential theory there is a need to let the opera-

(¹) Our terminology is justified by the fact: if E was finite dimensional g would be the classical Newtonian convolution kernel with respect to Lebesgue's measure associated to the Euclidean structure of H.

tors P_t and G_λ act not only on function spaces but, by duality, on spaces of measures too. Namely, for any measure μ on $(E, \mathcal{B}_E)$ and for any non negative numbers t and λ we define the measures μP_t and μG_λ by:

$$[\mu P_t](A) = \int_E p_t(x,A)\ \mu(dx) \qquad A \in \mathcal{B}_E,$$

$$[\mu G_\lambda](A) = \int_E g_\lambda(x,A)\ \mu(dx) \qquad A \in \mathcal{B}_E.$$

The potential kernels g_λ satisfy the so called "unicity of mass principle".

Proposition 5:

If μ is such that μG_λ is locally finite (*), *then μG_λ determines μ.*

Proof:

Let μ and ν be measures on $(E, \mathcal{B}_E)$ such that μG_λ and νG_λ are locally finite, and such that $\mu G_\lambda = \nu G_\lambda$. Classical results in the finite dimensional case (see for example [13.Proposition 1.1 p.54]) imply that μ and ν coincide on a system of cylindrical sets which is stable with respect to finite intersections and which generates the σ-field $\mathcal{B}_E$. This forces $\mu = \nu$. ■

The following "domination principle" is taken from [14]. Rost's result is in fact a refinement of a former one of G.A.Hunt for Markov processes in locally compact spaces (see [9.Proposition 7.3]). Its proof applies to the present situation.

Proposition 6:

Let μ and ν be finite measures on $(E, \mathcal{B}_E)$ with locally finite potentials μG_λ and νG_λ. If $A \in \mathcal{B}_E$ is such that $\mu G_\lambda \leq \nu G_\lambda$ on A and if W_A is defined by:

$$W_A = \inf \left\{ t>0;\ \int_0^t \mathbb{1}_A(X_u)du > 0 \right\},$$

then we have:

$$\left(W_A = 0 \quad \mathbb{P}_\mu\text{-a.s.} \right) \implies \left(\mu G_\lambda \leq \nu G_\lambda \quad \text{on } E \right).$$

Another principle of classical potential theory which holds in the present situation is the so called "balayage principle".

(*) A measure μ on $(E, \mathcal{B}_E)$ is said to be locally finite if $\mu(A) < +\infty$ for all bounded Borel subsets A of E.

Proposition 7:

If U is an open subset of E and if μ is a measure on $(E, \mathcal{B}_E)$, there exists a measure μ_U (called balayage of μ on U) whose support is contained in the closure of U and such that:

$$\mu_U G \leqslant \mu G \qquad \text{on } E$$

$$\mu_U G = \mu G \qquad \text{on } U.$$

Proof:

As usual let us set $T = T_{U^c}$ where A^c denotes the complementary set of A and define the measure μ_U by:

$$\mu_U(A) = \mathbb{P}_\mu\{X_T \in A\} \qquad A \in \mathcal{B}_E. \blacksquare$$

Remark 8:

As it is stated, the above proposition requires unnecessarily restrictive assumptions and asserts weaker properties than it should. Indeed, from the definition of μ_U it is clear first, that A need not be open in E and second, that the support of μ_U can be further specified. Nevertheless we believe that stated that way, Proposition 7 should admit a proof based only on analytical properties of the kernels, without any appeal to the probabilistic properties of the process (see for example [4] for such a proof in the locally compact case).

We would like to end this survey with one more problem.
Even though our potential kernels satisfy most of the fundamental principles of classical potential theory, their behavior is very pathological and this prevents us from using the powerful machinery of probabilistic potential theory.

In the present infinite dimensional setting we do not see what should be the relevant notions of Newtonian capacity and energy because the usual duality theory is not available. To illustrate this last assertion, let us check that the mere condition known as "Meyer's assumption (L)" (see [11]) is not satisfied.

Proposition 8:

For each fixed $\lambda > 0$, there is no σ-finite measure on $(E, \mathcal{B}_E)$ with respect to which all the measures $g_\lambda(x,.)$ for $x \in E$ are absolutely continuous.

Proof:

Once more let us choose $x \in E \setminus H$ and let L_x be given by Proposition 1. For each $\alpha>0$ we set:

$$A_\alpha = \alpha x + L_x .$$

Since $\gamma(L_x)=1$ it follows from (3.1) that:

$$p_t(\alpha x, A_\alpha) = 1$$

for all $t>0$ and $\alpha>0$. Consequently, for all $\alpha>0$ we have:

$$g_\lambda(\alpha x, A_\alpha) > 0.$$

So, there is no finite measure which can charge all of the uncountably many Borel sets A_α which are disjoint. The proof is complete.∎

Acknowledgments:

I would like to thank Simone Chevet and Jacques Berruyer for numerous discussions during the preparation of this talk.

REFERENCES

[1] C.BORELL: Random linear functionals and subspaces of probability one. Ark. Mat. 14 (1976) 79-92.

[2] R.CARMONA: Potentials on Abstract Wiener Space. J. Functional Anal. 26 (1977) 215-231.

[3] R.CARMONA: Tensor product of Gaussian measures. in "Vector Space Measures and Applications I" Proc. Dublin 1977 ed. R.M.Aron and S.Dineen Lect. Notes in Math. #644 p.96-124.

[4] J.DENY: Noyaux de convolution de Hunt et noyaux associés à une famille fondamentale. Ann. Inst. Fourier 12 (1962) 643-667.

[5] X.FERNIQUE: Intégrabilité des vecteurs gaussiens. C.R.Acad.Sci. Paris, ser.A 270 (1970) 1698-1699.

[6] V.GOODMAN: A Liouville theorem for abstract Wiener spaces. Amer. J. Math. 95 (1973) 215-220.

[7] L.GROSS: Abstract Wiener Spaces.
Proc. Fifth Berkeley Symp. on Math. Stat. and Prob. vol.II Univ. California Press (1967) p.31-42.

[8] L.GROSS: Potential Theory on Hilbert Space.
J. Functional Anal. 1 (1967) 123-181.

[9] G.A.HUNT: Markoff Processes and Potentials I.
Ill. J. Math. 1 (1957) 44-93.

[10] H.H.KUO: Gaussian Measures in Banach Spaces.
Lect. Notes in Math. #463 (1975).

[11] P.A.MEYER: Processus de Markov.
Lect. Notes in Math. #26 (1967).

[12] M.A.PIECH: Regularity of the Green operator on abstract Wiener space.
J. Differential Equat. 12 (1972) 353-360.

[13] S.C.PORT and C.J.STONE: Brownian Motion and Classical Potential Theory.
Academic Press (1978).

[14] H.ROST: Die Stoppverteilungen eines Markoff-Prozesses mit lokalendlichem Potential. Manuscripta Math. 3 (1970) 321-330.

MULTIPARAMETER PROCESSES AND VECTOR-VALUED PROCESSES

S.D. CHATTERJI

§1. Introduction.

If $\{\xi_{s,t}\}$, $(s,t) \in D \subset \mathbb{R}^2$, is some real-valued stochastic process on some probability space (Ω,Σ,P) parametrized by points of a subset of the plane, then clearly

$$X_s(\omega) = \{t \to \xi_{s,t}(\omega) \ , \ t \in D_s\}$$

(where $D_s = \{t \in \mathbb{R} \mid (s,t) \in D\}$ = s-section of D) defines a vector-valued process $\{X_s\}$, $s \in I$, where I is the projection of D on the s-axis and the values of X_s are in some space E_s of functions on D_s. This simple remark has been used in a previous paper [3] to derive a number of results on multiparameter processes. The purpose of this paper is to give one further application of the idea.

Let $\{\xi_{s,t} \ , \ \Sigma_{s,t}\}$, $(s,t) \in \mathbb{R}^2_+$, be a two-parameter martingale with $(s,t) \to \Sigma_{s,t}$ an increasing, right-continuous family of σ-algebras under the usual ordering of $\mathbb{R}^2_+$ i.e. $(s,t) \leq (s',t')$ if $s \leq s'$ and $t \leq t'$. The problem is to prove the existence of a modification of the process $\{\xi_{s,t}\}$ which is regular i.e. has a.s. right continous path with left hand limits. It is clear from the discrete two-parameter case that one must assume that ξ_{st} belong to $L^1 \log L^1$ class (cf. [1],[2],[3]) and that $\Sigma_{s,t}$'s should satisfy the (F4) condition of [2] p.113 i.e. $\Sigma_{s,\infty} = \sigma\{\Sigma_{s,t} \ , \ t \geq 0\}$ and $\Sigma_{\infty,t} = \sigma\{\Sigma_{s,t} \ , \ s \geq 0\}$ be conditionally independent given $\Sigma_{s,t}$. This latter condition can be written analytically as $S_s T_t = T_t S_s = E_{s,t}$ where S_s, T_t, $E_{s,t}$ are the conditional expectation operators given the σ-algebras $\Sigma_{s,\infty}$,

$\Sigma_{\infty,t}$ and $\Sigma_{s,t}$ respectively. The existence of such modifications were unknown until quite recently even in the case of bounded $\xi_{s,t}$ (cf. [2] p.119). Only a few months ago, Prof. Cairoli indicated to me that M. Bacry in Paris has obtained a proof (in a preprint) of the complete result. In §2 we give a slightly stronger result in a more restricted situation.

§2. <u>The main result.</u>

We keep the notations introduced above.

<u>Theorem.</u>

Let $\{\xi_{s,t}, \Sigma_{s,t}\}_{s \geq 0,\, t \geq 0}$ be a martingale with the σ-algebras $\Sigma_{s,t}$ satisfying the conditions of §1 and $E\{|\xi_{s,t}| \log^+ |\xi_{s,t}|\} < \infty$. If for each $s \geq 0$, the martingale $\{\xi_{s,t}\}_{t \geq 0}$ has a modification with continuous sample paths, then there is a modification $\{\eta_{s,t}\}$ of $\{\xi_{s,t}\}$ such that for all $s_0 \geq 0$ and compact subinterval $I \subset [0,\infty[$,

$$\lim_{\substack{s \to s_0 \\ s > s_0}} \sup_{t \in I} |\eta_{s,t}(\omega) - \eta_{s_0,t}(\omega)| = 0 \qquad \text{a.s.}$$

and

$$\lim_{\substack{s \to s \\ s < s}} \sup_{t \in I} |\eta_{s,t}(\omega) - \eta_{s_0,t}(\omega)| = 0 \qquad \text{a.s.}$$

<u>Proof</u>: Let us suppose without loss of generality that the $\xi_{s,t}$ are such that for each dyadic rational s $(s \in \Delta)$, $t \to \xi_{s,t}(\omega)$ is continuous for all ω and that s,t vary in some fixed compact interval $I = [0, N]$. Clearly the process $\{X_s\}_{s \in I \cap \Delta}$ is a $C(I)$-valued martingale (where $X_s(\omega) : I \to \mathbb{R}$ given by $X_s(\omega)(t) = \xi_{s,t}(\omega)$) with $E\{||X_N||\} < \infty$ where $||\cdot||$ is the sup-norm in $C(I)$. The integrability of $||\cdot||$ follows from the $L^1 \log L^1$ condition. Now it is a well-known fact that any Banach-valued process $\{X_s\}_{s \in I \cap \Delta}$ is such that a.s., the right

and left limits of X_{s}, $s \in \Delta$, exist at any $s_0 \in I$.

Defining Y_{s_0} to be the right hand limit of X_s as $s \to s_0$ through values of $s > s_0$ and $s \in I \cap \Delta$ we get a regular process $\{Y_s\}_{s \in I}$ with values in $C(I)$. Clearly $\eta_{s,t}$ defined by $\eta_{s,t}(\omega) = Y_s(\omega)(t)$ is a modification of $\xi_{s,t}$ and satisfies the conditions of the theorem Q.E.D.

Remarks: The continuity condition on $t \to \xi_{s,t}(\omega)$ was used to assure that the r.v's X_s are strongly measurable; to carry out the above proof, all that is necessary is that the maps $X_s : \Omega \to B(I)$, ($B(I)$ = the space of bounded functions on I under the sup norm) have separable range. It is doubtful that the above proof would work in the general case. It also seems highly improbable that the modifications would have, in general, such uniform right and left limits as in our theorem.

References.

[1] Cairoli, R.
Une inégalité pour martingales à indices multiples et ses applications. Sém. de Prob. IV, Univ. de Strasbourg, Lecture Notes in Math. No.124, 1-23, Springer-Verlag, Berlin (1970).

[2] Cairoli, R. and J. Walsh.
Stochastic integrals in the plane. Acta Math. Vol.134, 111-183 (1975).

[3] Chatterji, S.D.
Vector-valued martingales and their applications. Probability in Banach spaces, Oberwolfach 1975. Lecture Notes in Maths. No.526, 33-51, Springer-Verlag, Berlin (1976).

S.D. Chatterji
Professeur, Dépt. de Math.
Ecole Polytechnique Fédérale
de Lausanne
61, av. de Cour
CH-1007 LAUSANNE
Suisse

ON GEOMETRY OF ORLICZ SPACES

by

Z.G. Gorgadze, V.I.Tarieladze

Abstract. In terms of the function Φ it is established when Orlicz space L_Φ does not contain l_∞^n uniformly and when it has some type or cotype.

We study some geometrical properties of Orlicz spaces. Namely using the results of [1] we establish when the given Orlicz space does not contain l_∞^n uniformly and when it has some type or cotype.

Let X be a real Banach space, X^* the dual space and let $(\varepsilon_k)_{k\in N}$ be the sequence of independent random variables with $P[\varepsilon_k = 1] = P[\varepsilon_k = -1] = 1/2$ (the Bernoulli, or the Rademacher sequence). A Banach space X is said to be a space of type p, $0 < p \leq 2$, if there exists a constant $c > 0$ such that for each finite collection $x_1,\dots,x_n$ of elements of X there holds the inequality

$$E\left\|\sum_{k=1}^{n} x_k\,\varepsilon_k\right\|_X^p \leq c\sum_{k=1}^{n}\|x_k\|_X^p$$

Here and below E denotes the mathematical expectation,

It is clear that every Banach space has type $p, 0 < p \leq 1$.

A Banach space X is said to be a space of cotype q $2 < q < \infty$, if there exists a constant $c' > 0$ such that for each finite collection $x_1,\dots,x_n$ of elements of X there holds the inequality

$$E\left\|\sum_{k=1}^{n} x_k\ \varepsilon_k\right\|_X^q \geq c'\sum_{k=1}^{n}\|x_k\|_X^q .$$

If X is of type p, $1 < p \leq 2$, then X^* is of cotype $p' = p/p-1$. Denote by l_∞^n the R^n with the maximum-norm. We shall say that a Banach space X contains l_∞^n uniformly if for each $\varepsilon > 0$ and any integer n there exists an injective linear operator $J:\ l_\infty^n \to X$ such that $\|J\|\|J^{-1}\| < 1+\varepsilon$. X does not contain l_∞^n uniformly if and only if it has certain cotype q ([2]).

Let (T, Σ, ϑ) be a positive σ-finite measure space and $\Phi:\ R^+ \to R^+$ denotes a convex continious non-decreasing and vanishing at zero function. For measurable function $x: T \to R$ define

$$\rho_\Phi(x) = \int_T \Phi(|x(t)|)d\,\vartheta(t)$$

and denote $L_\Phi = L_\Phi(T,\Sigma,\vartheta)$ the collection of all measurable functions x with $\rho_\Phi(\lambda x) < \infty$ for some $\lambda > 0$. L_Φ is a vector space. Moreover, L_Φ is Banach space under the norm

$$\|x\|_\Phi = \inf\{\lambda > 0 : \rho_\Phi(x/\lambda) \leq 1\}$$

and this space is said to be Orlicz space.

By Δ_2 we denote the family of functions Φ that satisfy the so-called Δ_2 condition (i.e. $\Phi(2u) \leq c\,\Phi(u)$ for some $c > 0$ and every $u \in R^+$). If $\Phi \in \Delta_2$, then $x \in L_\Phi$ if and only if $\rho_\Phi(x) < \infty$.

Theorem 1. If $\Phi \in \Delta_2$, then the Orlicz space $L_\Phi(T,\Sigma,\vartheta)$ does not contain l^n_∞ uniformly.

Proof. Let $(\gamma_k)_{k\in N}$ be a sequence of independent standard Gaussian random variables. It is sufficient to show that if a series $\sum_k x_k$, $(x_k)_{k\in N} \subset L_\Phi$, unconditionally converges in L_Φ then the sequence $(\|x_k\|_\Phi)_{k\in N}$ can not tend to zero arbitrarily slowly ([3]). Let us show that if the series $\sum_k x_k$ is unconditionally convergent then $\sup_n E\left\|\sum_{k=1}^n x_k\gamma_k\right\|_\Phi < \infty$. From this it follows that $\sup_k \|x_k\gamma_k\|_\Phi < \infty$ a.s. and thus $\sup_k \|x_k\| \log^{1/2}(k+1) < \infty$ (see [4], p. 72).

From the theorem 1 of [1] it follows that if the series $\sum_k x_k$ unconditionally converges in L_Φ then $(\sum_k x_k^2)^{1/2} \in L_\Phi$.
Now

$$E\left\|\sum_{k=1}^n x_k\gamma_k\right\|_\Phi \leq 1 + E\rho_\Phi\left(\sum_{k=1}^n x_k\gamma_k\right) =$$

$$= 1 + \int_T E\Phi\left|\left(\sum_{k=1}^n x_k(t)\gamma_k\right|\right)d\,\vartheta(t) \leq$$

$$\leq 1 + c\int_T \Phi\left(\left(\sum_{k=1}^\infty x_k^2(t)\right)^{1/2}\right)d\vartheta(t) < \infty .$$

Here the inequalities $\|x\|_\Phi \leq 1 + \rho_\Phi(x)$ and $E\,\Phi(|\delta\,\gamma_1| \leq c\,\Phi(|\delta|)$ are used, first of which follows from convexity of Φ and the second one can be easily proved using the Δ_2 condition. This completes the proof.

Note that if $\Phi \notin \Delta_2$ then the theorem 1 is not valid: it is well-known that Orlicz sequence space l_Φ when $\Phi \notin \Delta_2$ contains a subspace which is isomorphic to c_o. We note also that in case of separable L_Φ in the proof of theorem 1 we can also use the result of [5].

Theorem 2. a) Let $2 \le q < \infty$ and the function $u \longrightarrow \Phi(u^{1/q})$ is concave. Then $L_\Phi(T,\Sigma, \theta)$ is the space of cotype q.

b) Let $1 < p \le 2$, the function $u \longrightarrow \Phi(u^{1/p})$ is convex and $\Phi \in \Delta_2$. Then $L_\Phi(T,\Sigma, \theta)$ is the space of type p.

Proof. a) We have to show that if the series $\sum_k x_k \varepsilon_k$, $(x_k)_{k\in N} \subset L_\Phi$ is convergent a.s. then $\sum_k \|x_k\|_\Phi^q < \infty$.

If the series $\sum_k x_k \varepsilon_k$ converges a.s. in L_Φ then $(\sum_k x_k^2)^{1/2} \in L_\Phi$ (see the mentioned above theorem 1 from [1]), i.e.

$$\int_T \Phi(\lambda(\sum_k x_k^2(t))^{1/2}) d\theta(t) < \infty$$

for some $\lambda > 0$. On the other hand we have the inequality

$$\|x\|_\Phi \le 1/\lambda \ (1 + \rho_\Phi(\lambda x)).$$

Let now $(\alpha_k)_{k\in N}$ be a sequence of positive numbers such that $\sum_k \alpha_k^r = 1$, where $r = q/q-1$. Let us show that $\sum_k \alpha_k \|x_k\|_\Phi < \infty$.

We have

$$\lambda \sum_k \alpha_k \|x_k\|_\Phi = \sum_k \|x_k/\alpha_k^{r-1}\| \alpha_k^r \le$$

$$\le \sum_k \alpha_k^r (1 + \int_T \Phi(\lambda |x_k(t)|/\alpha_k^{r-1}) d\theta(t)) =$$

$$= 1 + \int_T \sum_k \alpha_k^r \Phi(\lambda(|x_k(t)|^q/\alpha_k^{q(r-1)}) d\theta(t) \le$$

$$\le 1 + \int_T \Phi(\lambda(\sum_k |x_k(t)|^q)^{1/q}) d\theta(t) \le$$

$$\le 1 + \int_T \Phi(\lambda(\sum_k x_k^2(t))^{1/2}) d\theta(t) < \infty .$$

Here we exploit the concavity of the function $u \longrightarrow \Phi(u^{1/q})$.

b) Let $(x_k)_{k \in N} \subset L_\Phi$ and $\sum \|x_k\|_\Phi^p = 1$. Let us show that the series $\sum_k x_k \gamma_k$ converges a.s. in L_Φ. For this it is sufficient to show that $(\sum_k x_k^2)^{1/2} \in L_\Phi$ (see the above theorem 1 and theorem 3 from [1]). Denote $z_k = x_k/\|x_k\|_\Phi$, $\alpha_k = \|x_k\|_\Phi$. We have

$$\int_T \Phi\left(\left(\sum_k x_k^2(t)^{1/2}\right)d\theta(t) \leq \int_T \Phi\left(\sum_k |x_k(t)|^p\right)^{1/p}\right)d\theta(t) =$$

$$= \int_T \Phi\left(\left(\sum_k \alpha_k^p |z_k(t)|^p\right)^{1/p}\right)d(t) \leq$$

$$\leq \sum_k \alpha_k^p \int_T \Phi(|z_k(t)|)d\theta(t) = \sum \alpha_k \rho_\Phi(z_k) \leq$$

$$\leq \sum_k \alpha_k^p = \sum \|x_k\|_\Phi^p < \infty$$

Here we use the convexity of the function $u \longrightarrow \Phi(u^{1/p})$ and the following property : if $\|x\|_\Phi \leq 1$ then $\rho_\Phi(x) \leq 1$. This concludes the proof.

In the case $\Phi(u) = |u|^r$, $1 \leq r < \infty$, we obtain that $L_r(T,\Sigma,\theta)$ is the space of cotype max $(2,r)$ and type min $(2,r)$

Note that in statement b) of the theorem the condition $\Phi \in \Delta_2$ is necessary. Indeed, if $\Phi(u) = \exp\{u^2\} - 1$ $(\Phi \notin \Delta_2)$ then obviously the function $u \longrightarrow \Phi(u^{1/2})$ is convex but L_Φ is not the space of type 2. Moreover, it contains a subspace isomorphic to c_0.

The functions Φ, Ψ are said to be equivalent $(\sim)$ whenever there are positive constans c_1, c_2, k_1, k_2 such that the inequality $c_1 \Phi(k_1 u) \leq \Psi(u) \leq c_2 \Phi(k_2 u)$ holds.

If $\Phi \sim \Psi$ then $L_\Phi = L_\Psi$ and the norms $\|\cdot\|_\Phi$ and $\|\cdot\|_\Psi$ are equivalent. If the function $u \longrightarrow \Phi(u)/u^p$ is non-decreasing (resp. non-increasing) then there exists Ψ, $\Psi \sim \Phi$, such that the function $u \longrightarrow \Psi(u^{1/p})$ is convex (resp. concave) (cf. [6]). In this contex particularly we see that if $\Phi(u) = u^2 \log(u+1)$ tzen L_Φ has type 2, has cotype q for every $q > 2$ (but does not have cotype 2).

The another conditions under which l_Φ has type p or cotype q are given in [7].

References

[1] S.A. Chobanjan, Z.G. Gorgadze, V.I.Tarieladze, Gaussian covariances in Banach sulattices of L_0. (in Russian) Dokl.AN SSR, 241, 3(1978), 528-531; Soviet Math. Dokl., 19, 4 (1978), 885-888.

[2] B.Maurey, G.Pisier, Caracterization d'une classe d'espaces de Banach par des series aleatoires vectorielles, C.R.Acad. Sci.Paris, 277 (1973), 687-690.

[3] S.A. Rakov, On Banach spaces for which Orlicz's theorem does not hold. (in Russian), Mat.Zamet. 14,1 (1973), 101-106.

[4] N.N. Vakhania, Probability distributions in Banach spaces. (in Russian), Metzniereba, Tbilisi 1971.

[5] Z.G. Gorgadze, V.I. Tarieladze, Gaussian measuries in Orlicz spaces, Soobsc.AN Gruz.SSR 74, 3 (1974) 557-559.

[6] W. Matuszewska, W.Orlicz, On certain properties of φ-functions, Bull.Acad.Polon.Sci.Ser.Math.Astronom.Phys. 8, 7 (1960), 439-443.

[7] T.Figiel, J.Lidenstrauss, V.D. Milman, The dimension of almost spherical sections of convex bodies, Acta Mathematica, 139, 1-2 (1977), 53-94.

Tbilisi State University, Tbilisi, USSR
Academy of Sciences of the Georgian SSR,
Computing Center
Tbilisi, USSR

THE GENERALIZED DOMAIN OF ATTRACTION OF SPHERICALLY SYMMETRIC STABLE LAWS ON $\mathbb{R}^d$

by

Marjorie G. Hahn[1] and Michael J. Klass[2]

1. Introduction.

Let $X, X_1, X_2, \ldots$ be i.i.d. d-dimensional random vectors with law $\mathcal{L}(X)$ and n^{th} partial sum S_n . Following Hahn (1979), we say X is in the generalized domain of attraction (GDOA) of a law $\mathcal{L}(Z)$ if there exist affine transformations T_n such that

$$\mathcal{L}(T_n S_n) \rightarrow \mathcal{L}(Z) \quad .$$

If the affine transformations can be replaced by linear transformations we say $X \in G_L$DOA of $\mathcal{L}(Z)$ and if the linear transformations $T_n = a_n I$ for some constants a_n where I is the identity transformation, then we say $X \in$ DOA of $\mathcal{L}(Z)$.

All possible limit laws $\mathcal{L}(Z)$ corresponding to a Z which is full (equivalently nondegenerate), i.e. the support of Z is not concentrated on any (d-1)-dimensional subspace, have been characterized by Michael Sharpe (1969). They are called the operator-stable laws and are a strictly larger class than the stable laws.

Actually, even for a stable law the GDOA $\supsetneq$ DOA . The following examples illustrate this fact by exhibiting simple but natural situations in which norming by affine transformations rather than constants is essential.

[1,2] Supported in part by NSF Grants MCS-78-02417-A01 and MCS-75-10376-A01 respectively.

Example 1. Let Y_α be a symmetric stable random variable with characteristic function $e^{-|t|^\alpha}$, $0 < \alpha \leq 2$. Let U and V be independent symmetric random variables $\in DOA(\mathcal{L}(Y_\alpha))$ with 1-dimensional norming constants a_n and b_n respectively. If $a_n/b_n \to 0$, then $X = Ue_1 + Ve_2 \notin DOA$ of a full law $\mathcal{L}(Z)$ because norming the partial sums by any constants either causes degeneracy or the nonexistence of a weak limit in some direction. However, if linear transformations T_n are defined by $T_n x = (\langle x,e_1\rangle/a_n)e_1 + (\langle x,e_2\rangle/b_n)e_2$ then $\mathcal{L}(T_n S_n) \to \mathcal{L}(Z_\alpha)$ where $Z_\alpha = Y_\alpha e_1 + Y'_\alpha e_2$ where Y'_α is an independent copy of Y_α.

In the above example Z_α is symmetric stable but not spherically symmetric stable unless $\alpha = 2$. In fact, for $0 < \alpha < 2$, there are no d-dimensional spherically symmetric random vectors with independent marginals. To see this, let θ be a unit vector and note that if $Z = \sum_{i=1}^{d} Y_{\alpha,i} e_i$ where $Y_{\alpha,i}$ are i.i.d. $\mathcal{L}(Y_\alpha)$ then

$$Ee^{it\langle Z,\theta\rangle} = \exp(-|t|^\alpha \sum_{i=1}^{d} |\langle\theta,e_i\rangle|^\alpha)$$

which is not independent of θ unless $\alpha = 2$.

There are situations in which rotations as well as componentwise norming are needed.

Example 2. Let U,V,X and Z be as in Example 1. For $0 < \varphi < \pi/2$, let $\theta_\varphi = (\cos\varphi)e_1 + (\sin\varphi)e_2$. Define

$$X_\varphi = \langle X,\theta_\varphi\rangle e_1 + \langle X,\theta_{\varphi+\pi/2}\rangle e_2 .$$

Componentwise norming the partial sums for X_φ yields a degenerate weak limit concentrated on the line at angle $\varphi - \pi/2$. In order to obtain the full limit law $\mathcal{L}(Z)$, it is necessary to first rotate back $\varphi (\text{mod } \pi/2)$ and then norm componentwise. The appropriate norming operators are thus

$$T_n x = (\langle x, \theta_{-\varphi}\rangle / a_n) e_1 + (\langle x, \theta_{-\varphi + \pi/2}\rangle / b_n) e_2 .$$

Finally, a varying coordinate system may, in fact, be required.

<u>Example 3</u>. Let Z be standard multivariate normal. Choose a random variable $U \in DOA(N(0,1))$ for which there exist a set A and two increasing sequences of integers j_n and j_n' with the properties that

(i) $X = Ue_1 + UI_{(U \in A)} e_2 \in GDOA(Z)$;

(ii) if $a_n(\theta)$ are appropriate 1-dimensional norming constants for $\langle X, \theta\rangle$ then $a_{j_n}(e_1)/a_{j_n}(e_2) \to 1$ while $a_{j_n'}(e_1)/a_{j_n'}(e_2) \to \infty$.

In this case, $T_{j_n'} x = (\langle x, e_1\rangle / a_{j_n'}(e_1)) e_1 + (\langle x, e_2\rangle / a_{j_n'}(e_2)) e_2$ while if $\tau_1 = (e_1 + e_2)/\sqrt{2}$ and $\tau_2 = (-e_1 + e_2)/\sqrt{2}$ then $T_{j_n} x = (\langle x, \tau_1\rangle / a_{j_n}(\tau_1)) e_1 + (\langle x, \tau_2\rangle / a_{j_n}(\tau_2)) e_2$. Notice $a_{j_n}(\tau_2)/a_{j_n}(\tau_1) \to 0$. However, $a_{j_n}(\tau_1) \sim a_{j_n}(e_1)$.

This is discussed in further detail in Example 2 of Hahn and Klass (1980) when $\alpha = 2$.

Example 5 in section 4 illustrates the need to use a varying coordinate system when $0 < \alpha < 2$.

The main purpose of this paper is to prove the following theorem which gives necessary and sufficient conditions for a random vector X to be in the GDOA of a full spherically symmetric random variable Z which, as we will see, must necessarily be spherically symmetric stable.

Theorem 1. Let $X, X_1, X_2, \ldots$ be i.i.d. full d-dimensional random vectors. If $E\|X\| < \infty$, assume $EX = \vec{0}$. Let $S_n = X_1 + \ldots + X_n$. Then there exists a full spherically symmetric random vector Z, linear transformations T_n and vectors v_n such that

$$\mathcal{L}(T_n(S_n - v_n)) \to \mathcal{L}(Z)$$

iff

(I) there exists $0 < \alpha \leq 2$ such that

$$\lim_{t \to \infty} \sup_{\|\theta\|=1} \left| \frac{t^2 P(\langle X,\theta\rangle > t)}{E(\langle X,\theta\rangle^2 \wedge t^2)} - \frac{2-\alpha}{4} \right| = 0 ;$$

(II) there exist orthonormal bases $\{\theta_{n1}, \ldots, \theta_{nd}\}_{n \geq 1}$ such that

$$\lim_{n \to \infty} \sup_{\|\theta\|=1} \left| a_n^2(\theta) / \left(\sum_{j=1}^{d} \langle \theta, \theta_{nj}\rangle^2 a_n^2(\theta_{nj}) \right) - 1 \right| = 0$$

where $a_n(\theta) = \sup\{a : c^{-1} nE(\langle X,\theta\rangle^2 \wedge a^2) \geq a^2\}$ for some $c > 0$;

(III) if $m_n(\theta) \equiv nE\langle X,\theta\rangle I_{(|\langle X,\theta\rangle| \leq a_n(\theta))}$, then

$$\lim_{n \to \infty} \sup_{\|\theta\|=1} \left| m_n(\theta) - \sum_{j=1}^{d} \langle \theta, \theta_{nj}\rangle m_n(\theta_{nj}) \right| / a_n(\theta) = 0 .$$

Condition (III) is implied by (I) and (II) when $0 < \alpha \leq 2$ and $\alpha \neq 1$.

Moreover, whenever (I) - (III) hold, Z is symmetric stable of index α . Thus for all unit vectors θ , $Ee^{it\langle Z,\theta\rangle} = e^{\tilde{c}|t|^{\alpha}}$ where

$$\tilde{c} = \begin{cases} \frac{c}{2}\frac{\Gamma(3-\alpha)}{\alpha-1}\cos\frac{\pi\alpha}{2} & \text{if } 0 < \alpha < 1 \text{ or } 1 < \alpha < 2 \\ \frac{-c\pi}{4} & \text{if } \alpha = 1 \end{cases}$$

for some $c > 0$. When the constant c determining $\tilde{c}$ is the same as that in $a_n(\theta)$, the linear transformations T_n and vectors v_n may be chosen to satisfy

$$\text{(IV)} \qquad \begin{aligned} T_n\theta_{nj} &= \theta_{nj}/a_n(\theta_{nj}) \\ \langle v_n,\theta_{nj}\rangle &= m_n(\theta_{nj}) \end{aligned}$$

for $j = 1,\ldots,d$.

Remark 1. If (II) holds for some $c > 0$ then it holds for all $c > 0$ and (III) remains valid in this context as well (see (12) and (19)). When Z is the standard multivariate normal these conditions are equivalent to those obtained in Hahn and Klass (1980) since (I) and $\alpha = 2$ imply (II) and (III). However, when $0 < \alpha < 2$, new and interesting complications arise. Besides a condition on the ratios of tail probabilities to "minimized second moments", a certain regularity of the norming constants is required. See Example 4 in section 4.

As noted in the theorem, the centering condition is automatically satisfied if $0 < \alpha \leq 2$ and $\alpha \neq 1$.

2. Necessary and sufficient conditions for norming sequences of random vectors by linear transformations.

The proof of Theorem 1 is our main goal. However, we first require some basic results concerning the weak convergence of sequences of linear transformations applied to sequences of d-dimensional random vectors. In this section we gather these more general results before specializing to partial sum sequences in the next section.

As a preliminary, recall that the Prohorov distance ρ between two probability measures P and Q in $\mathbb{R}^d$ is defined by

$$\rho(P,Q) = \inf\{\epsilon > 0: P(F) \leq Q(F^\epsilon) + \epsilon, \ \forall\, F \text{ closed}\}$$

$$= \inf\{\epsilon > 0: Q(F) \leq P(F^\epsilon) + \epsilon, \ \forall\, F \text{ closed}\}$$

where $F^\epsilon = \{x \in \mathbb{R}^d : \inf_{y \in F} ||x - y|| < \epsilon\}$.

This metric metrizes weak convergence.

Our first lemma shows that weak convergence in $\mathbb{R}^d$ is equivalent to weak convergence of all 1-dimensional projections at a uniform rate.

<u>Lemma 1</u>. Let Z_n , $n \in \mathbb{N}$ and Z be d-dimensional random vectors. Then

$$\mathcal{L}(Z_n) \to \mathcal{L}(Z) \quad \text{iff} \quad \lim_{n\to\infty} \sup_{||\theta||=1} \rho(\mathcal{L}(\langle Z_n,\theta\rangle), \mathcal{L}(\langle Z,\theta\rangle)) = 0 \quad .$$

<u>Proof</u>: Sufficiency is an immediate consequence of the Cramer-Wold device (see Billingsley ((1968), Theorem 7.7, p. 49)).

For necessity we appeal to a consequence of a lemma due to Rao, namely, if $\mathcal{L}(Z_n) \to \mathcal{L}(Z)$ then

$$(1)\qquad \lim_{n\to\infty} \sup_{\|\theta\|=1} \left| E\, e^{it\langle Z_n\theta\rangle} - E\, e^{it\langle Z,\theta\rangle} \right| = 0$$

for any $t \in \mathbb{R}$, (see Billingsley ((1968), Problem 8, p. 17) for Rao's lemma and Hahn and Klass (1978), §2 for the above consequence). Now suppose there exist a $\delta > 0$, integers $n_k \to \infty$, and unit vectors ψ_{n_k} such that

$$(2)\qquad \rho(\mathcal{L}(\langle Z_{n_k}, \psi_{n_k}\rangle), \mathcal{L}(\langle Z, \psi_{n_k}\rangle)) \geq \delta .$$

Since the unit ball is compact, there is also a unit vector ψ and a subsequence n_k' of n_k such that $\psi_{n_k'} \to \psi$. In view of (1) we observe that

$$E\, e^{it\langle Z_{n_k'}, \psi_{n_k'}\rangle} \to E\, e^{it\langle Z,\psi\rangle} .$$

Therefore

$$\lim_{n\to\infty} \rho(\mathcal{L}(\langle Z_{n_k'}, \psi_{n_k'}\rangle), \mathcal{L}(\langle Z,\psi\rangle)) = 0 .$$

Since we also have

$$\lim_{n\to\infty} \rho(\mathcal{L}(\langle Z, \psi_{n_k'}\rangle), \mathcal{L}(\langle Z,\psi\rangle)) = 0 ,$$

the triangle inequality yields

$$\lim_{n\to\infty} \rho(\mathcal{L}(\langle Z_{n_k'}, \psi_{n_k'}\rangle), \mathcal{L}(\langle Z,\psi_{n_k'}\rangle)) = 0$$

which contradicts (2). ///

Throughout the sequel we assume that $\mathcal{L}(Z)$ is full on $\mathbb{R}^d$. Given a sequence of d-dimensional random vectors V_n, we are ultimately interested in when there exist affine transformations T_n such that $\mathcal{L}(T_n V_n) \to \mathcal{L}(Z)$. For the moment, however, we will confine our attention to T_n linear. Lemma 1 suggests that V_n must satisfy a number of 1-dimensional conditions which we proceed to identify.

Lemma 2. Let V and W be d-dimensional random vectors. Let Q be an invertible linear transformation on $\mathbb{R}^d$ with adjoint Q^* whose inverse is $(Q^*)^{-1}$. Then, for any unitary operator U on $\mathbb{R}^d$,

$$(3) \quad \sup_{\|\theta\|=1} \rho(\mathcal{L}(\langle UQV,\theta\rangle),\mathcal{L}(\langle W,\theta\rangle))$$

$$= \sup_{\|\theta\|=1} \rho(\mathcal{L}(\langle V,\theta\rangle/\|(Q^*)^{-1}\theta\|),\mathcal{L}(\langle W,U(Q^*)^{-1}\theta/\|(Q^*)^{-1}\theta\|\rangle)) .$$

Proof. Note that Q is invertible iff Q^* is invertible. Suppose the lemma holds for $U = I$, the identity. Then since $\|((UQ)^*)^{-1}\theta\| = \|U((Q^*)^{-1}\theta)\| = \|(Q^*)^{-1}\theta\|$, the lemma holds for arbitrary unitary operators. Hence, we may assume $U = I$ and put $b(\theta) = \|(Q^*)^{-1}\theta\|$.

Observe that $\{\theta: \|\theta\| = 1\} = \{(Q^*)^{-1}\theta/b(\theta): \|\theta\| = 1\}$ and also $\langle QV,(Q^*)^{-1}\theta\rangle = \langle V,Q^*(Q^*)^{-1}\theta\rangle = \langle V,\theta\rangle$. Therefore,

$$\sup_{\|\theta\|=1} \rho(\mathcal{L}(\langle QV,\theta\rangle),\mathcal{L}(\langle W,\theta\rangle))$$

$$= \sup_{\|\theta\|=1} \rho(\mathcal{L}(\langle QV,(Q^*)^{-1}\theta\rangle/b(\theta)),\mathcal{L}(\langle W,(Q^*)^{-1}\theta/b(\theta)\rangle))$$

$$= \sup_{\|\theta\|=1} \rho(\mathcal{L}(\langle V,\theta\rangle/b(\theta)),\mathcal{L}(\langle W,(Q^*)^{-1}\theta/b(\theta)\rangle)) .$$

///

If W is spherically symmetric, Lemma 2 acquires a particularly nice form due to the fact that $U(Q^*)^{-1}\theta/\|(Q^*)^{-1}\theta\|$ is a unit vector.

Corollary 1. If W is spherically symmetric then

$$(4)\qquad \sup_{\|\theta\|=1} \rho(\mathcal{L}(\langle UQV,\theta\rangle),\mathcal{L}(\langle W,\theta\rangle)) = \sup_{\|\theta\|=1} \rho(\mathcal{L}(\langle V,\theta\rangle/b(\theta)),\mathcal{L}(\langle W,\theta\rangle))$$

where

$$(5)\qquad b(\theta) = \|(Q^*)^{-1}\theta\| .$$

Glancing ahead, when V is replaced by a partial sum S_n and UQ by a linear operator $T_n = U_nQ_n$, the function $b_n(\theta) = \|(Q_n^*)^{-1}\theta\|$ may be thought of as a 1-dimensional norming constant for the random variable $\langle S_n,\theta\rangle$. Thus, it is particularly important to notice that $b(\theta)$ assumes the following characteristic form.

Lemma 3. Let L be an invertible linear operator on $\mathbb{R}^d$. Let $b(\theta) = \|L\theta\|$. Then there exists an orthonormal basis $\theta_1,\ldots,\theta_d$ for $\mathbb{R}^d$ such that

$$(6)\qquad b^2(\theta) = \sum_{j=1}^{d} \langle\theta,\theta_j\rangle^2 b^2(\theta_j) .$$

Conversely, if $\theta_1,\ldots,\theta_d$ is an orthonormal basis for $\mathbb{R}^d$ and L is a linear transformation determined by the relations $L\theta_j = b_j\theta_j$, for $1 \le j \le d$, then $\|L\theta\|^2 = \sum_{j=1}^{d} \langle \theta, \theta_j\rangle^2 b_j^2$.

Proof. Assume $b(\theta) = \|L(\theta)\| = (\langle L^*L\theta,\theta\rangle)^{1/2}$. Since $B = L^*L$ is self-adjoint, the Spectral theorem implies the existence of an orthonormal basis $\theta_1,\ldots,\theta_d$ such that $B\theta = \sum_{j=1}^{d} \lambda_j\langle\theta,\theta_j\rangle\theta_j$. Hence $b^2(\theta) = \langle B\theta,\theta\rangle = \sum_{j=1}^{d} \lambda_j\langle\theta,\theta_j\rangle^2$. Putting $\theta = \theta_j$ identifies λ_j as $b^2(\theta_j)$.

The converse is a straight-forward calculation. ///

As a final preliminary we recall the polar decomposition theorem for invertible linear transformations on $\mathbb{R}^d$ (see e.g. Halmos (1958), p. 169).

Polar Decomposition. Let L be an invertible linear transformation on $\mathbb{R}^d$. Then there exist an orthonormal basis $\{\theta_1,\ldots,\theta_n\}$, a transformation D which is diagonal in this basis and a unitary transformation U such that $L = U \bullet D$.

For example let $D = \sqrt{L^*L}$ so that $D\theta_j = \|L\theta_j\|\theta_j$ and $U^{-1} = D \bullet L^{-1}$ so that $U\theta_j = L\theta_j/\|L\theta_j\|$, for $1 \leq j \leq d$.

Synthesizing the above preliminary results we obtain a characterization of the feasibility of matrix norming an arbitrary sequence of random vectors to get a full limit distribution.

Theorem 2. Let Z , V_1 , V_2 ,... be d-dimensional random vectors. Assume Z is full. Suppose there are linear operators T_n on $\mathbb{R}^d$ such that $\mathcal{L}(T_nV_n) \to \mathcal{L}(Z)$. Then there exist sequences of orthonormal bases $\{\theta_{n1},\ldots,\theta_{nd}\}_{n\geq 1}$, positive constants $\{b_{n1},\ldots,b_{nd}\}_{n\geq 1}$, unitary transformations U_n and diagonal transformations D_n such that $T_n = U_n \bullet D_n$,

$$D_n^{-1}\theta_{nj} = b_{nj}\theta_{nj}$$

and

(7) $$\lim_{n\to\infty} \sup_{\|\theta\|=1} \rho(\mathcal{L}(\langle V_n,\theta\rangle/b_n(\theta)),\mathcal{L}(\langle Z,U_nD_n^{-1}\theta/\|D_n^{-1}\theta\|\rangle)) = 0 ,$$

where $b_n(\theta) = (\sum_{j=1}^{d} \langle\theta,\theta_{nj}\rangle^2 b_{nj}^2)^{1/2}$.

Conversely, if such orthonormal bases, positive constants, and diagonal and unitary transformations exist then, letting $T_n = U_n \bullet D_n$,

$$\mathcal{L}(T_nV_n) \rightarrow \mathcal{L}(Z) .$$

Proof. ($\Rightarrow$) Since Z is full, the image of T_n cannot be contained in any $d-1$ dimensional subspace of $\mathbb{R}^d$ if n is sufficiently large. Hence there exists n_o such that T_n is invertible for $n \geq n_o$. For simplicity we take $n_o = 1$.

The polar decomposition implies the existence of orthonormal bases $\{\theta_{n1},\ldots,\theta_{nd}\}_{n\geq 1}$, positive constants $c_{n1},\ldots,c_{nd}$, diagonal transformations D_n and unitary transformations U_n such that $T_n = U_n \bullet D_n$ and $D_n\theta_{nj} = c_{nj}\theta_{nj}$. Define $b_{nj} = 1/c_{nj}$ and notice that $D_n = D_n^*$. If $b_n(\theta) \equiv \|D_n^{-1}\theta\| = (\sum_{j=1}^{d} \langle\theta,\theta_{nj}\rangle^2 b_{nj}^2)^{1/2}$ then Lemmas 2 and 3 imply that

$$\lim_{n\to\infty} \sup_{\|\theta\|=1} \rho(\mathcal{L}(\langle V_n,\theta\rangle/b_n(\theta)),\mathcal{L}(\langle Z,U_nD_n^{-1}\theta/\|D_n^{-1}\theta\|\rangle))$$

$$= \lim_{n\to\infty} \sup_{\|\theta\|=1} \rho(\mathcal{L}(\langle T_nV_n,\theta\rangle),\mathcal{L}(\langle Z,\theta\rangle)) = 0 \qquad \text{by Lemma 1.}$$

The converse direction is a direct consequence of Lemma 2.

///

Remark 2. If f_n is a sequence of functions then

$$\lim_{n\to\infty} \sup_{\|\theta\|=1} |f_n(\theta)| = 0 \text{ iff } \lim_{n\to\infty} |f_n(\psi_n)| = 0 \text{ for every sequence}$$

of unit vectors ψ_n. Consequently, a condition such as (7) can be verified by considering the asymptotic behavior along all sequences of unit vectors rather than using the supremum over the unit sphere. We will use this fact repeatedly.

Corollary 2. Suppose Z is spherically symmetric. Then there exist linear transformations T_n such that $\mathcal{L}(T_nV_n) \to \mathcal{L}(Z)$ iff there exist sequences of orthonormal bases $\{\theta_{n1},\ldots,\theta_{nd}\}_{n\geq 1}$ and positive constants $\{b_{n1},\ldots,b_{nd}\}_{n\geq 1}$ such that

$$(8) \qquad \lim_{n\to\infty} \sup_{\|\theta\|=1} \rho(\mathcal{L}(\langle V_n,\theta\rangle/b_n(\theta)),\mathcal{L}(\langle Z,\theta\rangle)) = 0$$

where

$$(9) \qquad b_n(\theta) = \Big(\sum_{j=1}^{d} \langle\theta,\theta_{nj}\rangle^2 b_{nj}^2\Big)^{1/2} .$$

Proof: For necessity let U_n, D_n, $\{\theta_{n1},\ldots,\theta_{nd}\}_{n\geq 1}$, and $\{b_{n1},\ldots,b_{nd}\}_{n\geq 1}$ be as in Theorem 2. Then (8) follows immediately from (7), the spherical symmetry of Z and the fact that $U_nD_n^{-1}\theta/\|D_n^{-1}\theta\|$ is a unit vector.

Sufficiency follows from Theorem 2 upon defining $D_n\theta_{nj} = \theta_{nj}/b_{nj}$ and $U_n = I$. In this case $T_n = D_n$. ///

Remark 3. The procedure for constructing the norming linear transformations for the sequence V_n when Z is spherically symmetric can be paraphrased as follows: At stage n, a preferred orthonormal basis is selected for the random vector V_n and then componentwise norming occurs along that orthonormal basis. This is precisely what D_n does. Furthermore, convergence is still preserved if unitaries are composed with the D_n .

3. Application to sums of i.i.d. random vectors.

Thus far we have obtained a general characterization of the feasibility of operator norming a sequence of random vectors V_n to obtain a limit distribution Z . We are now ready to consider the problem of characterizing the GDOA of a spherically symmetric operator-stable law, with Theorem 1 as our goal.

Problem: Suppose $X, X_1, X_2, \ldots$ are i.i.d. d-dimensional random vectors with partial sums $S_n = X_1 + \ldots + X_n$. Let Z be a spherically symmetric full random vector on $\mathbb{R}^d$. Find necessary and sufficient conditions for the existence of vectors v_n and linear transformations T_n such that

$$\mathcal{L}(T_n(S_n - v_n)) \to \mathcal{L}(Z) \quad .$$

Since Z is spherically symmetric, let Y be a random variable with $\mathcal{L}(Y) = \mathcal{L}(\langle Z,\theta\rangle)$ for all unit vectors θ . By Corollary 2, such T_n and v_n exist iff

(10) there are functions $b_n(\theta)$ of the special form (9) such that for any sequence of unit vectors ϕ_n ,

$$\mathcal{L}(\langle S_n - v_n, \phi_n\rangle / b_n(\phi_n)) \to \mathcal{L}(Y) \quad .$$

For $\phi_n \equiv \theta$, classical theory shows Y must be stable of index α for some $0 < \alpha \leq 2$. Replacing θ by $-\theta$, it follows that Y is symmetric stable. Thus, the only spherically symmetric d-dimensional operator-stable distributions are the symmetric stables. We will write the 1-dimensional symmetric stable of index α as Y_α to identify the index (characteristic exponent).

The characteristic function of Y_α is of the form $e^{\tilde{c}|t|^\alpha}$ where there is a positive constant c such that

$$\tilde{c} = \begin{cases} \frac{c}{2}\frac{\Gamma(3-\alpha)}{\alpha-1}\cos\frac{\pi\alpha}{2} & \text{if } 0<\alpha<1 \text{ or } 1<\alpha\leq 2 \\ \frac{-c\pi}{4} & \text{if } \alpha = 1 \ . \end{cases}$$

(Note: our c is $2/\alpha$ times the one appearing in Feller (1971) p. 570.)

The following necessary and sufficient conditions for (10) are an immediate consequence of applying 1-dimensional triangular array theorems to $X_{nj\theta_n} = \langle X_j,\theta_n\rangle / b_n(\theta_n)$. (See, for example, Gnedenko and Kolmogorov (1968), p. 116, 84).

Lemma 4. For any sequence of unit vectors θ_n

$$\mathcal{L}(\langle S_n - v_n, \theta_n\rangle / b_n(\theta_n)) \rightarrow \mathcal{L}(Y_\alpha)$$

iff for any $\epsilon > 0$,

(11) (i) $$\lim_{n\to\infty} nP(\langle X,\theta_n\rangle > \epsilon b_n(\theta_n)) = \frac{c}{4}(2-\alpha)\epsilon^{-\alpha}$$

$$\lim_{n\to\infty} nP(\langle X,\theta_n\rangle < -\epsilon b_n(\theta_n)) = \frac{c}{4}(2-\alpha)\epsilon^{-\alpha} \ ;$$

(ii) $$\lim_{\epsilon\downarrow 0^+}\ \limsup_{n\to\infty}\left|\frac{n}{b_n^2(\theta_n)}\operatorname{Var}\langle X,\theta_n\rangle I_{(|\langle X,\theta_n\rangle|\leq\epsilon b_n(\theta_n))} - \eta\right| = 0$$

$$\text{where } \eta = \begin{cases} c & \text{if } \alpha = 2 \\ 0 & \text{if } 0<\alpha<2 \end{cases} ;$$

(iii) $$\lim_{n\to\infty} nE(\langle X,\psi_n\rangle/b_n(\psi_n))I_{(|\langle X,\psi_n\rangle|\le b_n(\psi_n))} - (\langle v_n,\psi_n\rangle/b_n(\psi_n)) = 0 .$$

The validity of Lemma 4 is independent of whether or not $b_n(\cdot)$ assumes the special form of (9).

Remark 4. Due to the monotonicity of $P(\langle X,\theta\rangle > y)$ and $P(\langle X,\theta\rangle < -y)$ in y for each θ, (11)(i) is equivalent to

(11)(i') For every $0 < \underline{\varepsilon} \le \varepsilon_n \le \bar{\varepsilon} < \infty$,

$$\lim_{n\to\infty} |nP(\langle X,\psi_n\rangle > \varepsilon_n b_n(\psi_n)) - \tfrac{c}{4}(2-\alpha)\varepsilon_n^{-\alpha}| = 0$$

$$\lim_{n\to\infty} |nP(\langle X,\psi_n\rangle < -\varepsilon_n b_n(\psi_n)) - \tfrac{c}{4}(2-\alpha)\varepsilon_n^{-\alpha}| = 0 .$$

In order to proceed we require additional information on the $b_n(\theta)$ appearing in the last lemma.

Lemma 5. Let X and $b_n(\theta)$ be as in Lemma 4. Then

(12) $$\lim_{n\to\infty} \sup_{\|\theta\|=1} \sup_{1\le\gamma\le 2} \left|\frac{b_{[n\gamma]}(\theta)}{b_n(\theta)} - \gamma^{1/\alpha}\right| = 0$$

where $[r]$ denotes the integer part of r;

(13) for any $\delta < 1/\alpha$, $$\lim_{n\to\infty} \inf_{\|\theta\|=1} n^{-\delta} b_n(\theta) = \infty .$$

Proof. (12): Using ψ_n at "time" $2n$ in (11)(i),

$$\lim_{n\to\infty} 2nP(\langle X,\psi_n\rangle > \varepsilon b_{2n}(\psi_n)) = \tfrac{c}{4}(2-\alpha)\varepsilon^{-\alpha}$$

or, equivalently,

$$\lim_{n\to\infty} nP(\langle X,\psi_n\rangle > \epsilon b_{2n}(\psi_n)) = \frac{c}{4}(2-\alpha)(2^{1/\alpha}\epsilon)^{-\alpha} .$$

Since $\lim_{n\to\infty} nP(\langle X,\psi_n\rangle > 2^{1/\alpha}\epsilon b_n(\psi_n)) = \frac{c}{4}(2-\alpha)(2^{1/\alpha}\epsilon)^{-\alpha}$, it follows from (11)(i') that $\lim_{n\to\infty} b_{2n}(\psi_n)/b_n(\psi_n) = 2^{1/\alpha}$. A slight refinement of this argument yields (12).

<u>(13)</u>: If (13) fails, there exists $\delta < 1/\alpha$, a subsequence $n' \uparrow \infty$, and unit vectors $\psi_{n'}$ such that

$$\lim_{n'\to\infty} (n')^{-\delta} b_{n'}(\psi_{n'}) < \infty .$$

Take $2^{\delta - 1/\alpha} < q < 1$. According to (12), there exists $n_o \geq 1$ such that for any $1 \leq j/n \leq 2$, $n \geq n_o$ and unit vector θ,

$$b_j(\theta)/b_n(\theta) \geq q(j/n)^{1/\alpha} .$$

Let $n' \geq n \geq n_o$ and let $k \geq 0$ satisfy $n2^k \leq n' < n2^{k+1}$. Then

$$\begin{aligned} b_{n'}(\theta)/b_n(\theta) &= (b_{n'}(\theta)/b_{2^k n}(\theta)) \prod_{i=1}^{k} (b_{2^i n}(\theta)/b_{2^{i-1}n}(\theta)) \\ &\geq q(n'/2^k n)^{1/\alpha}(q2^{1/\alpha})^k . \end{aligned}$$

(Thus, for each fixed θ, $b_{n'}(\theta) \to \infty$.) Hence,

$$\begin{aligned} (n')^{-\delta} b_{n'}(\psi_{n'}) &\geq b_{n_o}(\psi_{n'}) n_o^{-1/\alpha} q^{k+1} (n')^{-\delta + 1/\alpha} \\ &> b_{n_o}(\psi_{n'}) n_o^{-1/\alpha} (n'/n)^{\delta' - 1/\alpha} (n')^{-\delta + 1/\alpha} , \end{aligned}$$

for some $\delta' > \delta$.

The right side tends to infinity provided $\liminf_{n' \to \infty} b_{n_o}(\psi_{n'}) > 0$.

Thus to obtain a contradiction, it suffices to show that for all n sufficiently large, $\inf_{\|\theta\|=1} b_n(\theta) > 0$. If in fact this fails to hold, there is a subsequence $n' \to \infty$ and unit vectors $\psi_{n'}$ such that $b_{n'}(\psi_{n'}) \to 0$. By extracting a further subsequence we may suppose $\psi_{n'}$ converges to some ψ. By weak convergence, and (11)(i),

$$0 < P(\langle X,\psi\rangle > 0) \leq \liminf_{n \to \infty} P(\langle X,\psi_{n'}\rangle > b_{n'}(\psi_{n'})) = 0 ,$$

which gives a contradiction. Hence n_o may be chosen so large that $\inf_{\|\theta\|=1} b_n(\theta) > 0$, verifying (13). ///

One of the consequences of (13) is that the squares of truncated first moments are of lower order than truncated second moments in (11)(ii). Consequently, the variance condition implies a useful "minimized second moment" condition.

<u>Lemma 6</u>. If (11)(i) and (11)(ii) hold then

$$(14) \quad \lim_{\epsilon \downarrow 0^+} \limsup_{n \to \infty} \sup_{\|\theta\|=1} \left| nE((\langle X,\theta\rangle/b_n(\theta))^2 \wedge \epsilon^2) - \eta \right| = 0$$

$$\text{where } \eta = \begin{cases} c & \text{if } \alpha = 2 \\ 0 & \text{if } 0 < \alpha < 2 \end{cases} .$$

<u>Proof</u>. (14) is an immediate consequence of (11)(i), (11)(ii) and the following fact, to be verified below:

$$(15) \quad \lim_{n \to \infty} \sup_{\|\theta\|=1} (E\langle X,\theta\rangle I_{(|\langle X,\theta\rangle| \leq \epsilon b_n(\theta))})^2 / E\langle X,\theta\rangle^2 I_{(|\langle X,\theta\rangle| \leq \epsilon b_n(\theta))} = 0.$$

For $\alpha = 2$, (15) is a consequence of $EX = \vec{0}$, and the Equivalence lemma of Hahn and Klass (1980). Now suppose $0 < \alpha < 2$. There exist $t_n \to \infty$ such that $t_n = o(\inf_{||\theta||=1} n^{-1/2} b_n(\theta))$.

If $p_n = P(||X|| > t_n)$ then $p_n \to 0$. Now

$$(E\langle X,\theta\rangle I_{(|\langle X,\theta\rangle| \le \epsilon b_n(\theta))})^2$$

$$\le 2t_n^2 + 2(E\langle X,\theta\rangle I_{(t_n < |\langle X,\theta\rangle| \le \epsilon b_n(\theta))})^2$$

$$\le 2t_n^2 + 2P(t_n < |\langle X,\theta\rangle| \le \epsilon b_n(\theta)) E\langle X,\theta\rangle^2 I_{(t_n < |\langle X,\theta\rangle| \le \epsilon b_n(\theta))}$$

(by Cauchy-Schwarz)

$$\le 2t_n^2 + 2p_n E\langle X,\theta\rangle^2 I_{(|\langle X,\theta\rangle| \le \epsilon b_n(\theta))} .$$

Using (11)(i), there exists $\delta > 0$ (depending on ϵ) and n_o such that for $n \ge n_o$ and for all unit vectors θ,

$$\delta b_n^2(\theta)/n \le E\langle X,\theta\rangle^2 I_{(|\langle X,\theta\rangle| \le \epsilon b_n(\theta))} .$$

Then since $t_n^2 = o(\inf_{||\theta||=1} b_n^2(\theta)/n)$, (15) holds. ///

Remark 5. This lemma allows the $b_n(\theta)$ to be replaced by canonical norming constants $a_n(\theta)$ defined as follows: let $a_y(\theta)$ be the largest real satisfying the implicit relation

$$(16) \qquad a_y^2(\theta) = c^{-1} y E(\langle X,\theta\rangle^2 \wedge a_y^2(\theta)) .$$

To see that this replacement can be made recall that for any random variable Y,

$$(17) \qquad E(Y^2 \wedge t^2) = 2\int_0^t yP(|Y| > y)dy .$$

In view of (11)(i'), (14), and (17) we may conclude that for any $0 < \underline{\varepsilon} \le \bar{\varepsilon} < \infty$,

$$(18) \quad \lim_{n\to\infty} \sup_{\substack{\|\theta\|=1 \\ \underline{\varepsilon} \le \varepsilon_{n\theta} \le \bar{\varepsilon}}} |nE((\langle X,\theta\rangle/b_n(\theta))^2 \wedge \varepsilon_{n\theta}^2) - c\varepsilon_{n\theta}^{2-\alpha}| = 0 .$$

Finally, we observe from (16) and (18) that

$$(19) \quad \lim_{n\to\infty} \sup_{\|\theta\|=1} |a_n(\theta)/b_n(\theta) - 1| = 0 .$$

Gathering our results thus far, we can prove Theorem 1.

Proof of necessity in Theorem 1.

Suppose the linear transformations T_n and vectors v_n of the theorem exist. Then (10) holds and implies (11)(i)-(iii). In order to obtain (I), take any $t_n \to \infty$ and unit vectors θ_n. Let

$$k_n = \max\{k: b_n(\theta_n) \le t_n\} .$$

According to (12), $\varepsilon_n \equiv t_n/b_{k_n}(\theta_n) \to 1$. Hence, using (18) and then (11)(i'),

$$\frac{t_n^2 P(\langle X,\theta_n\rangle > t_n)}{E(\langle X,\theta_n\rangle^2 \wedge t_n^2)} = \frac{k_n \varepsilon_n^2 P(\langle X,\theta_n\rangle > \varepsilon_n b_{k_n}(\theta_n))}{k_n E((\langle X,\theta_n\rangle/b_{k_n}(\theta_n))^2 \wedge \varepsilon_n^2)}$$

$$\sim c^{-1}\varepsilon_n^{\alpha} k_n P(\langle X,\theta_n\rangle > \varepsilon_n b_{k_n}(\theta_n))$$

$$\to (2-\alpha)/4 .$$

Consequently, (I) holds.

Recalling (7) and (10) there exist orthonormal bases $\{\theta_{n1},\ldots,\theta_{nd}\}_{n\geq 1}$ and positive constants $b_{nj} = b_n(\theta_{nj})$ such that (19) holds with $b_n^2(\theta) = \sum_{j=1}^{d} <\theta,\theta_{nj}>^2 b_{nj}^2 = \|(T_n^*)^{-1}\theta\|$ where $a_n(\theta)$ is defined by (16) for some $c > 0$. Thus Condition (II) holds.

To prove (III), note that by (11)(iii),

$$\lim_{n\to\infty} \sup_{\|\theta\|=1} |<v_n,\theta> - m_n(\theta)|/a_n(\theta) = 0 .$$

Therefore,

$$\begin{aligned} <v_n,\theta> &= \sum_{j=1}^{d} <v_n,\theta_{nj}><\theta,\theta_{nj}> \\ &= \sum_{j=1}^{d} m_n(\theta_{nj})<\theta,\theta_{nj}> + \sum_{j=1}^{d} o(a_n(\theta_{nj}))<\theta,\theta_{nj}> . \end{aligned}$$

By Cauchy-Schwarz,

$$\left|\sum_{j=1}^{d} a_n(\theta_{nj})<\theta,\theta_{nj}>\right| = \sqrt{d}\left(\sum_{j=1}^{d} a_n^2(\theta_{nj})<\theta,\theta_{nj}>^2\right)^{\frac{1}{2}} \sim \sqrt{d}\, a_n(\theta)$$

for n large, independent of θ. Thus (III) holds. ///

The proof of sufficiency must be separated into two cases.

Proof of sufficiency in Theorem 1.

Case 1. $(\alpha = 2)$ If condition (I) holds then, by Hahn and Klass (1980), there exist linear transformations T_n such that $\mathcal{L}(T_n(S_n - v_n)) \to N(\vec{0},I)$, with $v_n = 0$. Moreover, the T_n have a special form. In particular, there exists an orthonormal basis $\theta_{n1},\ldots,\theta_{nd}$ such that, letting $a_n(\theta)$ satisfy (16) with $c = 1$, $a_n(\theta_{nj}) = \|(T_n^*)^{-1}\theta_{nj}\|$. Therefore, letting $b_n(\theta) = \|(T_n^*)^{-1}\theta\|$ and using (19) we find that condition (II) holds. Furthermore, by the Equivalence lemma of Hahn and Klass (1980), $\lim_{n\to\infty} \sup_{\|\theta\|=1} \|m_n(\theta)/a_n(\theta)\| = 0$. Consequently (III) holds. Observe that (II) and (III) are automatic consequences of (I).

Case 2. ($0 < \alpha < 2$). Fix $0 < \alpha < 2$ and let X be a d-dimensional random vector satisfying conditions (I) - (III). Let T_n and v_n be as defined in (IV). By (II) the $a_n(\theta)$ defined in (16) satisfy (9) as required by (10) with $a_n(\theta)$ replacing $b_n(\theta)$. Thus, according to (10), sufficiency reduces to verifying (11)(i)-(iii) with $a_n(\theta)$ replacing $b_n(\theta)$.

Let $R(t,\theta) = t^2 P(|\langle X,\theta\rangle| > t)/E(\langle X,\theta\rangle^2 \wedge t^2)$. Note that for each fixed θ, $\log(E(\langle X,\theta\rangle^2 \wedge t^2))$ is differentiable for all t's which are not atoms of the distribution of $|\langle X,\theta\rangle|$, (see (17)). It is therefore absolutely continuous, with

$$\frac{\partial}{\partial t} \log E(\langle X,\theta\rangle^2 \wedge t^2) = 2t^{-1}R(t,\theta) \quad \text{a.s.}$$

Integrating this for t between $a_n(\theta)$ and $\varepsilon a_n(\theta)$ and using (I),

$$\lim_{n\to\infty} \sup_{\|\theta\|=1} \left| \frac{E(\langle X,\theta\rangle^2 \wedge (\varepsilon a_n(\theta))^2)}{E(\langle X,\theta\rangle^2 \wedge a_n^2(\theta))} - \varepsilon^{2-\alpha} \right| = 0 . \tag{20}$$

Hence, for any sequence of unit vectors ψ_n,

$$\begin{aligned}\lim_{n\to\infty} nP(|\langle X,\psi_n\rangle| > \varepsilon a_n(\psi_n)) &= \lim_{n\to\infty} \frac{n(\varepsilon a_n(\psi_n))^2 P(|\langle X,\psi_n\rangle| > \varepsilon a_n(\psi_n))}{\varepsilon^2 c^{-1} nE(\langle X,\psi_n\rangle^2 \wedge a_n^2(\psi_n))} \\ &= \lim_{n\to\infty} \frac{(\varepsilon a_n(\psi_n))^2 P(|\langle X,\psi_n\rangle| > \varepsilon a_n(\psi_n))}{c^{-1}\varepsilon^{\alpha} E(\langle X,\psi_n\rangle^2 \wedge (\varepsilon a_n(\psi_n))^2)} \\ &= \frac{c}{2}(2-\alpha)\varepsilon^{-\alpha} .\end{aligned}$$

By (I), $P(\langle X,\psi_n\rangle > \varepsilon a_n(\psi_n)) \sim P(\langle X,-\psi_n\rangle > \varepsilon a_n(\psi_n))$ $= P(\langle X,\psi_n\rangle < -\varepsilon a_n(\psi_n))$, therefore (11)(i) follows.

Using (20) and the definition of $a_n(\theta)$,

(21) $$\lim_{\varepsilon\downarrow 0^+} \limsup_{n\to\infty} \sup_{\|\theta\|=1} nE((\langle X,\theta\rangle/a_n(\theta))^2 \wedge \varepsilon^2) = 0 .$$

Therefore (11)(ii) holds.

Finally (III) implies (11)(iii), completing the proof of sufficiency. ///

Remark 6. When $\alpha \neq 1$ we observe that, letting $v_n = 0$, condition (11)(iii) always holds. We omit the elementary though slightly tedious proof. Thus, condition (III) is always trivially satisfied when $\alpha \neq 1$. We do not know whether (III) follows from (I) and (II) if $\alpha = 1$. On the other hand, when $0 < \alpha < 2$, (II) is independent of conditions (I) and (III), not a consequence. Example 4 illustrates this point.

Remark 7. Another proof of sufficiency can be obtained by using (I) and a uniform version of Feller (1971), Theorem 2(ii) p. 283, to obtain the existence of slowly varying functions $L_\theta(t)$ such that

(22) for every $x > 0$,

$$\lim_{t\to\infty} \sup_{\|\theta\|=1} |(L_\theta(t)/L_\theta(tx)) - 1| = 0$$

and

(23) $$\lim_{t\to\infty} \sup_{\|\theta\|=1} \left|(E\langle X,\theta\rangle^2 I_{(|\langle X,\theta\rangle|\leq t)}/\alpha t^{2-\alpha}L_\theta(t)) - 1\right| = 0 .$$

From this it is easy to deduce that

(24) for every $\epsilon > 0$,

$$\lim_{n\to\infty} \sup_{\|\theta\|=1} \left|(nL_\theta(\epsilon a_n(\theta))/a_n^\alpha(\theta)) - \frac{c}{2}\right| = 0$$

which further elucidates the behavior of the norming constants $a_n(\theta)$.

4. Examples.

If $\alpha = 2$, condition (II) is automatically implied by (I). However, if $0 < \alpha < 2$ it is necessary to assume (II).

Example 4. Fix $0 < \alpha < 2$. Let U be a symmetric random variable such that

$$P(|U| > y) = y^{-\alpha} \quad \text{if} \quad y \geq 1 .$$

Let the random vector (X,Y) be independent of U and concentrated on the points $(1,1)$, $(-1,-1)$, $(1,0)$ and $(-1,0)$ each with probability $1/4$. Let $Z = U(X,Y)$. Z automatically satisfies (III) by symmetry. We will show that Z satisfies (I) but not (II) and hence is not in the GDOA of a spherically symmetric stable.

By symmetry, if $\theta \neq e_2$, $\|\theta\| = 1$, and $t \geq \sqrt{2}$,

$$\begin{aligned} P(\langle Z,\theta\rangle > t) &= \tfrac{1}{2}(P(U > t/|\langle\theta,e_1\rangle + \langle\theta,e_2\rangle|) + P(U > t/|\langle\theta,e_1\rangle|)) \\ &= t^{-\alpha}c_\theta/4 \end{aligned}$$

where $C_\theta = |\langle\theta,e_1\rangle + \langle\theta,e_2\rangle|^\alpha + |\langle\theta,e_1\rangle|^\alpha$. For $\theta = e_2$, $P(\langle Z,e_2\rangle > t) = \frac{1}{2}P(U > t) = t^{-\alpha}C_{e_2}/4$. Furthermore, (**uniformly in θ**),

$$E(\langle Z,\theta\rangle^2 \wedge t^2) = 2\int_0^t uP(|\langle Z,\theta\rangle| > u)du$$

$$\sim \int_0^t u^{1-\alpha}C_\theta du = t^{2-\alpha}C_\theta/(2-\alpha) .$$

Consequently, (**uniformly in θ**),

$$\frac{t^2P(\langle Z,\theta\rangle > t)}{E(\langle Z,\theta\rangle^2 \wedge t^2)} \sim \frac{t^{2-\alpha}C_\theta/4}{t^{2-\alpha}C_\theta/(2-\alpha)} = (2-\alpha)/4$$

verifying (I).

To see that (II) fails, notice that (uniformly in θ),

$$a_n^2(\theta) = c^{-1}nE(\langle X,\theta\rangle^2 \wedge a_n^2(\theta)) \sim c^{-1}na_n^{2-\alpha}(\theta)C_\theta/(2-\alpha)$$

which implies

$$a_n(\theta) \sim (c^{-1}nC_\theta/(2-\alpha))^{1/\alpha} .$$

Condition (II) therefore requires the existence of an orthonormal basis (φ_1,φ_2) such that for all unit vectors θ,

$$(|\langle\theta,e_1\rangle + \langle\theta,e_2\rangle|^\alpha + |\langle\theta,e_1\rangle|^\alpha)^{2/\alpha}$$

$$= \langle\theta,\varphi_1\rangle^2(|\langle\varphi_1,e_1\rangle + \langle\varphi_1,e_2\rangle|^\alpha + |\langle\varphi_1,e_1\rangle|^\alpha)^{2/\alpha}$$

$$+ \langle\theta,\varphi_2\rangle^2(|\langle\varphi_2,e_1\rangle + \langle\varphi_2,e_2\rangle|^\alpha + |\langle\varphi_2,e_1\rangle|^\alpha)^{2/\alpha} .$$

This is impossible so (II) fails.

The following example illustrates the need to use a varying coordinate system.

Example 5. Fix $0 < \alpha < 2$. Let $V, V_1, V_2, \ldots$ be i.i.d. positive stable random variables of index $\alpha/2$ such that $P(V > x) \sim x^{-\alpha/2}$ as $x \to \infty$. Let $Y, \widetilde{Y}, Y_1, \widetilde{Y}_1, \ldots$ be standard normal variables. All of the above random variables are assumed to be independent.

For any real ψ let U_ψ be the unitary matrix

$$U_\psi = \begin{pmatrix} \cos\psi & \sin\psi \\ -\sin\psi & \cos\psi \end{pmatrix} .$$

Fix $0 < \sigma < \infty$ with $\sigma \neq 1$. Define a random vector $\vec{Y} = (Y, \sigma\widetilde{Y})$ and similarly define $\vec{Y}_i$. Let $\{c_n\}$ be a sequence of constants such that

$$0 = c_0 < c_1 < c_2 < \ldots$$

and

$$\lim_{n\to\infty} c_{n+1}/c_n = \infty .$$

Let $\{\theta_n\}$ be a sequence of reals which is dense in $[0,2\pi]$ (mod 2π) and has the further property that $\theta_{n+1} - \theta_n \to 0$ (e.g. $\theta_n = \ln n$). Define $\vec{X} = V^{\frac{1}{2}} \sum_{k=0}^{\infty} U_{\theta_k} \vec{Y} I_{(c_k < V^{\frac{1}{2}} \leq c_{k+1})}$. Define $\vec{X}_i$ similarly.

Let $\vec{Z}$ be a random vector on $\mathbb{R}^2$ such that for all unit vectors $\vec{\theta}$,

$$\mathcal{L}(\langle\vec{Z},\vec{\theta}\rangle) = \mathcal{L}(V^{\frac{1}{2}}Y) ,$$

which is stable of index α. Then letting $j_n = \inf\{j: c_j \geq n^{1/\alpha}\}$ and letting $D = \begin{pmatrix} 1 & 0 \\ 0 & \sigma^{-1} \end{pmatrix}$,

$$\mathcal{L}(D \cdot U_{\theta_{j_n}}^{-1}((\vec{X}_1 + \ldots + \vec{X}_n)/n^{1/\alpha})) \to \mathcal{L}(\vec{Z}) \quad .$$

Thus, $\vec{X} \in \mathrm{GDOA}(\mathcal{L}(\vec{Z}))$ where $\vec{Z}$ is a spherically symmetric stable of index α .

Proof. We record a few basic facts. The norms we will be using for operators refer to L_2 norms. So in particular,

(25) $$\|U_{\theta+\delta} - U_\theta\| = \|U_\delta - I\| \leq 4|\delta| \quad .$$

Let $\Delta_n = (\theta_{j_n-1} - \theta_{j_n})^2$.

Define an increasing sequence $t_n \to \infty$ in such a way that both

(26) $$\Delta_n t_n^{2/\alpha - 1} \to 0$$

and

(27) $$(nt_n)^{1/\alpha} \leq c_{j_n+1} \quad .$$

For this sequence,

(28) $$nP(V^{1/2} > (nt_n)^{1/\alpha}) \to 0 \quad .$$

Also,

(29) $$EVI_{(V \leq x)} \sim \alpha(2-\alpha)^{-1}x^{1-\alpha/2} \quad \text{as} \quad x \to \infty \quad .$$

We will replace $\sum \vec{X}_i$ by an equivalent sequence. It will be obvious that this new sequence converges weakly to $\vec{Z}$. Clearly,

$$P\left(\sum_{i=1}^{n} \vec{X}_i I_{(V_i^{\frac{1}{2}} > (nt_n)^{1/\alpha})} \neq 0\right) \leq nP(V^{\frac{1}{2}} > (nt_n)^{1/\alpha}) \to 0 \ .$$

Also,

$$P\left(\left\|\sum_{i=1}^{n} \vec{X}_i I_{(V_i^{\frac{1}{2}} \leq c_{j_n-1})} / n^{1/\alpha}\right\| > \epsilon\right)$$

$$\leq (\epsilon n^{1/\alpha})^{-2} E\left\|\sum_{i=1}^{n} \vec{X}_i I_{(V_i^{\frac{1}{2}} \leq c_{j_n-1})}\right\|^2$$

$$= \epsilon^{-2} n^{1-2/\alpha} E\left(V \sum_{k=0}^{j_n-2} \|U_{\theta_k} \vec{Y}\|^2 I_{(c_k < V^{\frac{1}{2}} \leq c_{k+1})}\right)$$

$$= \epsilon^{-2} n^{1-2/\alpha} \sum_{k=0}^{j_n-2} E\|U_{\theta_k} \vec{Y}\|^2 EVI_{(c_k < V^{\frac{1}{2}} \leq c_{k+1})}$$

$$= \epsilon^{-2} n^{1-2/\alpha} E\|\vec{Y}\|^2 EVI_{(V \leq c_{j_n-1}^2)}$$

$$\sim \epsilon^{-2}(1+\sigma^2) n^{1-2/\alpha} \alpha(2-\alpha)^{-1} c_{j_n-1}^{2-\alpha}$$

$$= o(n^{1-2/\alpha} n^{(2-\alpha)/\alpha})$$

$$= o(1) \ .$$

Hence,

$$P\left(\left\|\sum_{i=1}^{n} \vec{X}_i - \sum_{i=1}^{n} \vec{X}_i I_{(c_{j_n-1} < V_i^{1/2} \leq (nt_n)^{1/\alpha})}\right\| / n^{1/\alpha} > \epsilon\right) \to 0 \ .$$

Next we note that

$$P\Big(\Big\| \sum_{i=1}^{n} \vec{X}_i I_{(c_{j_n-1}<V_i^{\frac{1}{2}}\le (nt_n)^{1/\alpha})} - \sum_{i=1}^{n} (U_{\theta_{j_n}} \vec{Y}_i) V_i^{\frac{1}{2}} I_{(c_{j_n-1}<V_i^{\frac{1}{2}}\le (nt_n)^{1/\alpha})} \Big\| / n^{1/\alpha} > \varepsilon\Big)$$

$$\le (\varepsilon n^{1/\alpha})^{-2} E \Big\| \sum_{i=1}^{n} V_i^{\frac{1}{2}} \sum_{k=j_n-1}^{j_n} (U_{\theta_k} - U_{\theta_{j_n}}) \vec{Y}_i I_{(c_k<V_i^{\frac{1}{2}}\le c_{k+1}\wedge (nt_n)^{1/\alpha})} \Big\|^2$$

$$= \varepsilon^{-2} n^{1-2/\alpha} EV \sum_{k=j_n-1}^{j_n} \| (U_{\theta_k} - U_{\theta_{j_n}}) \vec{Y} \|^2 I_{(c_k<V^{\frac{1}{2}}\le c_{k+1}\wedge (nt_n)^{1/\alpha})} .$$

Note that $E\| (U_{\theta_k} - U_{\theta_{j_n}}) \vec{Y} \|^2 \le \| U_{\theta_k} - U_{\theta_{j_n}} \|^2 E \| \vec{Y} \|^2 \le 16 \Delta_n (1 + \sigma^2)$.

Using independence, the preceding probability is at most

$$16(1+\sigma^2)\varepsilon^{-2}\Delta_n n^{1-2/\alpha} EV I_{(c_{j_n-1}<V^{\frac{1}{2}}\le (nt_n)^{1/\alpha})}$$

$$\sim 16\alpha(2-\alpha)^{-1}(1+\sigma^2)\varepsilon^{-2}\Delta_n n^{1-2/\alpha}(nt_n)^{2/\alpha-1} \to 0 \quad \text{by choice of } t_n.$$

Finally, using our initial arguments,

$$P\Big(\Big\| \sum_{i=1}^{n} (U_{\theta_{j_n}} \vec{Y}_i) V_i^{\frac{1}{2}} I_{(c_{j_n-1}<V_i^{\frac{1}{2}}\le (nt_n)^{1/\alpha})} - \sum_{i=1}^{n} (U_{\theta_{j_n}} \vec{Y}_i) V_i^{\frac{1}{2}} \Big\| > \varepsilon n^{1/\alpha}\Big) \to 0 .$$

Hence,

$$P\Big(\Big\| \sum_{i=1}^{n} \vec{X}_i - U_{\theta_{j_n}} \sum_{i=1}^{n} \vec{Y}_i V_i^{\frac{1}{2}} \Big\| > \varepsilon n^{1/\alpha}\Big) \to 0$$

and therefore

$$P\Big(\Big\| D \cdot U_{\theta_{j_n}}^{-1} \Big(\sum_{i=1}^{n} X_i / n^{1/\alpha}\Big) - D\Big(\sum_{i=1}^{n} \vec{Y}_i V_i^{\frac{1}{2}} / n^{1/\alpha}\Big) \Big\| > \varepsilon\Big) \to 0 .$$

Finally, note that

$$\mathcal{L}(D(\sum_{i=1}^{n} \vec{Y}_i V_i^{\frac{1}{2}}/n^{1/\alpha})) = \mathcal{L}(D(\vec{Y}V^{\frac{1}{2}})) = \mathcal{L}((Y,\tilde{Y})'V^{\frac{1}{2}}) = \mathcal{L}(\vec{Z}) ,$$

whence $\mathcal{L}(D \cdot U_{\theta_{j_n}}^{-1} ((\vec{X}_1 + \dots + \vec{X}_n)/n^{1/\alpha})) \to \mathcal{L}(\vec{Z})$.

The fact that $\vec{Z}$ is spherically symmetric of index α results from calculating the characteristic function of $\vec{Z}$ and utilizing the known forms for the characteristic function of $\vec{Y}$ and the Laplace transform of V $(g(s) = e^{-bs^{\alpha/2}})$. For $d = 1$, this is noted in Feller (1971) and for $d > 1$ it was recorded by Wolfe (1975). ///

References

Billingsley, P. (1968). Convergence of Probability Measures. Wiley.

Feller, W. (1971). An Introduction to Probability Theory and its Applications, v. II, second edition. Wiley.

Gnedenko, B. V. and Kolmogorov, A. N. (1968). Limit Distributions for Sums of Independent Random Variables. Addison-Wesley.

Hahn, M. (1979). The generalized domain of attraction of a Gaussian law on Hilbert space. Lecture Notes in Math., 709, 125-144.

Hahn, M. and Klass, M. (1980). Matrix normalization of sums of i.i.d. random vectors in the domain of attraction of the multivariate normal. Annals of Probability, April, 1980.

Hahn, M. and Klass, M. (1979). The multi-dimensional central limit theorem for arrays normed by affine transformations. (Preprint).

Halmos, P. R. (1958). Finite-dimensional Vector Spaces, second edition. Van Nostrand Co.

Sharpe, M. (1969). Operator-stable probability distributions on vector groups. Trans. of Amer. Math. Soc. 136, 51-65.

Wolfe, S. J. (1975). On the unimodality of spherically symmetric stable functions. J. of Multivariate Analysis 5, 236-242.

M. G. Hahn
Department of Mathematics
Tufts University
Medford, MA 02155
USA

M. J. Klass
Department of Statistics
University of California
Berkeley, CA 94720
USA

A CLASS OF CONVOLUTION SEMI-GROUPS OF MEASURES ON A LIE GROUP *)

by

A.Hulanicki (Wrocław)

The aim of this article is to propose an investigation of a class of semi-groups of measures on a Lie group which seems to be a natural generalization of the class of stable semi-groups, if the group is $\underline{R}^r$. The definitions, propositions and problems concerning this class are naturally formulated in terms of the infinitesimal generators of the semi-groups.

We try to make the whole story self-contained: only elementary informations concerning Lie groups and the theory of one-parameter semi-groups as presented in [11] are the prerequisites. To do so we include here shortcuts through the paper of G.A.Hunt [6]. This is our section 2 in which we make an utmost use of a recent aricle of M.Duflo [1]. The second part of the proof of proposition 2.4 is due to A.Iwanik. Section 3 contains probabilistic interpretation of the operation of taking sum of the infinitesimal generators of two semi-groups. It comprises two theorems, the first can be deduced from a theorem by T.G. Kurtz [8], but we give a shorter direct proof adapted to our situation, the latter has been implicitly used in [4]. In section 4 the decay of the measures in a semi-group at infinity is studied. For this a long list of historical references should perhaps be included, but we refer the reader only to [1]. Section 5 is new but easy, the idea of the proofs of lemmas 5.1, 5.2 presented here is due to T.Pytlik. Section 6, finally, contains the definition of the class of the semi-groups of measures mentioned at the begining. Some properties of them are deduced in the general setting of an arbitrary Lie group. Also some problems and partial results the proofs of which are not included are presented.

The author is grateful for many helpful comments on the subject of this paper to Tomasz Byczkowski, Paweł Głowacki, Anzelm Iwanik, Tadeusz Pytlik and Czesław Ryll-Nardzewski.

---

*) A summary of this paper has been presented at the conference Probability Theory on Vector Spaces II. The author is very grateful to Professor Weron for his kind invitation to the conference and his hospitality.

1. Preliminaries. Let G be a Lie group, $\underline{g}$ its Lie algebra. We identify $\underline{g}$ with the differential operators X of order one which commute with the right translations via the exponential map, i.e.

$$Xf(x) = \frac{d}{dt}f(\exp tX\cdot x)\Big|_{t=0} .$$

The exponential map is a local diffeomorphism of an open neighbourhood of zero in $\underline{g}$ onto an open neighbourhood U of the unit element e in G.

Let $X_1,\dots,X_r$ be a basis in $\underline{g}$. We introduce the coordinates in U:

$$U \ni x = \exp(x_1X_1+\dots+x_rX_r)$$

and adjusting the length of $X_1,\dots,X_r$, if necessary, we write

$$U_0 = \left\{ \exp(x_1X_1+\dots+x_rX_r) : |x_j|<1 \right\}$$

with $U_0 \subset U$ and $x = (x_1,\dots,x_r)$. Also

$$|x| = (\sum_j x_j^2)^{1/2} .$$

For X in $\underline{g}$ we define also a left-invariant operator

$$X^* : C^\infty(G) \longrightarrow C^\infty(G)$$

by

$$X^*f(x) = \frac{d}{dt} f(x\cdot \exp tX)\Big|_{t=0} .$$

For a function f on G we write

$${}_af(x) = f(a^{-1}x) \quad \text{and} \quad f_a(x) = f(xa) \quad , a,x \in G .$$

For a multi-index $n = (n_1,\dots,n_r)$, $n_j \in \underline{Z}^+$ and a chosen basis $X_1,\dots,X_r$ in $\underline{g}$ we write

$$X^{*n} = X_1^{*n_1}\dots X_r^{*n_r} .$$

Let $|n| = n_1+\dots+n_r$. We define for $k = 0,1,\dots,\infty$

$$C^k = \left\{f : X^{*n}f \in C(G) \text{ for } |n|<k+1\right\} .$$

Two subspaces of C^k are of particular interest for us

$$C^k_\infty = \left\{f \in C^k : \lim_{x\to\infty} X^{*n}f(x) \text{ exists for } |n|<k+1\right\}$$

$$C^k_0 = \left\{f \in C^k_\infty : \lim_{x\to\infty} X^{*n}f(x) = 0 \text{ for } |n|<k+1\right\}$$

For $k < \infty$ we equip C_∞^k with a Banach space norm

$$\|f\|_{C_\infty^k} = \sum_{|n| \leq k} \|X^{*n} f\|_{C(G)} .$$

Of course C_o^k is a closed subspace of C_∞^k .

Let

$$D_j f(x) = \frac{\partial}{\partial x_j} f(\exp(x_1 X_1 + \ldots + x_r X_r)), \quad x \in U_o .$$

Then

$$D_i f(x) = \sum_j a_{ij}(x) X_j^* f(x), \text{ where } a_{ij}(x) \in C^\infty(G)$$

and

$$a_{ij}(e) = \delta_{ij} .$$

Hence

$$D_i D_j f(x) = \sum_{k,l} a_{ik}(x)\, a_{jl}(x) X_k^* X_l^* f(x)$$

$$+ \sum_{k,l} a_{ik}(x) (X_k^* a_{jl}(x)) X_l^* f(x) .$$

Consequently, by Taylor's formula, for $f \in C^2$ such that

$$f(e) = X_j^* f(e) = 0 \quad \text{for } j = 1, \ldots, r$$

(1.1) we have

$$f(x) = \frac{1}{2} \sum_{i,j} X_i^* X_j^* f(e) x_i x_j + o(|x|^2) \text{ as } a \longrightarrow 0 .$$

The differential of the left-invariant Haar measure on G is denoted by dx. We shall consider the spaces $L^p(G)$, $1 \leq p < \infty$ with respect to this measure. By $M(G)$ we denote the Banach $*$-algebra of bounded measures on G and by $P(G)$ we denote the subspace of $M(G)$ of non-negative measures μ such that $\mu(G) \leq 1$.

2. Semi-groups of measures. By a semi-group of measures in P(G) we mean a semi-group $\{\mu_t\}_{t>0}$ such that

for every $t > 0$ $\mu_t \in P(G)$

(2.1) $$\mu_{s+t} = \mu_s * \mu_t$$

$$\lim_{t \to 0} \| \mu_t * f - f\|_{C_\infty} = 0 \quad \text{for all f in } C_\infty .$$

Since for every X in $\mathfrak{g}$ the operator X^* commutes with the operator

(2.2) $$f \longrightarrow \mu_t * f ,$$

the semi-group $\{\mu_t\}_{t>0}$ defines a strongly continuous semi-group of operators on C_∞^k for every $k = 0, 1, \ldots, \infty$, which preserves C_o^k .

From now on the semi-group $\{\mu_t\}_{t>0}$ is to be viewed as a one-parameter semi-group of contractions on various spaces of functions on G and one of the basic tools is going to be the Hille-Yoshida theorem.

Let A be the infinitesimal generator of $\{\mu_t\}_{t>0}$ on C^k_∞, i.e.

$$(2.3)\qquad Af = \lim_{t\to 0} \frac{1}{t} (\mu_t * f - f),$$

the domain $D^k(A)$ of A being the set of functions f in C^k_∞ such that the limit (2.5) exists in the norm of C^k_∞. For every k and every function f in C^k_∞, for $s > 0$ the function

$$(2.4)\qquad g = \int_0^s \mu_t * f\, dt$$

belongs to $D^k(A)$. Thus, if $D(A) = D^0(A)$, then the cone of positive functions in $D(A) \cap C^k_\infty$ is dense in the cone of positive functions in C^k_∞.

We are going to consider the following linear form

$$(2.5)\qquad F: D(A) \ni f \longrightarrow (Af)(e) \in \underline{C}.$$

<u>Proposition 2.1.</u> The linear form as defined by (2.5) is a continuous functional on C^2_∞.

In the proof we use the following easy to prove and well-known

<u>Lemma.</u> Let B be a Banach space, C a cone in B and C_0 a dense convex subset of C. Then for every $\varepsilon > 0$ and a finite set of functionals $\varphi_1,\dots,\varphi_r$ in B' and a f in C there is a g in C_0 such that $\langle f, \varphi_j\rangle = \langle g, \varphi_j\rangle$, $j = 1,\dots,r$ and $\|f-g\|_B < \varepsilon$.

<u>Proof of proposition 2.1.</u> Let

$$M = \{f \in C^2_\infty : f(e) = (X^*_j f)(e) = 0, \quad j=1,\dots,r\}.$$

By the lemma, $D(A) \cap M = M_0$ is dense in M and, of course, M is of finite co-dimension in C^2_∞. Another application of the lemma shows that there exists a function φ such that

$$\varphi \in D(A) \cap C^2_\infty$$

$$(2.6)\qquad \varphi(e) = (X^*_j\varphi)(e) = 0, \quad (X^*_i X^*_j \varphi)(e) = 2\delta_{ij}, \quad i,1=1,\dots,r$$

$$\varphi(x) > 0 \quad \text{for } x \neq e$$

$$\lim_{x\to\infty} \varphi(x) > 0.$$

Hence, by (1.1), for a constant $c > 0$

$$c\varphi(x) \geq |x|^2 \quad \text{for} \quad x \in U_o \ .$$

Suppose now that $f_n \in M_o$ and $\lim_{n\to\infty} \|f_n\|_{C^2_\infty} = 0$. Then for every $\varepsilon > 0$, by (1.1), it follows that for n large enough

$$|f_n(x)| \leq \varepsilon \, |x|^2 \leq \varepsilon \, c\varphi(x) \quad \text{for} \quad x \in U_o$$

and, for n still larger,

$$|f_n(x)| \leq \varepsilon \, \varphi(x) \quad \text{for all } x \text{ in } G \ .$$

Thus, since $\varphi \in D(A)$, we have

$$|<f_n, F>| = \lim_{t\to 0} \frac{1}{t} \, | \int f_n(x^{-1}) d\mu_t(x)|$$

$$\leq \lim_{t\to 0} \frac{1}{t} \int \varphi \, (x^{-1}) d\mu_t(x) = <\varphi, F>$$

which shows that F extends continously to M and, since M has finite co-dimension in C^2_∞ and $D(A) \cap C^2_\infty$ is dense in C^2_∞, proposition 2.1 follows.

We say that a distribution $F \in C^\infty_c(G)$, is <u>dissipative</u> if For every real function f in $C^\infty_c(G)$ such that

(2.7) $\max\{f(x) : x \in G\} = f(e)$ we have $<f, F> \leq 0$,

<u>Corollary 2.2</u> . The linear form (2.5) definies a dissipative distribution.

<u>Proof.</u> By proposition 2.1 , (1.5) defines a functional on C^2_∞ and so the limit

$$\lim_{t\to 0} \frac{1}{t} \, (\mu_t * f(e) - f(e)) = <f, F>$$

exists for all f in $C^2_\infty \supset C^\infty_c(G)$. Consequently, since $\mu_t \in P(G)$ implies $\mu_t * f(e) \leq \max\{f(x) : x \in G\}$, corollary 2.2 follows.

An easy application of the Riesz theorem yields

<u>Proposition 2.3</u>. If F is a dissipative distribution, then for every neighbourhood V of e

$$F = F_V + \mu_V \ ,$$

where F_V is a distribution supported by V and $\mu_V \in M(G)$.

With a dissipative distribution F we associate an operator A_F such that

$$(2.8)\qquad \begin{aligned} &D(A_F) = C_c^\infty(G) \\ &A_F f = F^* * f \ , \ \text{i.e.}\ (A_F f)(x) = \langle f_x \ , F \rangle \ . \end{aligned}$$

Proposition 2.4. Let F be a dissipative distribution. There exists a unique semi-group of measures $\{\mu_t\}_{t>0}$ in P(G) such that the infinitesimal generator of $\{\mu_t\}_{t>0}$ on C_0 is the closure of the operator defined by (2.8).

Proof. We note first that since F is dissipative, for $\lambda>0$ and every real function f we have

$$\|\lambda f\|_{C_0} \le \|\lambda f - A_F f\|_{C_0} \ .$$

Moreover, we check that the range of $\lambda - A_F$ is dense in C_0. In fact, suppose that

$\langle \lambda f - A_F f \ , \mu \rangle = 0$ for all f in $C_c^\infty(G)$ and a μ in M(G).

Then

$$(2.9)\qquad 0 = \langle \lambda f - F^* * f \ , \mu \rangle = \langle \lambda \delta_e - F^* \ , \mu * f^\sim \rangle \ .$$

Let a real function f in $C_c^\infty(G)$ be such that

$$\mu * f^\sim(e) = \max\{\mu * f^\sim(x) : x \in G\} \ge 0 \ .$$

Then, since F is dissipative, (2.9) shows that $\mu * f^\sim(e) = 0$. Translating f on the left and multiplying by -1, if necessary, we see that $\mu * f^\sim = 0$ for all real f in $C_c^\infty(G)$, whence $\mu = 0$.

Now we apply the Yoshida-Hille theorem which shows that the closure of A_F generates a strong semi-group of contractions $\{T_t\}_{t>0}$ on C_0 which commute with right translations (since so does A_F). Therefore $T_t f = \mu_t * f$, where $\{\mu_t\}_{t>0}$ is a semi-group of measures such that $\|\mu_t\|_{M(G)} \le 1$. We easily check that, since F is dissipative, for $\lambda>0$ the operator $(\lambda - A_F)^{-1}$ maps non-negative functions onto non-negative functions and, consequently, so does T_t. This proves that $\{\mu_t\}_{t>0}$ is a semi-group of measures in P(G).

Suppose now that $\{\nu_t\}_{t>0}$ is a semi-group of measures in P(G) such that

$$(2.10)\qquad \lim_{t\to 0} \frac{1}{t}(\nu_t * f(e) - f(e)) = \langle f \ , F \rangle \qquad \text{for } f \in C_c^\infty(G).$$

Let B be the infinitesimal generator of $\{\nu_t\}_{t>0}$. We are going to show that B coincides with the closure of A_F which will prove the equality $\nu_t = \mu_t$.

First we verify the following well-known fact.

(2.11) $$D(B) = \{f \in C_o : \lim_{t \to 0} \frac{1}{t}(\nu_t * f(x) - f(x)) = g(x) \ , \ g \in C_o\}.$$

In fact, let B' be the operator defined by the point-wise limit in (2.11), the domain of B' being the right hand side of (2.11). Of course $D(B') \supset D(B)$. But for every $\lambda > 0$ the operator $\lambda - B'$ is one-to-one on $D(B')$ because if for $f \in D(B')$ $\max\{f(x): x \in G\} = f(x_o) > 0$, then, by the definition of B', we have $B'f(x_o) \leq 0$, whence $(\lambda - B')f(x) = 0$ is impossible. But $\lambda - B$ maps $D(B)$ onto C_o, and so $D(B') = D(B)$.

Now translating on the right, (2.10) yields

(2.12) $$\lim_{t \to 0} \frac{1}{t}(\nu_t * f(x) - f(x)) = F^* * f(x)$$

for all x in G and $f \in C_c^\infty(G)$. Of course $F^* * f \in C_o$, whence $D(\overline{A}_F) \subset D(B') = D(B)$, i.e. $\overline{A}_F \subset B$. But since again for $\lambda > 0$ $\lambda - \overline{A}_F$ maps $D(\overline{A}_F)$ onto C_o and is equal to $\lambda - B$ on $D(\overline{A}_F)$, the latter being one-to one on D(B), we have $D(\overline{A}_F) = D(B)$, which completes the proof of proposition 2.4.

The followig theorem summarizes the above considerations.

<u>Theorem 2.5</u>. Let G be a Lie group and let $\{\mu_t\}_{t>0}$ be a semi-group of measures in P(G). Let A be the infinitesimal generator of $\{\mu_t\}_{t>0}$ on C_o. Then $C_c^\infty(G) \subset D(A)$, $\langle f, F \rangle = Af(e)$ defines a dissipative distribution on $C_c^\infty(G)$ and $Af = F^* * f$ for $f \in C_c^\infty(G)$. Conversely, every dissipative distribution F defines a unique semi-group of measures in P(G) and the infinitesimal generator A of this semi-group is of the form $Af = F^* * f$, $f \in C_c^\infty(G)$.

<u>Remarks</u>.

(a) It is easy to see that proposition 2.3 implies that a dissipative distribution which is bounded on C_∞^2 defines an operator A_F as in (2.8) which maps C_∞^2 into C_∞. Thus, by proposition 2.1 we see that not only $C_c^\infty(G) \subset D(A)$ but also $C_\infty^2 \subset D(A)$.

(b) Riesz theorem used in the proof of proposition 2.3 does not give an effective method of constructing the measure μ_V.

The following argument shows that μ_V is the restriction of a non-negative measure μ defined on $G\setminus\{e\}$ such that

(2.13) $\quad < f,\mu > \; = \lim_{t\to 0} \frac{1}{t} < f, \mu_t >, \quad e \notin \operatorname{supp} f$.

Let φ be the function defined by (2.2). Then, for $f \in C_c^\infty(G)$, $f\varphi \in D(A)$ and $f\varphi(e) = 0$. Therefore the limit

$$\lim_{t\to 0} \frac{1}{t} \int f(x)^{-1})\varphi(x^{-1})d\mu_t(x) = A(f\varphi)(e) = \; < f,\mu_\varphi >$$

exists and, clearly,

(2.14) $\quad | < f, \mu_\varphi >| \leq \|f\|_{C_o} \lim_{t\to 0} \frac{1}{t} \int \varphi(x^{-1})du_t(x)$.

Inequality (2.14) shows that μ_φ is a non-negative bounded measure. If now $f \in C_c^\infty(G)$ and $e \in \operatorname{supp} f$, then $\varphi(x)^{-1}$ is bounded on $\operatorname{supp} f$ and we see that μ of (2.13) is given by

$$< f, \mu > \int f(x^{-1})\varphi(x^{-1})du_t(x) \; .$$

Proposition 2.6. A semi-group of measures in $P(G)$ $\{\mu_t\}_{t>0}$ consists of probability measures if and only if the infinitesimal generator A of $\{\mu_t\}_{t>0}$ on C_∞ anihilates constant functions.

Proof. Let K be the one-dimensional subspace of C_∞ consisting of constant functions. The semi-group of operators $f \longrightarrow \mu_t * f$ preserves K and it is the semi-group consisting of the identity operator on K, if and only if its infinitesimal generator on K is zero.

Proposition 2.7. If $\{\mu_t\}_{t>0}$ is a semi-group of measures in $P(G)$, then for every $1 < p < \infty$

$$L^p(G) \ni f \longrightarrow \mu_t * f \in L^p(G)$$

is a strongly continuous semigroup of contractions. Let A be the infinitesimal generator of it, then $C_c^\infty(G) \subset D(A)$.

Proof. By theorem 2.5, for $f \in C_c^\infty(G)$ the limit

$$\lim_{t\to 0} \frac{1}{t} (\mu_t * f(x) - f(x))$$

exists uniformly with respect to x in G. By proposition 2.3 and theorem 2.5 , it is equal to

$$F^* * f(x) = F_V^* * f(x) + \mu_V^* * f(x) \; .$$

Thus

$$\mu_t * f(x) - f(x) = \int_0^t \mu_s * F_V^* * f(x)ds + \int_0^t \mu_s * \mu_V^* * f(x)ds,$$

whence, since $F_V^* * f \in C_c^\infty(G)$,

$$\| \mu_t * f - f \|_{L^p(G)} \leq t \| F_V^* * f \|_{L^p(G)} + \|\mu\|_{M(G)} \|f\|_{L^p(G)} ,$$

which completes the proof.

3. Sums of the infinitesimal generators. Let $\{\mu_t(0)\}_{t>0}$ and $\{\mu_t(1)\}_{t>0}$ be two semi-groups of measures in $P(G)$ and let $A(0)$ and $A(1)$ be their infinitesimal generators (on C_o), respectively. By theorem 2.5, the operator $\frac{1}{2}(A(0)+A(1))$ defined at least on C_o^2 is the infinitesimal generator of a semi-group of measures $\{\mu_t\}_{t>0}$ in $P(G)$. Two theorems of this section describe the semi-group $\{\mu_t\}_{t>0}$ in terms of the semi-groups $\{\mu_t(0)\}_{t>0}$ and $\{\mu_t(1)\}_{t>0}$. Both have very natural probabilistic interpretation.

Let

$$D = \{w=(w_1,w_2,\dots) : w_j \in \{0,1\}\} .$$

We equip D with the direct product measure m of countably many copies of the measure assigning the value 1/2 to each of the points 0 and 1.

For a positive integer k we define a map

$$D \ni w \longrightarrow w^k \in D$$

by

$$(w^k)_j = w_{j+k} .$$

The following proposition is a version of the ordinary strong law of large numbers.

Proposition 3.1. There exists a subset M of D such that $m(M)=1$ and for every $w \in M$, non-negative integer k, non-negative real numbers s,t with $t > 0$ and a vector valued function φ on the two-point space $\{0,1\}$ we have

$$\lim_{n\to\infty} \frac{1}{[nt]} \sum_{j=1}^{[nt]} \varphi(w_{j+[sn]}) = \lim_{n\to\infty} \frac{1}{[nt]} \sum_{j=[ns]+1}^{[ns]+[nt]} \varphi(w_j) = \frac{1}{2}(\varphi(0)+\varphi(1)).$$

Theorem 3.2. For every w in M and every f in C_o we have

$$\lim_{n\to\infty} \langle f, \mu_{\frac{1}{n}}(w_1) * \dots * \mu_{\frac{1}{n}}(w_{[tn]}) \rangle = \langle f, \mu_t \rangle$$

Proof. For a w in D we define

(3.1) $$\mu_t(w,n) = \mu_{\frac{1}{n}}(w_1)*\dots*\mu_{\frac{1}{n}}(w_{[nt]})$$

Clearly $\mu_t(w,n) \in P(G)$.

For a f in C_0^2 we have the following estimate

(3.2) $$\| \mu_t(w,n)*f-f\|_{C_0} \leq t \max\{\|A(0)f\|_{C_0}, \|A(1)f\|_{C_0}\}+o(1/n)$$

as $n \to \infty$. In fact,

(3.3) $$\mu_t(w,n)*f-f= \sum_{j=1}^{[nt]} \mu_{\frac{1}{n}}(w_1)*\dots*\mu_{\frac{1}{n}}(w_{j-1})*(\mu_{\frac{1}{n}}(w_j)*f-f)$$

$$= t\frac{1}{[nt]}\sum_{j=1}^{[nt]} \mu_{\frac{1}{n}}(w_1)*\dots*\mu_{\frac{1}{n}}(w_{j-1})*A(w_j)f+o(1/n)$$

which implies (3.2).

Since $\mu_t(w,n) \in P(G)$, for every f in C_0 we have

(3.4) $$\lim_{t\to 0} \limsup_{n\to\infty} \| \mu_t(w,n)*f-f\|_{C_0} = 0 \quad \text{uniformly in } w \in D.$$

It follows immediately from (3.1) that

(3.5) $$\|\mu_{s+t}(w,n)*f - \mu_s(w,n)*\mu_t(w^{[ns]},n)*f\|_{C_0} \to 0 \text{ as } n\to\infty .$$

For a fixed w in M (of proposition 3.1) let n_k be an increasing sequence of integers such that

(3.6) $$\mu_t(w^{[n_k s]},n_k)$$

convergences *-weakly to a measure $p_{t,s}$. Passing to a subsequence if necessary, we may assume that (3.6) is convergent for all rational t and s. It follows from (3.4) and (3.5) that sequence (3.6) is convergent for all real t and s. We write

$$p_t = p_{t,0}$$

We are going to verify

(3.7) $$\frac{d}{dt}p_{t,s} *f \Big|_{t=0} = Af, \quad \text{for } f \in C_c^\infty(G),$$

where

$$A = \frac{1}{2}(A(0) + A(1)) .$$

Let $\varepsilon > 0$. For t small and n_k large enough and all j between $[n_k s]+1$ and $[n_k s]+[n_k t]$, by (3.4) we have

$$\| \mu_{\frac{1}{n_k}}(w_{[n_k s]+1})*\ldots*\mu_{\frac{1}{n_k}}(w_{j-1})*A(w_j)f - A(w_j)f\|_{C_0} < \varepsilon$$

Hence, by (3.3),

$$\| \tfrac{1}{t}(\mu_t(w^{[n_k s]}, n_k)*f-f) - Af\|_{C_0}$$

$$= \| \frac{1}{n_k t} \sum_{j=[n_k s]+1}^{[n_k s]+[n_k t]} \mu_{\frac{1}{n_k}}(w_{[n_k s]+1})*\ldots*\mu_{\frac{1}{n_k}}(w_{j-1})*A(w_j)f - Af\|_{C_0} +$$

$$+ \left(\frac{1}{n_k t}\right) \leq \| \frac{1}{n_k t} \sum_{j=[n_k s]+1}^{[n_k s]+[n_k t]} A(w_j)f - Af\|_{C_0} + o\left(\frac{1}{n_k t}\right) + \varepsilon$$

Consequently, for $t \to 0$ and $n_k t \to \infty$, by proposition 3.1, we obtain

$$\frac{d}{dt} p_{t,s}*f \Big|_{t=0} = \lim_{t\to 0} \lim_{n_k \to \infty} \tfrac{1}{t}(\mu_t(w^{[n_k s]}, n_k)*f-f) = Af$$

and so (3.7) is proved.

Now passing to the limit in (3.5) as $n_k \to \infty$ we get

$$p_{s+t} = p_s * p_{t,s}$$

Hence, by (3.7), for f in C_0^2 we get

$$(3.8) \quad \frac{d}{ds} p_s*f = \frac{d}{dt} p_{t+s}*f \Big|_{t=0} = p_s* \frac{d}{dt} p_{t,s}*f \Big|_{t=0} = p_s*Af .$$

Let $\{\mu_t\}_{t>0}$ be the semi-group of measures the infinitesimal generator of which is A. For f in C_0^2 and $t > 0$, by (3.8) we have

$$(p_t-\mu_t)*f = \int_0^t \frac{d}{ds}(p_s*\mu_{t-s}*f)ds$$

$$= \int_0^t p_s*A(\mu_{t-s}*f) - p_s*A(\mu_{t-s}*f)\, ds = 0 ,$$

which proves that $p_t = \mu_t$ for all t.

Since, as it has just turned out, the limit

$$*w\text{-}\lim_{n_k\to\infty} \mu_t(w,n_k) = \mu_t$$

does not depend on the sequence, if only $w \in M$ and the limit exists, the theorem is proved.

An immediate corollary of theorem 3.2 is the following

Theorem 3.3 If $(\mu_t(0))_{t>0}$, $(\mu_t(1))_{t>0}$ and $(\mu_t)_{t>0}$ are semigroups of measures in $P(G)$ such as in theorem 3.2, then

$$*w\text{-}\lim_{n\to\infty} [\tfrac{1}{2}(\mu_{\frac{t}{n}}(0) + \mu_{\frac{t}{n}}(1))]^{*n} = \mu_t .$$

Proof. For f in C_o we have

$$\lim_{n\to\infty} < [\tfrac{1}{2}(\mu_{\frac{t}{n}}(0) + \mu_{\frac{t}{n}}(1))]^{*n}, f >$$

$$= \lim_{n\to\infty} \frac{1}{2^n} \sum_{|w|\leq n} < \mu_{\frac{t}{n}}(w_1)*\dots*\mu_{\frac{t}{n}}(w_n), f >$$

$$= \lim_{n\to\infty} \frac{1}{2^{[nt]}} \sum_{|w|\leq[nt]} < \mu_{\frac{1}{n}}(w_1)*\dots*\mu_{\frac{1}{n}}(w_{[nt]}), f >$$

$$= \int_D < \mu_t, f > dm(w) = < \mu_t, f > .$$

4. Decay at infinity . Let G be a locally compact group. We say that a function φ on G is submultiplicative, if

φ is locally bounded

$\varphi(x^{-1}) = \varphi(x), \quad \varphi(x) \geq 1$

$\varphi(xy) \leq \varphi(x)\varphi(y)$.

We say that a submultiplicative function is a polynomial weight, if

$$w(xy) \leq C(w(x) + w(y)) .$$

Example. Let G be compactly generated and let $U = U^{-1}$ be a compact set of generators of G, then

$$\tau_U(x) = \min\{ n: x \in U^n\}$$

is subadditive and

$$w(x) = (1 + \tau_U(x))^a \qquad a > 0$$

is a polynomial weight.

Also it is not difficult to prove that every submultiplicative function on G is dominated by $e^{C\tau_U(x)+C}$.

The following theorem gives a simple criterium for a semi-group of measures in P(G) to integrate a submultiplicative function. Here the semi-group is considered as a semi-group of contractions on $L^1(G)$.

Theorem 4.1. Let $\{\mu_t\}_{t>0}$ be a semi-group of measures in P(G). Let A be the infinitesimal generator of $\{\mu_t\}_{t>0}$ on $L^1(G)$ and let φ be a submultiplicative function on G. Suppose that for a single function f in D(A) with $0 \neq f \geq 0$ such that $< f,\varphi > < +\infty$ we have $< |Af|, \varphi > = C < +\infty$. Then for a constant k

$$< \mu_t, \varphi > < k \quad \text{for} \quad t \in (0,1]$$

Proof. We note first that if ϕ is a submultiplicative function, then for every non-negative f such that $<f,\phi> < \infty$ we have

$$(4.1) \qquad < \tilde{f},\phi^{-1} > \phi(x) \leq \phi * f(x) \leq < \tilde{f},\phi > \phi(x)$$

In fact, since

$$(4.2) \qquad \phi(xy) \leq \phi(x)\phi(y) \quad \text{and} \quad \varphi(y^{-1}) = \varphi(y)$$

we have $\phi(x) \leq \phi(xy^{-1})\phi(y)$, whence $\phi(x) \leq \phi(xy)\phi(y)$ and so

$$(4.3) \qquad \phi(x)\phi(y)^{-1} \leq \phi\,(xy)\ .$$

Thus, since

$$\phi * f(x) = \int \phi(y)f(y^{-1}x)dy = \int \phi(xy)\tilde{f}(y)dy$$

and f is non-negative, a simple application of (4.2) and (4.3) yields (4.1)

Now for the function φ and an arbitrary positive integer m we define

$$\varphi_m(x) = \min\{\, m, \varphi(x)\,\}\ .$$

Clearly, φ_m is submultiplicative and bounded. We define

$$h_m(t) = < \mu_t * f, \varphi_m > = < \mu_t, \varphi_m * \tilde{f} >\ .$$

We note that since $f \in D(A)$, $Af \in L^1(G)$ and

$$\frac{d}{dt}\mu_t * f = \mu_t * Af\ .$$

Hence

$$h_m'(t) = < \mu_t * Af, \varphi_m > = < \mu_t, \varphi_m * (Af)^{\sim} >$$

and consequently, by (4.1),

$$|h_m'(t)| \leq < \mu_t, \varphi_m * |(Af)^{\sim}| > = < \mu_t * |Af|, \varphi_m >$$
$$\leq < \mu_t, \varphi_m > < |Af|, \varphi_m >$$
$$\leq < \mu_t, \varphi_m * f^{\sim} > < f, \varphi_m^{-1} >^{-1} < |Af|, \varphi_m > .$$

But $\sup \{ < f, \varphi_m >^{-1} : m = 1,2,\ldots \} = < f, \varphi^{-1} >^{-1} = c_1$.
By assumption $< |Af|, \varphi_m > \leq < |Af|, \varphi_m > = $. Consequently,

$$| h_m'(t) | \leq c_1 C\, h_m(t)$$

and so

$$h_m(t) \leq h_m(o)\, e^{tc_2}$$

i.e.

$$< \mu_t * f, \varphi_m > \leq < f, \varphi_m > e^{tc_2} .$$

Using (4.1) again we get

$$< \mu_t, \varphi_m > \leq < f, \varphi_m^{-1} >^{-1} < \mu_t * f, \varphi_m >$$
$$\leq c_1 < f, \varphi\, e^{tc_2} ,$$

whence

$$< \mu_t, \varphi > \leq c_3\, e^{tc_2} ,$$

which completes the proof of theorem 4.1.

For a submultiplicative function φ we define a Banach space

$$L^1_\varphi = \{ f \in L^1(G) : \int f(x) | \varphi(x) dx = \|f\|_{L^1_\varphi} < \infty \} .$$

<u>Proposition 4.2.</u> Given a submultiplicative function φ. Suppose that for a subset M of $D(A)$ dense in L^1_φ and containing non-zero non-negative functions we have $< |Af|, \varphi > < +\infty$ for f in M. Then the semi-group of measures $\{\mu_t\}_{t>0}$ in $P(G)$ whose infinitesimal generator is A defines by left convolutions a strongly continuous semi-group of operators on L^1_φ and M is contained in the domain of A in L^1_φ .

<u>Proof.</u> By theorem 4.1, we have $< \mu_t, \varphi > < k$ for $t \in [0,1]$
Hense, since

$$|< \mu_t * f, \varphi >| \leq < \mu_t, \varphi >| < f, \varphi > | \quad f \in L^1_\varphi$$

the operators

$$L^1_\varphi \ni f \longrightarrow \mu_t * f \in L^1_\varphi$$

are uniformly bounded for $t \in (0,1]$. Thus it is sufficient to prove

(4.4) $\lim_{t\to 0} \langle |\mu_t * f - f|, \varphi \rangle\ 0$ for $f \in M$.

Let, as before, $\varphi_m(x) = \min\{m, \varphi(x)\}$. For f in M we have

$$\mu_t*f-f = \int_0^t \frac{d}{ds}(\mu_s*f)ds = \int_0^t \mu_s*Af\ ds ,$$

where the integral is the Riemann integral of $L^1(G)$ valued functions. Consequently,

$$\langle |\mu_t*f-f|, \varphi_m\rangle \leq \int_0^t \langle \mu_s*|Af|, \varphi_m \rangle\ ds$$

$$\leq t \sup\{ \langle \mu_s,\varphi_m \rangle : s\in(0,1], m=1,2,\dots,\}\langle |Af|\ \varphi\rangle$$

$$\leq tk \langle |Af|\ , \varphi \rangle \text{ for } t \in (0,1].$$

Thus

$$\langle |\mu_t*f-f|, \varphi \rangle \leq tk\langle |Af|, \varphi \rangle$$

which completes the proof of (4.4) and proposition 4.2 .

<u>5. Subordinated semi-groups.</u> Let $\{\mu_t\}_{t>0}$ be a semi-group of measures in P(G) and let A be the infinitesimal generator of it. Since for $0 < a < 1$ and $d > 0$ we have

$$\int_0^\infty t^{-a-1}(e^{-td}-1)dt = -d^a \int_0^\infty t^{-a-1}(e^{-t}-1)dt = -cd^a ,$$

we define the fractional power of A putting for f in $C_c^\infty(G)$

$$-|A|^a f = c^{-1} \int_0^\infty t^{-a-1}(\mu_t*f-f)dt .$$

It is immediate to see that

$$f \longrightarrow -|A|^a f(e)$$

is a dissipative distribution. The semi-group generated by it is denoted by $\{\mu_t^{(a)}\}_{t>0}$ and is called subordinated to $\{\mu_t\}_{t>0}$.

The aim of this section is to prove that if $\{\mu_t\}_{t>0}$ is a semi-group of measures in P(G) which defines a strongly continuous semi-group on L^1_w, where w is a polynomial weight, then the subordinated semi-group defines a strongly continuous semi-group on $L^1_{w^\alpha}$ for some $0 < \alpha \leq 1$. First we need three easy lemmas.

<u>Lemma 5.1.</u> Let $\varphi : \underline{R}^+ \longrightarrow \underline{R}^+$ be a function with the following properties

(5.1) $\varphi(s+t) \leq C(\varphi(s) + \varphi(t))$

(5.2) $\varphi(s+t) \leq \varphi(s)\ \varphi(t)$

(5.3) $\qquad \varphi(t) \leq K$ for $t \in (0,1]$.

Then for constants M and k we have

$$\varphi(t) \leq Mt^k \text{ for } t \geq 1.$$

Proof. First we notice that (5.1) implies

$$\varphi(2t) \leq 2C\varphi(t) ,$$

whence

$$\varphi(2^n t) \leq (2C)^n \varphi(t) .$$

Thus, if

$$l = \sum_{j=0}^{n} w_j 2^j, \quad w_j \in \{0,1\} ,$$

we have

$$\varphi(lt) = \varphi(\sum_{j=0}^{n} w_j 2^j t) \leq C(w_n(2C)^n \varphi(t) + \varphi(\sum_{j=0}^{n-1} w_j 2^j t)$$

$$\leq C^{n+1} \sum_{j=0}^{n} w_j 2^j \varphi(t) \leq C^2 C^{\log_2 l} l \, \varphi(t)$$

$$= C^2 \, l^{1+\log_2 C} \varphi(t) .$$

This by (5.2) and (5.3) completes the proof of lemma 5.1.

Lemma 5.2. Suppose w is a polynomial weight on G and $\{\mu_t\}_{t>0}$ is a semi-group of measures in $P(G)$ such that $< \mu_t, w > \leq K$ for $t \in (0,1]$. Then for constants M and k we have $< \mu_t, w > \leq Mt^k$.

Proof. We put $\varphi(t) = < \mu_t, w >$ and apply lemma 5.1.

Lemma 5.3. If the conclusion of lemma 5.2 holds, then for $0 < \alpha < 1$ we have

$$< \mu_t, w^\alpha > \leq M^\alpha t^{\alpha k} \text{ for } t \geq 1 .$$

Proof. Since $0 < \alpha \leq 1$ and $\mu_t \in P(G)$, we have

$$< \mu_t, w^\alpha > \int w^\alpha(x) d\mu_t(x) \leq (\int w(x) d\mu_t(x))^\alpha \leq M^\alpha t^{\alpha k} .$$

Theorem 5.4. Suppose that w is a polynomial weight, $\{\mu_t\}_{t>0}$ semi-group of measures in $P(G)$ which by left convolutions defines a strongly continuous semi-group of operators on L^1_w. Let for an $0 < a < 1$ $\{\mu_t^{(a)}\}_{t>0}$ be the subordinated semi-group. Then there exists an $0 < \alpha \leq 1$ such that for every non-zero, non-negative function f in the domain of the infinitesimal generator A on L^1_w f belongs to the domain of $-|A|^a$ and $< \||A|^a f| , w^\alpha > < +\infty$, which shows that f belongs to the domain of $-|A|^a$ in L^1_w .

Proof. In virtue of proposition 4.2 is suffices to show that

$$< \| A|^a f| , w^\alpha > < +\infty \; .$$

We have

$$c < \| A|^a f| , w^\alpha > \leq \int_0^\infty t^{-a-1} < |\mu_t * f - f| , w^\alpha > .$$

We estimate

$$\int_0^1 t^{-a-1} < |\mu_t * f - f| , w^\alpha > dt \leq \int_0^1 t^{-a-1} < |\mu_t * f - f|, w > dt$$

$$\leq c_1 \int_0^1 t^{-a} \, dt, \quad \text{since } f \text{ belongs to the domain of } A \text{ in } L^1_w .$$

Also

$$\int_1^\infty t^{-a-1} < |\mu_t * f - f| , w^\alpha > dt$$

$$\leq \int_1^\infty t^{-a-1} < \mu_t * f , w^\alpha > dt + \int_1^\infty t^{-a-1} < f, w > dt$$

$$\leq \int_1^\infty t^{-a-1} < \mu_t , w^\alpha > dt < f, w > + \int_1^\infty t^{-a-1} dt < f, w > \; .$$

But, since w is a polynomial weight, by lemma 5.3 we get

$$< \mu_t , w^\alpha > \leq M^\alpha t^{\alpha k} ,$$

so, taking $\alpha = (a-\varepsilon)k^{-1}$, we get

$$\int_1^\infty t^{-a-1} < \mu_t , w^\alpha > dt \leq M^\alpha \int_1^\infty t^{-1-\varepsilon} dt < \infty ,$$

which completes the proof of the theorem.

6. Stable semi-group. On $\underline{R}^r$ a semi-group $\{\mu_t\}_{t>0}$ of probability measures is (non-isotropically) stable, if the infinitesimal generator of it is of the form

$$- |A_1|^{a_1} - \dots - |A_k|^{a_k}$$

where $0 < a_j \leq 1$ and A_i is a partial Laplace operator

$$A_i = \frac{\partial^2}{\partial x_{i1}^2} + \dots + \frac{\partial^2}{\partial x_{ij_i}^2}$$

for a partition of the coordinates $x_{11}, \dots, x_{1j_1}, \dots, x_{k1}, \dots, x_{kj_k}$.

For a general (connected) Lie group let $X_1, \dots, X_k$ be some fixed elements in the Lie algebra. Then. of course,

$$C_c^\infty(G) \ni f \longrightarrow (X_1^2 + \dots + X_k^2) f(e) \in \underline{C}$$

is a dissipative distribution. The semi-group of measures $\{\mu_t\}_{t>0}$

in P(G) defined by it has the infinitesimal generator which is the closure of $X_1^2+\dots+X_k^2$ defined on C_∞^2 and, since it anihilates the constant functions, μ_t are probability measures for all $t > 0$. Such semi-groups are called Gaussian.

Definition. Let (S) be the smallest class of semi-groups of measures in P(G) which contains Gaussian semi-groups and is closed with respect of taking sums of the generators and subordination.

The following two properties of the semi-groups in (S) are easily deduced from what has been proved in sections 2-5.

(6.1) If $\{\mu_t\}_{t>0} \in (S)$, then $\mu_t(G) = 1$ for all $t > 0$.

(6.2) If $\{\mu_t\}_{t>0} \in (S)$, then for a non-trivial polynomial weight w we have $< \mu_t, w > < +\infty$. (μ_t have "fractional moments".)

Let us conclude with some problems and comments.

Problem 1. Suppose $\{\mu_t\}_{t>0} \in (S)$, is every μ_t absolutely continuous with respect to the Haar measure of a Lie subgroup of G ?

The answer is positive in case of Gaussian measures [10],[3] and also for semi-groups generated by $-|X^2|^a - |Y^2|^b$, $0 < a,b \leq 1$, where X,Y,Z is the basis of the Heisenberg Lie algebra with $[X,Y] = Z$, [2] .

Let (X,dx) be a locally compact space with a Radon measure dx. Consider a strongly continuous semi-group $\{T_t\}_{t>0}$ of operators defined on all $L^p(X)$, $1 \leq p < \infty$ such that $\|T_t f\|_{L^p} \leq \|f\|_{L^p}$, T_t is self-adjoint on $L^2(X)$, $T_t 1 = 1$ and $T_t f \geq 0$ for $f \geq 0$.

We say that the semi-group $\{T_t\}_{t>0}$ has tauberian property. if for a φ in $L^\infty(X)$

$$\lim_{x\to\infty} T_{t_0}\varphi(x) = a \quad \text{for} \quad t_0 > 0 \quad \text{implies} \quad \lim_{x\to\infty} T_{0+}\varphi(x) = a.$$

The tauberian property formulated above means that for the Markov process associated with the semi-group $\{T_t\}_{t>0}$ for every (unbounded) set M the following implication holds :

If the probability of comming from infinity to M in time t_0 exists and is equal to a, then the probability of comming from infinity to M in an arbitrarily small time is a.

Problem 2. Suppose $\{\mu_t\}_{t>0} \in (S)$. Does it have tauberian property (on G) ?

Some very partial answers have been obtained in [5]. Also a better functional calculus has been recently obtained by T.Pytlik [9] which shows a possibility of generalizations of the results in [5] to larger class of groups.

A semi-group of measures in P(G) is called <u>holomorphic</u>, if the map

$$\underline{R}^+ \ni t \longrightarrow \mu_t \in M(G)$$

extends holomorphically to

$$\{ z: \operatorname{Arg} z < \theta \} \ni z \longrightarrow \mu_z \in M(G).$$

It is easy to verify that holomorphic semi-groups have tauberian property.

<u>Problem 3.</u> Is every semi-group of class (S) holomorphic ?

This is true for Gaussian semi-groups [7]. Some partial results have been recently obtained by the author for semi-groups in (S) on the Heisenberg group.

<u>References</u>

[1] M.Duflo, Representations de semi-groupes de measures sur un groupe localement compact, Ann.Inst.Fourier, Grenoble 28 (1978), 225-249.

[2] Paweł Głowacki, A calculus of symbols and convolution semi-groups on the Heisenberg group, Studia Math.(to appear).

[3] A.Hulanicki, Commutative subalgebra of $L^1(G)$ associated with a subelliptic operator on a Lie group G, Bull.Amer.Math.Soc. 81 (1975), 121-124.

[4] A.Hulanicki, The distribution of energy in the Brownian motion in the Gaussian field and analytic-hypoellipticity of certain subelliptic operators on the Heisenberg group, Studia Math.56 (1976), 165-173.

[5] A.Hulanicki, A tauberian property of the convolution semi-group generated by $X^2 - |Y|$ on the Heinsenberg group, Preceedings of Symposia in Pure Math.35, Part 2 (1979), 403-405.

[6] G.A.Hunt, Semi-groups of Measures on Lie Groups, Trans.Amer. Math.Soc.81 (1956), 264-293.

[7] Jan Kisyński, Holomorphicity of semigroups of operators generated by sublaplacians on Lie groups, Lecture Notes in Math. Springer-Verlag.

[8] Thomas G.Kurtz, A random Trotter product formula, Proc.Amer. Math.Soc. 35 (1972), 147-154.

[9] T.Pytlik, Functional calculus on Beurling algebras, (to appear)

[10] D.Wehn, Some remarks on Gaussian distributions on a Lie group, Z.Wahrscheinlichkeitstheorie verw.Geb.30 (1974), 255-263.

[11] K. Yoshida, Functional Analysis, Berlin-Göttingen-Heidelberg: Springer (1965).

Institute of Mathematics
Polish Academy of Sciences
ul. Kopernika 18,
51-617 Wrocław
Poland

CONVERGENCE OF TWO-SAMPLE EMPIRICAL PROCESSES

T. Inglot

1. Introduction. Let (ξ_n) , (η_n) be independent sequences of independent identically distributed real random variables with the same distribution function $F(t)$. Let $F_n(t)$, $G_n(t)$ be empirical distribution functions of (ξ_n) and (η_n), respectively. Classical Kolmogorov's result states that for continuous F the sequence $\{n^{1/2} \sup_t |F_n(t) - F(t)|\}$ of real random variables converges in distribution to the so-called Kolmogorov's distribution. Smirnov proved two-sample analogue of this fact, namely, that the sequence $\{(nk/(n+k))^{1/2} \sup_t |F_n(t) - G_k(t)|\}$ converges in distribution to the Kolmogorov's distribution (in fact, he needed an additional assumption, that n/k is constant). However, the original proofs of these facts were rather complicated. Donsker showed (see [1] for the proof and references) that there is a convergence in distribution in Skorohod topology of the space $D[0,1]$ of one-sample empirical process $n^{1/2}(F_n(t) - F(t))$ to the Brownian Bridge process $W^o(F(t))$. The two-sample case was not covered by that considerations. Recently, Dudley ([4], section 6) proved that the sequence $\{(nk/(n+k))^{1/2} (F_n(t)-G_k(t))\}$ converges in distribution in $D[0,1]$ to the Brownian Bridge process $W^o(F(t))$. More precisely, he considered more general problem of empirical measures instead of empirical processes and the proof is based on the theorem of Wichura. Restricting ourselves to empirical processes only the present note provides an alternative proof of the same fact in the spirit of weak convergence methods of the space $D[0,1]$. The main idea of the proof is the use of a well-known characterization of Gaussian measures (see [5]). We also give some corollaries connected with two-sample problem.

2. Fundamental Lemma. Let $D = D[0,1]$ be the space of right-continuous functions on the unit interval with left limits at every point endowed with the Skorohod metric d_o (see [1]). $\mathcal{D}$ stands for Borel σ-algebra of subsets of (D,d_o). In the product $D \times D$ we always consider the product topology and the product σ-algebra.

Define on $D \times D$ a family of measurable transformations (called rotations) $\{T_{uv}\}$, where $u^2+v^2 = 1$, as follows: $T_{uv}(x,y) = (ux-vy,\ vx+uy)$, $x,y \in D$. Then we have :

(i) rotations are continuous on $D \times C \cup C \times D$, where, $C = C[0,1] \subset D$ is the space of continuous functions;

(ii) if $u_n \to u$, $v_n \to v$, then $T_{u_n v_n}$ converges to T_{uv} uniformly on bounded sets of $D \times D$.

A random element X of D is called symmetric Gaussian if all its finite-dimensional distributions are symmetric Gaussian. The following characterization is well-known (see [5]):

A random element X of D is Gaussian iff the distribution of (X_1, X_2) in $D \times D$ is invariant under all rotations, where X_1, X_2 are any independent copies of X.

<u>Lemma.</u> Let A be a directed set without final element and such that there exists an increasing sequence cofinal with A. Let $\{T_{u_\alpha v_\alpha}\}_{\alpha \in A}$ be any net of rotations and $\{X_\alpha\}_{\alpha \in A}$, $\{Y_\alpha\}_{\alpha \in A}$ be nets of random elements of D such that the net $\{(X_\alpha, Y_\alpha)\}$ converges in distribution to (X,Y) in $D \times D$, where X,Y are independent identically distributed Gaussian random elements concentrated on C ($P(X \in C) = 1$). Then the net $\{T_{u_\alpha v_\alpha}(X_\alpha, Y_\alpha)\}$ converges in distribution to (X,Y).

<u>Proof.</u> It suffices to show that for every increasing sequence $\{\alpha_n\}$ cofinal with A, a sequence $\{T_{u_{\alpha_n} v_{\alpha_n}}(X_{\alpha_n}, Y_{\alpha_n})\}$ converges to (X,Y). This will follow if every such sequence has further increasing subsequence converging to (X,Y).

Let $\{\alpha_n\}$ be an increasing sequence cofinal with A. Then there exists an increasing subsequence $\{\alpha'_n\}$ such that $u_{\alpha'_n} \to u$, $v_{\alpha'_n} \to v$ for some u,v. Obviously $(X_{\alpha'_n}, Y_{\alpha'_n}) \xrightarrow{D} (X,Y)$. So by (i) (ii), assumption $P(X \in C) = 1$ and Theorem 5.5 from [1] we infer that

$$T_{u_{\alpha'_n} v_{\alpha'_n}}(X_{\alpha'_n}, Y_{\alpha'_n}) \xrightarrow{D} T_{uv}(X,Y) .$$

It follows by the above characterization of Gaussian random elements that $T_{uv}(X,Y)$ has the same distribution as (X,Y). The proof is complete.

<u>3. Applications</u>. Suppose $\{\xi_n\}$, $\{\eta_n\}$ are sequences of independent identically distributed real random variables having the distribution function $F(t)$. Assume $F(0) = 0$, $F(1) = 1$. Let $F_n(t)$, $G_n(t)$ be the empirical distribution functions of $\{\xi_n\}$ and $\{\eta_n\}$, respectively, and $X_n(t) = n^{1/2}(F_n(t) - F(t))$,

$Y_n(t) = n^{1/2}(G_n(t) - F(t))$ the corresponding empirical processes.

If W^o is the Brownian Bridge in D then we write W_F^o for the random element defined by $W_F^o(t) = W^o(F(t))$.

The following theorem is a two-sample analogue of Theorem 16.4 in [1].

Theorem 1. If the sequences $\{\xi_n\}$ and $\{\eta_n\}$ are independent then

$$\left(\frac{nk}{n+k}\right)^{1/2}(F_n - G_k)\xrightarrow[(n,k)]{D} W_F^o .$$

Proof. We may assume that F is continuous since the general case may be treated exactly as in the proof of Theorem 16,4 in [1]. The sequences $\{X_n\}$and $\{Y_n\}$ are independent. Theorem 16.4 in [1] yields $X_n \xrightarrow{D} X$, $Y_n \xrightarrow{D} Y$, where X,Y are independent Gaussian random elements distributed as W_F^o. Therefore $(X_n,Y_k) \xrightarrow[(n,k)]{D} (X,Y)$. Put in Lemma $A =\{(n,k) : n \geq 1, k \geq 1\}$ and $u_{nk} = (k/(n+k))^{1/2}$, $v_{nk} = (n/(n+k))^{1/2}$. Then we have

$$T_{u_{nk}\, v_{nk}}(X_n,Y_k)=\left(\left(\frac{nk}{n+k}\right)^{1/2}(F_n-G_k),(n+k)^{1/2}\left(\frac{nF_n+kG_k}{n+k} - F\right)\right)\xrightarrow[(n,k)]{D}(X,Y)$$

Now the theorem follows immediately.

It is possible to obtain a random version of Theorem 1. Let $\{\mu_n\}$, $\{\theta_n\}$ be sequences of positive integer-valued random variables and $\{a_n\}$, $\{b_n\}$ be constants going to infinity as $n \to \infty$.

Theorem 2. Suppose sequences $\{\xi_n\}$, $\{\eta_n\}$ are independent and $\mu_n/a_n \xrightarrow{P} \theta_1$, $\vartheta_n/b_n \xrightarrow{P} \theta_2$ with θ_1,θ_2 positive constants. Then

$$\left(\frac{\mu_n\vartheta_k}{\mu_n+\vartheta_k}\right)^{1/2}(F_{\mu_n} - G_{\vartheta_k})\xrightarrow[(n,k)]{D} W_F^o .$$

Proof. Without loss of generality we may assume that F is continuous, $0 < \theta_1 = \theta_2 = \theta < 1$ and a_n, b_n are integers.

Let $D_D[0,1]$ be the space of right continuous functions defined on the unit interval with values in D, endowed with Skorohod metric (see [2] for details). Using the Invariance Principle for such spaces (see [2]) and similar argumentation as in the proof of Theorem 2 in [2] or Theorem 17.1 in [1] (but for two sequences $\{X_n\}$ and $\{Y_k\}$ jointly) we claim

$$(a_n{}^{1/2}(F_{\mu_n}-F),b_k{}^{1/2}(G_{\vartheta_k}-F))\xrightarrow[(n,k)]{D}(\theta^{-1/2}\,X,\ \theta^{-1/2}\,Y),$$

where X, Y are (as above) independent random elements of D distributed like W_F^o. Now, put in Lemma $A = \{(n,k): n \geq 1, k \geq 1\}$ and $u_{nk} = (b_k)/(a_n+b_k))^{1/2}$, $v_{nk} = (a_n/(a_n+b_k))^{1/2}$. Then

$$((\frac{a_n b_k}{a_n+b_k})^{1/2}(F_{\mu_n}-G_{\vartheta_k}),(a_n+b_k)^{-1/2}(a_n F_{\mu_n}+b_k G_{\vartheta_k}-(a_n+b_k)F))\xrightarrow[(n,k)]{D}\theta^{-1/2}(X,Y)$$

It is obvious that

$$(\frac{\mu_n\vartheta_k}{\mu_n+\vartheta_k})^{1/2}(\frac{a_n b_k}{a_n+b_k})^{-1/2}\xrightarrow[(n,k)]{P}\theta^{1/2} .$$

These two last facts give our assertion.

4. Renyi-Wang Statistics. Many known statistics, as it was indicated in [3], may be applied in Theorem 1 and 2 to obtain several limit theorems. As an example we consider Renyi-Wang statistics [7],[8] of the form

$$\varphi_{ab}(x,y) = \sup_{a\leq y(t)<b}\frac{|x(t)|}{y(t)},\ \varphi^+_{ab}(x,y) = \sup_{a\leq y(t)<b}\frac{x(t)}{y(t)},$$

$$\varphi^-_{ab}(x,y) = \inf_{a\leq y(t)<b}\frac{x(t)}{y(t)},$$

defined on $D \times D^+$, where $0 < a < b < \infty$ and D^+ denotes the subset of nonnegative, nondecreasing functions belonging to D (D^+ is a closed subset of D). To make the definition correct we take a convention that the supremum over the empty set is equal to zero.

Let C^+_{ab} denotes the subset of $C \cap D^+$ with the property that $y(t) = a$ for at most one value of t and the same for $y(t) = b$ for any $y \in C^+_{ab}$.

The following Proposition may be proved by a standard argument (so we omit the proof).

Proposition. The mappings $\varphi_{ab}, \varphi^+_{ab}, \varphi^-_{ab}$ are Borel measurable on $D \times D^+$ and are continuous on $C \times C^+_{ab}$.

Theorems 1 and 2 together with Proposition give the following two corollaries. First of them is a generalization of the result of Wang [8].

Corollary 1. Let $\{\xi_n\}, \{\eta_n\}, \{F_n\}, \{G_n\}$ be as in Theorem 1 and let $a < b$ be such that $F \in C^+_{ab}$. Then

$$(1)\quad \sup_{a\leq G_k(t)<b}(\frac{nk}{n+k})^{1/2}\frac{|F_n(t)-G_k(t)|}{G_k(t)}\xrightarrow[(n,k)]{D}\sup_{a\leq t<b}\frac{|W^o(t)|}{t},$$

$$(2)\quad \sup_{a\leq G_k(t)<b} \left(\frac{nk}{n+k}\right)^{1/2} \frac{F_n(t)-G_k(t)}{G_k(t)} \xrightarrow[(n,k)]{D} \sup_{a\leq t<b} \frac{W^o(t)}{t},$$

$$(3)\quad \inf_{a\leq G_k(t)<b} \left(\frac{nk}{n+k}\right)^{1/2} \frac{F_n(t)-G_k(t)}{G_k(t)} \xrightarrow[(n,k)]{D} \inf_{a\leq t<b} \frac{W^o(t)}{t}.$$

Proof. According to the Glivenko-Cantelli Theorem we have $G_k \xrightarrow{P} F$. It follows by Theorem 1 and Theorem 4.4 in [1] that

$$\left(\left(\frac{nk}{n+k}\right)^{1/2}(F_n-G_k),\ G_k\right) \xrightarrow[(n,k)]{D} (X,F),$$

where X is distributed like W_F^o and so concentrated on C (F is continuous). Since $F \in C^+_{ab}$ Proposition can be applied. Moreover

$$\sup_{a\leq F(t)<b} |W^o(F(t))|/F(t) = \sup_{a\leq t<b} |W^o(t)|/t,$$

by continuity of F (and similary for other statistics). This proves our corollary.

Recall that Renyi (see [6],[7]) found the limiting distribution in (1), (2),(3), for $b > 1$.

Corollary 2. Let $\{\xi_n\}$, $\{\eta_n\}$, $\{F_n\}$, $\{G_n\}$, a, b, F be as in Corollary 1 and let $\{\mu_n\}$, $\{\vartheta_n\}$, a_n, b_n be as in Theorem 2. Then

$$\sup_{a\leq G_{\vartheta_k}(t)<b} \left(\frac{\mu_n\vartheta_k}{\mu_n+\vartheta_k}\right)^{1/2} \frac{|F_{\mu_n}(t)-G_{\vartheta_k}(t)|}{G_{\vartheta_k}(t)} \xrightarrow[(n,k)]{D} \sup_{a\leq t<b} \frac{|W^o(t)|}{t}$$

and likewise for two other statistics.

Proof. By Theorem 1 of Csörgö [3] (see also [2]) we have $G_{\vartheta_k} \xrightarrow{P} F$. The rest of the proof is exactly as in Corollary 1.

References

[1] P.Billingsley, Convergence of probability measures, Wiley, 1967.

[2] T.Byczkowski, T.Inglot, The invariance principle for vector valued random variables with applications to functional random limit theorems, to appear in Lecture Notes in Statistics.

[3] S.Csörgö, On weak convergence of the empirical process with random sample size, Acta Sci. Math.Szeged. 36 (1974),17-25.

[4] R.M. Dudley, Central limit theorems for empirical measures, Ann.Prob.6 (1978), 899-929.

[5] X. Fernique, Integrabilite des vecteurs gaussiens, C.R. 270 (1970), 1698-1699.

[6] M. Fisz, Probability theory and mathematical statistics, (1963, third edition) John Wiley, New York.

[7] A. Renyi, On the theory of order statistics, Acta Math. Acad.Sci.Hung. 4 (1953), 191-227.

[8] S.J. Wang, On the limiting distribution of the ratio of two empirical distributions, Acta Math.Sinica 5 (1955).

Institute of Mathematics
Wrocław Technical University
50-370 Wrocław, Poland

V-DECOMPOSABLE MEASURES ON HILBERT SPACES

R. Jajte

0. INTRODUCTION

We are going to consider some classes of infinitely divisible probability measures on Hilbert spaces. The measures we shall deal with arise as the limit laws for some special arrays of Hilbert space-valued random variables. Let us begin with some notation. Let H be a separable real Hilbert space. For a probability measure m on H and a linear operator A acting in H, we define a measure Am by putting $(Am)(E) = m\{x \in H: Ax \in E\}$ for every Borel subset E of A. Let $V = \{V_a, a \in R^+\}$ denote a weakly continuous unitary representation in H of the multiplicative group $R^+ = (0,\infty)$ of positive numbers. Let us consider a representation $S = \{S_a, a \in R^+\}$ where $S_a = aV_a$ for all $a \in R^+$. The purpose of this paper is to introduce and characterize some classes of limit laws for the sums of the form

$$\eta_n = S_{a_n}\sum_{k=1}^{k_n} \xi_k + x_k ,$$

where $\{\xi_k\}$ is a sequence of H-valued independent random variables, and $\{x_k\} \subset H$. A natural modification of the definitions of stable, semi-stable and self-decomposable measures will lead us to the notions of V-stable, V-semi-stable and V-decomposable measures. In comparison with the classical situation, we consider the norming of sums of random variables by operators (instead of numbers). In our context, we obtain the classical case by considering the trivial representation $V = \{V_a \equiv I$ (identity)$\}$. In general, besides multiples aI of the unit operator I we also consider the rotations (given by the unitary operators) taken from the representation V. It should be mentioned here that not long ago K.Urbanik considered the operator-decomposable measures in a very general setting [26]. Namely, he considered the limit laws for the sums of the form

$$A_m(\xi_1 + \dots + \xi_n) + x_n ,$$

where ξ_j are independent Banach space-valued random variables, A_n - linear automorphisms of the Banach space. The author assumed that semi-group generated by the operators

$\{A_m A_n^{-1} : n = 1,2,\dots,m;\ m = 1,2,\dots\}$ is compact in the norm topology. Urbanik's approach was developed in Krakowiak's papers under the same assumption on the norming operators $\{A_n\}$. Let us remark that our results do not coincide with those of the works of Urbanik and Krakowiak. This follows from the fact that, in the case of the unbounded generator of the unitary group **V**, Urbanik's condition is not satisfied.

In the following three chapters we shall preserve all the notation introduced in this section.

1. V-stable measures

1.1. Definition. A probability measure μ on H is said to be V-stable if, for every pair a,b of positive numbers, there exist $c > 0$ and $x \in H$, such that

(1.1) $$S_a\mu * S_b\mu = S_c\mu * \delta(x) .$$

The operator-stable measures on R^N were introduced and examined by M.Sharpe [21] and later, by J.Kucharczak [10]. Recently, operator-stable measures on Banach spaces have been considered by W.Krakowiak [11]. In this chapter we are interested in some special cases of operator-stable measures on Hilbert spaces. Our results do not coincide with those of Krakowiak (see Introduction) We omit here the extensive bibliography concerning stable measures on Hilbert and Banach spaces. We shall refer to only a few papers when we use some techniques developed earlier.

From the above definition the following can easily be deduced.

1.2. If μ is V-stable, then, for any positive integer n, there exist $a_n > 0$ and $x_n \in H$, such that

$$\mu = \delta(x_n) * S_{a_n} \mu^n$$

Here, and in the next chapters, the following continuity lemma will be needed.

1.3. Lemma. Let μ, $\{\mu_n\}$ be probability measures on H , and let a, $\{a_n\}$ be positive numbers. Assume that $a_n \longrightarrow a$ and $\mu_n \longrightarrow \mu$ (weakly). Then $S_{a_n} \mu_n \longrightarrow S_a\mu$ (weakly).

Proof. It suffices to show that

(*) $$\liminf_n S_{a_n} \mu_n(U) \geq S_a\mu(U)$$

for every non-empty open subset U of H. Let U be such a set, and let $x \in S_a^{-1}U$. We choose $\varepsilon > 0$ such that $\|z\| < 2\varepsilon$ implies $S_a x + z \in U$. We find a positive integer n_o such that

$\| S_{a_n} x - S_a x \| < \varepsilon$ for $n \geq n_o$.

Put $M = \sup_n |a_n|$ and $Q = \{h \in H : \|h\| < \frac{\varepsilon}{M}\}$.

Then we have $Q + x \subset \bigcap_{n \geq n_o} S_{a_n}^{-1} U$.

Let T_k denote the interior of the set $\bigcap_{n \geq k} S_{a_n}^{-1} U$. Then the above reasoning shows that $S_a^{-1}U \bigcup_{k=1}^{\infty} T_k$, and, consequently

$$S_a\mu(U) \leq \lim_k \mu(T_k).$$

Since $\mu_n \longrightarrow$ and T_k are open, we have

$$\mu(T_k) \leq \lim_n \inf \mu_n(T_k) \quad \text{for all } k.$$

But $T_k \subset S_{a_n}^{-1} U$ for $n \geq k$, so we obtain

$$\mu_n(T_k) \leq S_{a_n} \mu_n(U) \quad \text{for } n \geq k$$

Thus we have

$$\lim_n \inf S_{a_n} \mu_n(U) \geq \lim_k \mu(T_k) \geq S_a\mu(U),$$

and the proof is completed.

Using the continuity lemma 1.3 systematically, by an easy modification (mutatis mutandis) of the proofs of lemmas 2 and 3 in [2], we obtain the following two results

1.4. __Lemma.__ If, for a measure μ, $a_n > 0$, $x_n \in H$, we have

$$\lim_n S_{a_n} \mu^n * \delta(x_n) = \vartheta ,$$

and ϑ is non-degenerate, then

(1.3) $\quad a_n \longrightarrow 0$

and

(1.4) $\quad \frac{a_{n+1}}{a_n} \longrightarrow 1$ as $n \longrightarrow \infty$.

1.5. __Lemma.__ Let the assumption of Lemma 1.4 be satisfied. Then there exist a $\lambda > 0$ and a function $z : \mathbb{R}^+ \times \mathbb{R}^+ \longrightarrow H$, such that

(1.5) $\quad \hat{\vartheta}(S_a^* x)\, \hat{\vartheta}(S_b^* x) = e^{i(z(a,b),x)} \cdot \hat{\vartheta}(S^*_{(a^\lambda + b^\lambda)^{1/\lambda}} x)$

for all $a,b > 0$. $\hat{\vartheta}$ denotes the Fourier transform of ϑ .

1.6. <u>Lemma</u>. If μ is a V-stable measure, than, for any positive integer n , there exists a vector $x_n \in H$ such that

(1.6) $\qquad \mu^n = S_{n^{1/\lambda}} \mu * \delta(x_n)$.

<u>Proof</u>. From (2.2) we always have

(1.7) $\qquad \mu^n = S_{c^n} \mu * \delta(z_n)$

for some $c_n > 0$ and $z_n \in H$. From (1.5) it easily follows (by induction, cf. [19] that c_n in (1.7) is $n^{1/\lambda}$ for $n = 1,2,\ldots$.

Now we are in a position to prove the following characterization of V-stable measures

1.7. <u>Theorem</u>. A functional $\varphi : H \longrightarrow C$ is the characteristic functional of a V-stable measure if and only if either

(1.8) $\qquad \varphi(y) = \exp\{i(a,y) - 1/2(Dy,y)\}$

or

(1.9) $\qquad \varphi(y) = \exp\{i(a,y) + \int K(x,y)\, N(dy)\}$,

where $a \in H$, D is an s-operator commuting with the unitary group $V = \{V_t, t \in R^+\}$, and there exists $0 < \lambda < 2$ such that

(1.10) $\qquad S_a M = a^\lambda M$ for every $a \in R^+$,

where M is defined by the formula

(1.11) $\qquad M(dx) = \dfrac{1 + \|x\|^2}{\|x\|^2} N(dx)$.

K is, as elsewhere, the Lévy-Khinchine kernel given by the formula

$$K(x,y) = \left(e^{i(x,y)} - 1 - \frac{i(x,y)}{1+\|x\|^2}\right) \frac{1+\|x\|^2}{\|x\|^2} .$$

The representation (1.8 - 9) is unique.

<u>Proof</u>. The above theorem is a natural generalization of the result obtained earlier by the author [2]. In the proof we shall follow the general idea of [2], but we shall also exploit the technique developed by B. Rajput [19].

Let μ be an S-stable probability measure on H.

To exclude the trivial complications, we assume that μ is non-degenarate. It follows from lemmas 1.2 and 1.4 that μ is infinitely divisible.
Writing the Lévy-Khinchine representation for the characteristic function $\hat{\mu}$ of μ, we have

$$\hat{\mu}(y) = \exp \rho(y) ,$$

where

$$\rho(y) = i(x_0,y) - 1/2(Dy,y)+\int (e^{i(x,y)}-1-\frac{i(x,y)}{1+||x||^2})M(dx), \tag{1.12}$$

where M is given by (1.11).
We shall now show that, for any $t > 0$, there exists a $x_t \in H$ such that

$$t\rho(y) = \rho(S^*_{t^{1/\lambda}} y) + i(x_t,y) , \tag{1.13}$$

where λ is given by lemma 1.5.
In fact, by (1.6), we have (1.13) for $t = n$ $(n = 1,2,\ldots)$.
For $t = \frac{k}{n}$, where k,n are positive integers, we have

$$\begin{aligned} n\rho(S^*_{(k/n)^{\lambda}} y) &= \rho(S^*_{n^{1/\lambda}(k/n)^{1/\lambda}} y) + i(\overline{x}_{k/n} , y) \\ &= k\rho(y) + i(x_{k/n} , y) \end{aligned}$$

for some $\overline{x}_{k/n}$, $x_{k/n} \in H$, so we have (1.13) for a rational t.
For an arbitraty $t > 0$, let $\{r_n\}$ be a sequence of positive rationals such that $r_n \to t$. By the continuity of ρ and of the group $\{S_a, \ a \in R^+\}$, we have

$$\rho(S^*_{r_n^{1/\lambda}} y) \to \rho(S^*_{t^{1/\lambda}} y) \quad \text{as } n \to \infty$$

Since $i(x_{r_n} , y) = r_n \rho(y) - \rho(S^*_{r_n^{1/\lambda}} y)$, by passing to the limit, we obtain (1.13) for all $t > 0$.
From the formulas (1.13) and (1.12), and by the uniqueness of the Lévy-Khinchine representation, we obtain

$$tD = t^{2/\lambda} V_{t^{1/\lambda}} DV^*_{t^{1/\lambda}} , \quad t > 0 , \tag{1.14}$$

and

$$S_{t^{1/\lambda}} M = t^{\lambda} M , \qquad t > 0 . \tag{1.15}$$

Then, from (2.14), by the unitarity of V_s, we obtain

(1.16) $(t - t^{2/\lambda})$ trace $D = 0$

so, either $D = 0$ or $\lambda = 2$.
Now, we shall show that in the case $\lambda \geq 2$ we have $M \equiv 0$.
In fact, let us fix same $0 < d < 1$, and put $Z_o = \{d \leq \|x\| < 1\}$
and $Z_n = \{S_{d^n} x : x \in Z_o\}$ for $n = 1,2,\ldots$, . If $M(Z_o) = 0$,
then $M(H - \{0\}) = 0$, i.e., $M \equiv 0$.
Assuming $M(Z_o) > 0$, we obtain

$$\int_{\|x\| \leq 1} \|x\|^2 M(dx) = \sum_0^\infty \int_{Z_n} \|x\|^2 M(dx) \geq$$

$$\geq \sum_0^\infty d^{2(n+1)} (S_{d^{-n}} M)(Z_o) = d^2 M(Z_o) \sum_0^\infty d^{(2-\lambda)n} = \infty ,$$

which contradicts the assumption that M is a Lévy-Khinchine spectral measure. From the formula (1.14), for $\lambda = 2$, we obtain

(1.17) $V_t D = D V_t$ for $t > 0$,

so, a Gaussian measure with the covariance operator D is V-stable if and only if (1.17) holds.
It is easy to check that the formula (1.9) describes the Fourier transform of a V-stable probability measure. The uniqueness of our representation follows immediately from the uniqueness of the Lévy-Khinchine-Varadhan representation of an infinitely divisible measure on a Hilbert space [15, 23].

2. V-decomposable measures

2.1. **Definition.** A probability measure μ on H is said to be V-decomposable if, for each $0 < a < 1$, there exists a probability measure μ_a such that

(2.1) $\mu = S_a \mu * \mu_a$,

One can easily prove that, if μ is non-degenerate and S_a-decomposable (in the sense of (2.1)) for some $a \in R^+$, then it must necessarily follow that $0 < a < 1$ (cf.[16], p.322).
The problem of characterizing self-decomposable measures ($V_a \equiv I$) on the real line has been solved by P.Lévy [15].

In 1968 K. Urbanik gave another characterization [22]. Urbanik's representation has been extended to R^N by himself [23] and to H by Kumar and Schreiber [14]. In 1972 K. Urbanik introduced and examined operator self-decomposable measures on Euclidean spaces [24] and not long ago, he characterized a large class of such measures on Banach spaces [26]. The works of K. Urbanik were continued by W. Krakowiak [11, 12] and J. Kucharczak [10].

We shall give the description of V-decomposable measures. It will be the Urbanik type representation for the characteristic function of a V-decomposable measure.

2.2. Theorem. A function $\varphi : H \to C$ is the characteristic functional of a V-decomposable measure on H if and only if it is of the form

$$\varphi(y) = \exp \{ i(a,y) - 1/2(Dy,y) + \int_{H\setminus\{0\}} L(x,y) \frac{M(dx)}{\log(1+\|x\|^2)} \} , \tag{2.2}$$

where $a \in H$, M is an arbitrary finite Borel measure on $H\setminus\{0\}$, D is an S-operator such that

$$D - t^2 V_t D V_t^{-1} \geq 0 , \quad 0 < t < 1 , \tag{2.3}$$

and the kernel $L(x,y)$ is given by the formula

$$L(x,y) = \int_0^1 (e^{i(S_t x,y)} - 1 - \frac{i(S_t x,y)}{1+t^2\| x\|^2}) \frac{dt}{t} . \tag{2.4}$$

In the case when the generator A of the group $\{U_t\}$, where $U_t = V_{e^t}$ $(t \in R)$ is bounded, D satisfies the condition (2.3) if and only if

$$2D + i(DA - AD) \geq 0 \tag{2.5}$$

(of course, in the description of D we use the natural complex extensions of H and of operators acting in H).

The representation (2.2) is unique. Before starting the proof, let us remark that one can show that V-decomposable measures are the limit laws for the uniformly infinitesimal sums of independent random variables. This can be done in a similar way as in [14], by applying the continuity lemma 1.3. In particular, it follows from it that V-decomposable measures are infinitely divisible.

<u>Proof of Theorem 2.2.</u>

The proof is based on the extreme-point method (Krein-Milman-Choquet Theorem [7, 1, 18]) adapted to the theory of infinitely divisible measures by Kendall [6], Johansen [5], Urbanik [22-26], and others.

Let us denote by U the closed unit ball in H, Σ denotes the unit sphere in H, $H_o = H \setminus \{0\}$ $U_o = U \cap H_o$.
$\mathbb{K} = U \times [0,\infty]$, where $[0,\infty]$ is a compactification of $\mathbb{R}^+ = (0,\infty)$.
Let us consider a mapping $\pi : H_o \longrightarrow \mathbb{K}$ given by the formula

(2.6) $\qquad \pi(x) = < \frac{S_1}{\|x\|} x \, , \, \|x\| > , \quad x \in H_o$

Let us endow U with the relative weak topology of H.
Then U and K become compact metric spaces, and π is a Borel automorphism from H_o onto $\Sigma \times \mathbb{R}_+$ (because Borel fields on H with respect to the norm and weak topologies coincide).

Let us denote by $\mathbb{M}(K)$ the set of all finite non-negative Borel measures on $\mathbb{K}$, and by $\mathbb{M}^o K)$ the subset of $\mathbb{M}(\mathbb{K})$, consisting of those measures which are concentrated on $\Sigma \times \mathbb{R}^+$.
Let $\mathbb{P}(\mathbb{K})$ be the set of all probability measures on $\mathbb{K}$ (with the topology of weak convergence). $\mathbb{P}(\mathbb{K})$ is a compact metric space (see, e.g., [17], p.45).
For $q = (x,r) \in K$ and $a \in [0, \infty]$, let us put $\|q\| = r$ and $aq = < x, ar >$.
Then we have

(2.7) $\qquad \|\pi(x)\| = \|x\|$
and
(2.8) $\qquad \pi(S_a x) = a\pi(x)$.

For a finite Borel measure m on $\mathbb{K}$, we put

(2.9) $\qquad I_m(E) = \int_E (1 + \|q\|^{-2}) \; m(dq)$.

Let us denote by $\mathbb{M}^*$ the set of all the measures m from $\mathbb{M}$ for which

(2.10) $\qquad I_m(E) - I_m(a^{-1}E) \geq 0$

for all $0 < a < 1$ and all E such that $\inf\{\|q\| : q \in E\} > 0$.

Let $\mathbb{P}^* \subset \mathbb{P}(K)$ denote the subset of $\mathbb{M}^*$, consisting of probability measures.
Obviously, $\mathbb{M}^*$ and $\mathbb{P}^*$ are convex. Moreover, $\mathbb{P}^*$ is metric and compact (in the weak topology).

Let us remark that the extreme points of $\mathbb{P}^*$ are concentrated on one of the following sets

(2.11) $\quad < x,0 >, < x, \infty >$ or $F_x = \{ < x,t > , t \in (0, \infty) \}$,

where x runs over U. This follows immediately from the fact that, if a Borel subset Z of $\mathbb{K}$ is invariant under dilatations $q \longrightarrow aq$, and $m \in M^*$, then the restriction $m|Z$ belongs to $\mathbb{M}^*$, too.
The extreme points of $\mathbb{P}^*$ concentrated on the sets F_x can be treated, in a natural way, as the measures on R^+ , so we can use the result of K. Urbanik [22]. Namely, every such extreme point (considered as a measure on R^+) is of the form

$$(2.12) \qquad m_c(dt) = \frac{2t\, I_{[o,c]}(t)}{\log(1+c^2)(1+t^2)}\, dt ,$$

where $I_{[o,c]}$ denotes the indicator of the interval $[o,c]$,
and c is an arbitrary fixed positive number.
Thus every extreme point concentrated on F_x is given by a pair $< x,c > \in \mathbb{K}$, where $x \in U$ is a "direction" of F_x , and $c \in R^+$.
This extreme point will be denoted by $m_{< x,c >}$.
Evidently, the extreme points of $\mathbb{P}^*$ concentrated on one-point sets $< x,0 >$ and $< x, \infty >$ (invariant under dilatations $a \longrightarrow aq$) are of the form

$$(2.13) \qquad m_{< x,0 >} = \delta_{< x, \infty >} \text{ and } m_{< x, \infty >} \text{ for } x \in U .$$

Thus we established a 1-1 mapping ρ between K and the set extr $\mathbb{P}^*$) of extreme points of $\mathbb{P}^*$. The mapping $\rho : K \longrightarrow \text{extr}(\mathbb{P}^*)$ is a Borel automorphism.

Let us now write the Lévy-Khinchine-Varadhan formula for an V-decomposable measure μ on H

$$(2.14) \qquad \hat{\mu}(y) = \exp \{i(a,y)-1/2(Dy,y)+ \int_{H_o} K(x,y)N(dx) ,$$

where $a \in H$, D is an S-operator, and N is a finite Borel measure on H_o . K is the Lévy-Khinchine kernel given by the formula

$$(2.15) \qquad K(x,y) = (e^{i(x,y)} - 1 - \frac{i(x,y)}{1+\|x\|^2})\, \frac{1+\|x\|^2}{\|x\|^2}$$

Writing the decomposition formula (2.1) in terms of the spectral decomposition (2.14), we see that N is a spectral measure of μ if and only if $N(\pi^{-1}) \in \mathbb{M}^*$.

By the Choquet theorem, we can write

$$(2.16)\qquad \int_{\mathbb{K}} f(n)N(\pi^{-1}dn) \int_{\mathbb{K}}\int_{\mathbb{K}} f(n)m_q(dn)\gamma(dq),$$

(where γ is a finite Borel measure on $\mathbb{K}$) for every continuous function f on $\mathbb{K}$. Evidently, $N(\pi^{-1})$ is concentrated on the set $\mathbb{K}_o = \pi(H_o)$ and, consequently, the measure γ is also concentrated on $\mathbb{K}_o$. Thus the measures m_q (for $q \varepsilon \mathbb{K}_o$) are also concentrated on $\mathbb{K}_o$, and we can write

$$(2.17)\qquad \int_{\mathbb{K}_o} f(n)N(\pi^{-1}dn) = \int_{\mathbb{K}_o}\int_{\mathbb{K}_o} f(n)m_q(dn)\gamma(dq)$$

for every bounded continuous function on $\mathbb{K}_o = \pi(H_o)$. It is easy to show that (1.14) holds also for every bounded continuous function f on $\mathbb{K}$.

Putting $\gamma(\pi\cdot) = m(\cdot)$ and $f(\pi\cdot) = g(\cdot)$, we can rewrite (2.17) in the form

$$(2.18)\qquad \int_{H_o} g(x)N(dx) = \int_{\mathbb{K}_o} | \int_0^a f\langle z,t\rangle \frac{tdt}{1+t^2}\,\frac{\gamma(d\langle z,a\rangle)}{\ln(1+a^2)} =$$

$$= \int_{H_o} | \int_0^{\|x\|} g(S_{\frac{t}{\|x\|}}\, x)\,\frac{t(dt)}{1+t^2}\,\frac{m(dx)}{\ln(1+\|x\|^2)}$$

by (2.12) and because, if $\pi : x \longrightarrow \langle z,a\rangle = q\ \varepsilon\ \mathbb{K}_o$, then $x = S_a z$ $z = S_{\frac{1}{\|x\|}}\, x$ and $\pi^{-1}\langle z,t\rangle = S_t\, z = S_t\, S_{\frac{1}{\|x\|}}\, x = S_{\frac{t}{\|x\|}}\, x$.

Let us now remark that,

$$(2.19)\qquad \int_0^{\|x\|} g(S_{\frac{t}{\|x\|}}\, x)\ \frac{t}{1+t^2}\, dt = \int_0^1 g(S_\tau\, x)\ \frac{\tau\|x\|^2\, d\tau}{1+\tau^2\|x\|^2}$$

(after the substitution $t = \tau\|x\|$).

Thus, putting the Lévy-Khinchine kernel (2.15) instead of g, we obtain (1.2). We shall now examine the covariance operator D (of the Gaussian part of (2.2)).

Let us take into consideration a unitary group $\{U_t,\ t \varepsilon R\}$, where $U_{\ln s} = V_s$ for $s > 0$. If this group is generated by a bounded, self-adjoint operator A, i.e., $U_t = e^{it A}$, $t \varepsilon R$ (after the natural extension to the complex space $\hat{H}$) we put $\ln t = \tau$.

Then the decomposition formula (2.1), written for the Gaussian part, gives

(2.20) $$D - t^2 V_t D V_t^{-1} \geq 0 .$$

This means that

(2.21) $$D - e^{2\tau}(I + i\tau A + o(\tau)) \; D(I - i\tau A + o(\tau)) \geq 0$$

Dividing by $(-\tau) > 0$ and passing to the limit as $\tau \to 0$, we obtain

(2.22) $$2D + i(DA - AD) \geq 0 .$$

In particular, for commuting operators A and D, (2.22) is satisfied. The condition can be written in a little more symmetric way as $B^* + B \geq 0$, where $B = D + iDA$.

Let us now assume that the condition (2.22) is satisfied, and put

(2.23) $$D_\tau = D - e^{2\tau} U_\tau D U_\tau^{-1}$$

and

(2.24) $$\varphi_x(t) = (D_t \, x, x) .$$

The routine computation shows that

(2.25) $$\frac{d}{dt} \varphi_x(t) = ((B + B^*) \, S^*_{e^t} \, x, \; S^*_{e^t} \, x) \geq 0.$$

But $\varphi_x(0) = 0$, so $\varphi_x(t) = (D_t \, x, x) \geq 0$.

This means that (2.20) is satisfied, so the Gaussian part is S-decomposable, which ends the proof.

The formula (1.2) describes the Fourier transform of an S-decomposable probability measure in H. In fact, writing the "Poissonian part" of (2.2) as the limit of integral sums and using the Prokhorov compactness criterion [17], we obtain, after routine calculations, that φ is a Fourier transform of an infinitely devisible measure μ. The Lévy's component μ_t of μ will then be of the form.

$$\hat{\mu}(y) = \exp \{ i(a_t, y) - \frac{t^2}{2} (V_t D V_t^{-1} y, y) + \int_{H \setminus \{0\}} L(x,y) \frac{m(dx)}{\ln(1+x^2)} \}, \; 0 < t < 1 ,$$

where

$$L(x,y) = \int_0^t \left(e^{i(S_\tau x,y)} - 1 - \frac{i(S_\tau x,y)}{1+\tau^2\|x\|^2}\right) \frac{d\tau}{\tau} ,$$

so μ is an S-decomposable measure.

The uniqueness of our representation can be proved in the same way as indicated, for example, in [23], p.147 (by taking the suitable functions f and g_f).

This completes the proof.

3. V-semi-stable measures

3.1. Definition. A probability measure μ on H is said to be V-semi-stable if it is a weak limit of measures of the form

$$(3.1) \qquad \mu = \lim_n S_{c_n} \theta^{K_n} * \delta(x_n),$$

where θ is a probability measure on H, $c_n > 0$, $x_n \in$ H, and (K_n) is an increasing sequence of positive integers such that $K_{n+1} / K_n \longrightarrow r < \infty$.

The class of semi-stable distributions on the real line $\mathbb{R}$ was introduced and examined by Kruglov [8]. Next, the same author extended his results to the case of a real Hilbert space [9]. He described the characteristic functions of semi-stable measures (for $S_a \equiv aI$) in the way analogous to that indicated in [2]. We shall follow the general ideas of [9] and [2].

3.2. Lemma. If μ is a non-degenerate V-semi-stable measure such that (3.1) holds, then

$$(3.2) \qquad \lim c_n = 0$$

and

$$(3.3) \qquad \lim \frac{c_{n+1}}{c_n} = d, \qquad 0 < d < \infty .$$

Proof. The proof of (3.2) is the same as in the case of V-stable measures. To prove (3.3), we assume that, for some subsequence (b_j) of (c_j), we have $\lim \frac{b_{j+1}}{b_j} = 0$.

Then

$$|\hat{\theta}((S^*_{b_j})^{-1} S_{b_{j+1}} S^*_{b_j}(x)|^{n_j} \longrightarrow |\hat{\mu}(0)| = 1$$

because

$$\| (S^*_{b_j})^{-1} S^*_{b_{j+1}} \| = \frac{b_{j+1}}{b_j} \|V_{b_j} V^{-1}_{b_{j+1}}\| \longrightarrow 0 \quad \text{as} \quad j \longrightarrow \infty.$$

On the other hand,

$$|\hat{\vartheta}((S^*_{b_j})^{-1} S^*_{b_{j+1}} S^*_{b_j} x|^{n_j} = |\hat{\vartheta}(S^*_{b_{j+1}} x)|^{n_{j+1} \frac{n_j}{n_{j+1}}} \longrightarrow$$
$$\longrightarrow |\hat{\mu}(x)|^{\frac{1}{r}},$$

where $r = \lim_k \frac{n_{k+1}}{n_k}$. Thus $|\hat{\mu}(x)| = 1$, and μ is degenerate (cf. [12], pp. 134–135).

In the same way we can prove that there is no subsequence $\{b_k\}$ of $\{c_k\}$ for which $\frac{b_{k+1}}{b_k} \longrightarrow \infty$. Let us now suppose that there are two limit points of the sequence $\frac{c_{k+1}}{c_k}$, say α and β. Let $\alpha < \beta$. Put $\gamma = \frac{\alpha}{\beta}$. Then we have

$$|\hat{\mu}(x)| = |\hat{\mu}(\alpha V^*_\alpha x)|^r = |\hat{\mu}(\beta V^*_\beta x)|^r$$

and, consequently,

$$|\hat{\mu}(x)| = |\hat{\mu}((\tfrac{\alpha}{\beta})^n V^n_{\frac{\beta}{\alpha}} x)| \longrightarrow 1 \quad \text{as} \quad n \longrightarrow \infty .$$

Thus μ would be degenerate. The lemma is proved.

3.3. <u>Theorem.</u> A functional $\varphi : H \longrightarrow C$ is a characteristic functional of a V-semi-stable measure μ if and only if either

(3.4) $$\varphi(y) = \exp\{i(a,y) - 1/2(Dy,y)\} ,$$

where D is the same as in Theorem 1.7,

or

(3.5) $$\varphi(y) = \exp\{i(a,y) + \int K(x,y)\, M(dx)\} ,$$

where $a \in H$, $K(x,y) = e^{i(x,y)} - 1 - \frac{i(x,y)}{1+\|x\|^2}$, and M is a semi-finite measure on $H \setminus \{0\}$, such that

(3.6) $$M\{x \in H : \|x\| > 1\} < \infty$$

and there exist two numbers $\alpha \in (0,2)$ and $0 < a \neq 1$ for which

(3.7) $$S_a \cdot M = a^\alpha M .$$

This representation is unique.

Proof. Let $\mu = \lim S_{c_k} \vartheta^{n_k} * \delta(x_k)$ for some $c_n > 0$, k_n and x_n as in Definition 3.1. By lemma 3.2, the limit (3.3) does exist. Let us first consider the case

$$(3.8) \qquad r = \lim_k \frac{n_{k+1}}{n_k} > 1.$$

Let us remark $\hat{\mu}$ satisfies the following condition

$$(3.9) \qquad |\hat{\mu}(x)| = |\hat{\mu}(dV_{\frac{1}{d}} x)|^r \cdot e^{i(x_o,x)}$$

with

$$(3.10) \qquad 0 < d = \lim \frac{c_{k+1}}{c_k} < 1 .$$

In fact, (3.9) follows immediately from the relationship

$$(3.11) \qquad \hat{\mu}(x) = \lim_k [\hat{\theta}(S^*_{c_{k+1}} x)]^{n_{k+1}} \cdot e^{i(\bar{x}_k,x)} =$$

$$= \lim_k [\hat{\theta}(S^*_{c_{k+1}} (S^*_{c_k})^{-1} S^*_{c_k} x)]^{n_k \frac{n_{k+1}}{n_k}} \cdot e^{i(\bar{\bar{x}}_k,x)} =$$

$$= |\hat{\mu}(dV_d^{-1} x)|^r \cdot e^{i(x_o,x)}$$

(for some $\bar{x}_k$, $\bar{\bar{x}}_k \in H$).

Contrary to (3.10), let us assume that $d \geq 1$.

If $d = 1$, then $[\hat{\mu}(y)]^r = \hat{\mu}(y)$ for $y \in H$.

Since $r > 1$, we obtain $\hat{\mu}(y) \equiv 1$, hence μ is degenerate which is impossible. Let $d > 1$. Then, by (3.9)

$$|\hat{\mu}(d^{-1}V_d y)| \leq |\hat{\mu}(y)|^r \leq |\hat{\mu}(y)| .$$

Thus, for every n, we would have

$$|\hat{\mu}(y)| \geq |\hat{\mu}(d^{-n} V_{d^n} y)| \longrightarrow 1 ,$$

which would imply again $\mu = \delta_x$ for some $x \in H$.

By lemma 3.2, every V-semi-stable measure is infinitely divisible. Let us write the Lévy-Khinchine-Varadhan formula the characteristic functional φ of the measure μ

$$(3.12) \qquad \varphi(y) = \exp [i(x,y) - 1/2(Dy,y) + \int K(x,y)\, M(dx)].$$

Writing the formula (3.9) in terms of (3.12), by the uniqueness of the representation (3.12), we obtain

(3.13) $$M = rS_dM = rdV_dM$$

and

(3.14) $$D = rd^2V_dDV_d^{-1} \quad .$$

The last formula gives $(1-rd^2)$ trace $D = 0$, so

(3.15) either $rd^2 = 1$ or $D = \theta$.

Let us put $\alpha = -\frac{\ln r}{\ln d}$, i.e., $rd^{\alpha} = 1$.

The formula (3.13) implies

(3.16) $$S_{d^k} M = d^{k\alpha}M$$

for all integers k.

Let us remark that, if $\alpha \geq 2$, then $M \equiv 0$ (the proof analogous as for V-stable measures (after the formula (1.16)).

Then, if $\alpha = 2$, we have

(3.17) $$\hat{\mu}(y) = \exp \{i(x_o,y) - 1/2(Dy,y)\} ,$$

i.e., μ is a Gaussian measure. The proof that D must commute with V is the same as for V-stable measures. If $\alpha \in (0,2)$, then we have

(3.18) $$\hat{\mu}(y) = \exp \{i(x_o,y) + \int K(x,y)\, M(dx)\}$$

with

$$M\{\|x\| > 1\} \leq \int_H \min(1, \|x\|^2)\, M(dx) < \infty \quad .$$

It remains to consider the case $r = 1$. By a rather standard reasoning, we can show that the measure μ is then V-stable. By Theorem 1.7, the conditions of our theorem hold, which ends the proof of necessity.

Now, let M be a semi-finite Borel measure on $H \setminus \{0\}$, satisfying the conditions (3.6) and (3.7). It is easy to show that M is then finite on the complement of every neighbourhood of zero in H. In fact, without loss of generality we may assume that $a > 1$. Then, for $\delta > 0$ and some n_o ,

$$\{x : \|x\| > \delta\} \subset \{x : \|x\| > (\tfrac{1}{a})^{n_o}\} = \{\|S_a^{n_o}x\| > 1\} .$$

Thus, for a sufficiently large n_o,

$$M\{\|x\| > \delta\} \leq M(S^{-1}_{a^{n_o}}\{\|x\| > 1\}) =$$

$$= (S_{a^{n_o}} M)\{\|x\| > 1\} = (a^{n_o})^{\alpha}\, M\{\|x\| > 1\} .$$

Let us define $Z_o = \{x : a \leq \|x\| < 1\}$, $Z_n = \{S_{d^n} x : x \in Z_o\}$ for $n = 1,2,\ldots$, and assume that $a \in (0,1)$ (if $a > 1$, then we can take a^{-1} instead of a).
After easy calculations we obtain

$$\int_{\|x\|<1} \|x\|^2 \, M(dx) \leq \sum_{0}^{\infty} a^{2n} M(Z_n) = M(Z_o) \sum_{0}^{\infty} a^{(2-\alpha)n} < \infty .$$

Thus M is a Lévy-Khinchine spectral measure, so φ, given by (3.5), is the characteristic functional of an infinitely divisible measure μ.
The representation (3.12) and property (3.7) of M lead us to

$$(3.19) \qquad \hat{\mu}(y) = e^{i(x_o,y)} [\hat{\mu}(S_a^* y)]^{a^{\frac{1}{\alpha}}}$$

for some $0 < a < 1$ and, consequently, to the more general formula

$$(3.20) \qquad \hat{\mu}(y) = e^{i(x_n,y)} [\hat{\mu}(S^*_{a^n} y)]^{a^{\frac{1}{n\alpha}}}$$

for $n = 1,2,\ldots$, and some $x_n \in H$, $a \in (0,1)$.
Putting $a^{-\alpha} = r > 1$, we find a sequence $\{n_j\}$ of positive integers, such that

$$(3.21) \qquad \lim_j \frac{n_j}{r^j} = 1 \quad \text{and} \quad n_j \leq r^j$$

and then, by (3.20) and (3.21), we have

$$(3.22) \qquad \hat{\mu}(y) = \lim_j [\hat{\mu}(S^*_{a^j} y)]^{n_j} \cdot e^{i(x_j,y)} .$$

It is sufficient to show that the sequence of measures $\{S_a^j \mu^{n_j} * \delta(x_j)\}$ is shift compact [15]. This follows immediately, by Prokhorov's theorem, from the inequalities

$$|\hat{\mu}(S^*_{a^j} y)|^{2n_j} \geq |\hat{\mu}(S^*_{a^j} y)|^{2r^j} = |\hat{\mu}(y)|^2$$

because then

$$(3.23) \qquad 1 - |\hat{\mu}(S^*_{a^j} y)|^{2n_j} \leq 1 - |\hat{\mu}(y)|^2 .$$

The uniqueness of the representation follows immediately from the uniqueness of the Lévy-Khinchine representation. This completes the proof.

3.4. Theorem. The class of characteristic functionals of V-semi-stable measures on H coincides with the class of all functions of the form (3.4) or of the form $e^{\phi(y)}$, where

$$(3.24) \qquad \phi(y) = \sum_{-\infty}^{\infty} d^{n\alpha} \int_{d \le \|x\| < 1} K(S_d^{-n} x, y)\, m(dx),$$

where $0 < d < 1$, $0 < \alpha \le 2$, m is an arbitrary finite Borel measure on the ring $\{d \le \|x\| < 1\}$, and K is the Lévy-Khinchine kernel, i.e.,

$$K(x,y) = e^{i(x,y)} - 1 - \frac{i(x,y)}{1+\|x\|^2} .$$

Proof. In view of Theorem 3.3, it suffices to show the formula (3.24) for the non-Gaussian case.

For the purpose, let us remark that, if M is a Lévy-Khinchine spectral measure of an S-semi-stable measure, then, for every M-integrable function f, putting $Z_o = \{d \le \|x\| < 1\}$, we have

$$\int_{H \setminus \{0\}} f(x) M(dx) = \sum_{-\infty}^{\infty} \int_{S_{d^n}^{-1} Z_o} f(x)\, M(dx) =$$

$$= \sum_{-\infty}^{\infty} \int_{S_{d^n}^{-1}} f(x) d^{n\alpha} (S_{d^{-n}} M)(dx) =$$

$$= \sum_{-\infty}^{\infty} \int_{Z_o} f(S_d^{-n} x)\, M(dx).$$

To obtain the formula (3.24), it suffices to put, for Borel subsets of Z_o,

$$m(E) = M(E) \quad \text{anf} \quad f(x) = K(x,y) .$$

From the above it follows that every Lévy-Khinchine spectral measure M of an S-semi-stable measure is of the form

$$(3.25) \qquad M(Z) = \sum_{-\infty}^{\infty} d^{n\alpha}\, m(S_d^n Z \cap Z_o)$$

for $Z \in$ Borel $(H \setminus \{0\})$, where m is a finite Borel measure on Z_o. To finish the proof, let us notice that every finite Borel measure on Z_o determines, by the formula (3.25), a Lévy-Khinchine measure M.

It suffices to show that

(3.26) $$\int_{\|x\|\leq 1} \|x\|^2 \, M(dx) < \infty$$

and

(3.27) $$d^{\alpha} M = S_d M .$$

We have

$$d^{\alpha}M(Z) = \sum_{-\infty}^{\infty} d^{(n+1)\alpha} \, m(S_d^n \, Z \cap Z_o) =$$

$$= \sum_{-\infty}^{\infty} d^{n\alpha} \, m(S_d^n \, (S_d^{-1} \, Z \cap Z_o)) = (S_d M)(Z),$$

so (3.27) holds.

The finiteness of the integral (3.26) follows easily from the convergence of the series

$$\sum_{0}^{\infty} d^{n\alpha} < \infty \quad \text{and} \quad \sum_{0}^{\infty} d^{n(2-\alpha)} < \infty \ .$$

This concludes the proof.

References

[1] G. Choquet, Le théoreme de représentation intégrale dans les ensembles convexes compact, Ann.Inst.Fourier 10 (1960),333-344,

[2] R. Jajte, On stable distributions in Hilbert spaces, Studia Math. 30 (1968), 63-71.

[3] R. Jajte, Semi-stable probability measures on R^N , Studia Math. 61 (1977), 29-39.

[4] R. Jajte, Semi-stable measures, Banach Center Publications, vol.5, 141-150.

[5] S. Johansen, An application of extreme-point methods to the representation of infinitely divisible distributions, Z.Wahrsch. und verw. Gebiete 5 (1966), 304-316.

[6] D.G. Kendall, Extreme-point methods in stochastic analysis, Z. Wahrsch. und verw. Gebiete 1 (1963), 295-300.

[7] M. Krein and D. Milman, On extreme points of regularly convex sets, Studia Math. 9 (1940), 133-138.

[8] V.M. Kruglov, On an extension of the class of stable distributions, Teor. Ver.Appl. 17 (1972), 723-732 (in Russian).

[9] V.M. Kruglov, On a class of limit laws in a Hilbert space, Lit. Mat. Sbornik 12 (1972), 85-88 (in Russian) .

[10] J. Kucharczak, Remarks on operator-stable measures, Coll.Math. 34 (1976), 109-119.

[11] W. Krakowiak, Operator-stable probability measures on Banach spaces, to appear.

[12] W. Krakowiak, Operator semi-stable probability measures on Banach spaces, to appear.

[13] A. Kumar and V. Mandrekar, Stable probability measures on Banach spaces, Studia Math. 42 (1972), 133-144.

[14] A. Kumar anb M.B. Schreiber, Self,decomposable probability measures on Banach spaces, Studia Math.53 (1975), 55-71.

[15] P. Lévy, Théorie de l'addition des variables aléatoires, Paris 1937.

[16] M. Loéve, Probability theory, New York, 1950

[17] K.R. Parthasarathy, Probability Measures in Metric Spaces, New York 1967.

[18] R.R. Phelps, Lectures on Choquet's theorem, Princenton, 1966.

[19] B.S. Rajput, A representation of the Characteristic Function of a Stable Probability Measure on Certain TV Spaces, J. Multivariate Analysis 6 (1976), 592-600.

[20] I. Ciszar, B. Rajput, A. Convergence of types theorem for probability measures on topological vector spaces with applications to stable laws, Z.Wahrschein, verw. Gebiete 36 (1976), 1-7.

[21] M. Sharpe, Operator-stable probability distributions on vector groups, Trans Amer.Math.Soc. 136 (1969), 51-65.

[22] K. Urbanik, A.representation of self-decomposable distributions Bull.Acad. Pol.Sci.Serie des.math.astronom, et phys. 16 (1968), 196-204.

[23] K. Urbanik, Self-decomposable probability measures on R^m, Applicationes Math. 10 (1969) 91-97.

[24] K. Urbanik, Lévy's probability measures on Euclidean spaces, Studia Math. 44 (1972), 119-148.

[25] K. Urbanik, Extreme-point method in probability theory, Probability Winter School - Karpacz 1975, Lecture Notes on Mathematics 472, 169-194.

[26] K. Urbanik, Lévy's probability measures on Banach spaces, Studia Math. 63 (1978), 283-308.

[27] S.R.S. Varadhan, Limit theorems for sums of independent random variables with values in a Hilbert space, Sankhya, the Indian Journal of Statistics 24 (1962), 213-238.

Institute of Mathematics
Łódź University
90-238 Ł ó d ź
Banacha 22
P o l a n d

ON STABILITY OF PROBABILITY MEASURES IN EUCLIDEAN SPACES

by
Zbigniew J. JUREK (Wrocław University)

1. INTRODUCTION

In the probability theory the class of stable distributions (measures) plays very important role. Let us recall that on stable measures (on the real line) one can see from two points. On the one hand as the limit distributions (see [2], p.176), and on the other hand as the measures satisfying an equation expressed in terms of a convolution and linear transformations (see [2], p.175). Thus in Euclidean spaces (also in linear topological spaces) there are two posibilities too. In the first case M.Sharpe in [28] (W.Krakowiak [15] for Banach space; see also K.Urbanik [29] and [31]) introduced the concept of operator-stable measures, and in the second case K.R.Parthasarathy and K.Schmidt in [24] introduced and examained the notion of stability measures with respect to groups of automorphisms. Moreover R.Jajte in [6] extended the class of operator-stable measures introducing semi-stable probability measures (but in the sequel we will use designation: operator-semistable measures). All these above classes of probability measures are subclasses of infinitely divisible measures.

As in the classical theory of stable measures ([2], pp.185-196) one can ask on the domains of attraction, the domains of normal attraction and moments for stable measures with respect to groups of automorphisms, or operator-stable or operator-semistable measures.

In this paper we shall give answer for these questions on domains of normal attraction and moments, if we consider probability measures on Euclidean spaces. In the case of arbitrary domains of attraction we present only partially answers. The contents of this paper are as follows: § 2 gives a short introduction the terminology and notations. § 3 is concerned with domains of normal attraction for stable measures with respect to the one-parameter groups. § 4 is devoted to the operator-stable measures and § 5 for operator-semistable measures. § 6 contains Lindeberg-Feller theorem and a characterization of domains of attraction for normal distribution. In § 7 we quote some results on convergence of types and Lévy's measures.

2. PRELIMINARIES AND NOTATIONS.

Let R^d denote d-dimensional Euclidean space with inner product $(\cdot,\cdot)$ and the norm $|\cdot|$. We write $P(R^d)$ for the set of probability measures on R^d, $\mu * \nu$ for the convolution of $\mu,\nu \in P(R^d)$ and δ_x $(x \in R^d)$ for measure concentrated at the point x. An element $\mu \in P(R^d)$ is called <u>infinitely divisible</u> if for any $n=2,3,\ldots$ there exists $\mu_n \in P(R^d)$ such that $\mu_n^{*n} = \mu$. Further μ is infinitely divisible if and only if the characteristic function $\hat{\mu}$ of μ is of the form

$$(2.1)\qquad \hat{\mu}(y) = \{\exp i(y,x_0) - \tfrac{1}{2}(Dy,y) + \int_{R^d\setminus\{0\}} [e^{i(y,x)} - 1 - \frac{i(y,x)}{1+|x|^2}] M(dx)\}$$

where x_0 is an element of R^d, D is real symmetric positive definite operator and M is σ-finite measure on $R^d\setminus\{0\}$, finite outside every neighbourhood of the zero and

$$(2.2)\qquad \int_{|x|\le 1} |x|^2 M(dx) < \infty .$$

The representation (2.1) is unique and in the sequel we will write $\mu = [x_0,D,M]$ if $\hat{\mu}$ have the representation (2.1).

By $\mu_n \Rightarrow \mu$ we shall denote the <u>weak convergence</u> of measures $\{\mu_n\}$ to a measure μ.

For moments of infinitely divisible measures we have the following

PROPOSITION 2.1 <u>Let</u> $\mu = [x_0,D,M]$. <u>Then for</u> $r>0$ $\int_{R^d} |x|^r d\mu(x) < \infty$ <u>if and only if</u> $\int_{|x|>1} |x|^r dM(x) < \infty$.

This is a partial case of Theorem 2 in [14] where were investigated the integrals with respect to infinitely divisible measures on Banach spaces.

A measure $\mu \in P(R^d)$ is called _full_ if its support is not contained in any $(d-1)$-dimensional hyperplane of R^d, and by $F(R^d)$ we denote the set of a full probability measures on R^d. We mention that $F(R^d)$ is an open subsemigroups of semigroup $P(R^d)$.

Given a linear operator A on R^d and $\mu \in P(R^d)$ by $A\mu$ we shall denote the probability measure defined by the formula $(A\mu)(E) = \mu(A^{-1}(E))$ for every Borel subset E of R^d. It is easy to check the equations for all linear operators A, B and measures μ, ν

$$A(B\mu) = (AB)\mu \ , \quad A(\mu * \nu) = A\mu * A\nu \ , \quad \widehat{A\mu}(y) = \hat{\mu}(A^* y) \ .$$

Moreover, the mapping $\langle A, \mu \rangle \to A\mu$ is jointly continuous where the space of linear bounded operators on R^d is provided with a norm topology and in $P(R^d)$ is given the topology of weak convergence.

K.Urbanik in [29] (see also P.Billingsley [1], M.Sharpe [28]), introduced the concept of decomposability semigroups of linear operators associated with the probability measure μ. Here we will use the following one. The semigroup _Inv_(μ) consists of all linear operators A in R^d for which the equality

$$\mu = A\mu \tag{2.3}$$

holds. Further we will widely exploited the following

PROPOSITION 2.2. _For_ $\mu \in P(R^d)$ _we have that_ $\mu \in F(R^d)$ _if and only if_ _Inv_(μ) _is compact group_.

The proof of this Proposition is simple consequence of Proposition 1.2. in [29] and Theorem 2 in [30]. It is worth to notice that some probabilistic properties of measures one can characterize by their decomposability semigroups (for example see [29] Theorem 5.1, [31]).

3. G-STABLE PROBABILITY MEASURES.

Let $GL(d,R)$ denote the general linear group, I the unit operator and G be a subgroup of $GL(d,R)$. A measure $\mu \in P(R^d)$ is called stable with respect to G (or shortly G-stable) if

$$\forall (A,B \in G)\ \exists (C \in G)\ \exists (x \in R^d) \quad A\mu * B\mu = C\mu * \delta_x \ . \tag{3.1}$$

According to Theorem 4.1 in [26] one can assume that the group is closed in $GL(d,R)$ and in view of Theorem 3.5 in [26] it is enough to consider only full G-stable measures. By Theorem 4.8 in [26] there exists a one parameter subgroup $G_1 = \{e^{tA}: t \in R\}$ of G such that μ is G_1-stable too. Finally, let us recall that every $\mu \in F(R^d)$ which is G-stable (and G is closed in $GL(d,R)$) can be decomposed into a convolution $\mu = \mu_1 * \mu_2$, where $\mu_1 = [x_1, D, 0]$, $\mu_2 = [x_2, 0, M]$ and μ_1, μ_2 are concentrated on subspaces R_1 and R_2 respectively, which are invariant under G and which have intersection $\{0\}$ (see [26], Theorem 5.3).

In this section we will consider only probability measures stable under one-parameter groups. For such measures we would like to propose the following definition of the domains of normal attraction.

Let $\mu \in F(R^d)$ be $G = \{e^{tA}: t \in R\}$ - stable and $\lambda > 0$ define the homomorphism c from G onto R^+ such that

$$c(e^{tA}) = e^{\lambda t} \tag{3.2}$$

and for $t \in R$ the following equation

$$\mu^{*c^{\lambda t}} = e^{tA}\mu * \delta_{x_t} \tag{3.3}$$

for some $x_t \in R^d$ holds good (see [26], Lemma 2.2 and proof of Theorem 4.8). We say that R^d-valued random vector X belongs to the domain of normal attraction of G-stable measure μ if there exists a sequence $\{a_n\} \subseteq R^d$ such that for any independent copies $X_1, X_2, \ldots$ of X we have

$$L(n^{-1/\lambda A}(X_1 + \ldots + X_n) + a_n) \Rightarrow \mu$$

as $n \to \infty$, i.e. in terms of probability distribution

(3.4) $$n^{-1/\lambda A}\nu^{*n} * \delta_{a_n} \Longrightarrow \mu$$

where $\nu = L(X)$. We use the term "normal" to stress that the norming operators are of the form $\{n^{-1/\lambda A}\}$.*)

Let us note that for $G = \{e^{tI}: t\in R\}$, a measure μ is G-stable if and only if μ is stable in classical sense. Moreover λ is the exponent of stable measure (see [24], Theorem 3.5). Thus our definition of normal attraction and classical notion of the domain of normal attraction are the same.

At first we will investigate the domain of attraction for full Gaussian G-stable measure (without Poissonian component; comp. [26] Theorem 5.3).

THEOREM 3.1. Let $G = \{e^{tA}: t\in R\}$ and $\mu = [x_0, D, 0]$ be full G-stable measure. Then ν belongs to the domain of normal attraction of μ if and only if ν has second moment and

$$(Dy,y) = \int_{R^d} (x-m,y)^2 \nu(dx)$$

where m is expectation value of ν.

Proof. From [26] Theorem 5.1. we have that $\mu = [x_0, D, 0]$ is G-stable if and only if for some $B \in GL(d.R)$

$$BAB^{-1} = \mathrm{diag}[\begin{pmatrix} \lambda/2 & d_1 \\ d_1 & \lambda/2 \end{pmatrix}, \ldots, \begin{pmatrix} \lambda/2 & d_k \\ d_k & \lambda/2 \end{pmatrix}, \ldots, \begin{pmatrix} \lambda/2 & 0 \\ 0 & \lambda/2 \end{pmatrix}]$$

where $\lambda > 0$ is the same as in (3.2)-(3.4), $d_k \in R$, and $DA = AD$.

The necessity. Of course we may assume that $x_0 = 0$. Let us suppose that for some sequence $\{a_n\} \subseteq R^d$ we have

(3.5) $$\mu_n \overset{df}{=} n^{-1/\lambda A}\nu^{*n} * \delta_{a_n} \Longrightarrow \mu = [0, D, 0] .$$

By (3.3), for $t\in R$

$$D = e^{t(A-\lambda/2\, I)} D e^{t(A^*-\lambda/2\, I)}$$

thus the one-parameter group $e^{t/\lambda(A/\lambda-\frac{1}{2}I)}$ belongs to the compact group $\mathrm{Inv}(\mu)$ (see Proposition 2.1). Further from equality

*) The definition does not depend on λ and A satisfying (3.3); cf. [13].

$$n^{-\frac{1}{2}I}\,\nu^{*n} = n^{(\frac{1}{\lambda}A-\frac{1}{2}I)}\,\mu_n * \delta_{b_n}$$

where $b_n = -n^{(1/\lambda\, A-\frac{1}{2}I)}(a_n)$, and from (3.5) we infer that the sequence of probability measures $\{n^{-\frac{1}{2}I}\circ\nu^{*n}\}$ is conditionally compact, because the sequence of operators $\{\exp[(1/\lambda\, A - \frac{1}{2}I)\lambda \log n]\}$ is conditionally compact. In view of [5], Remark 5.5 we conclude that $°\nu$ has finite second moment and the measure ν too. Then by [25] p.194 we have that

$$(3.6)\qquad \rho_n \overset{df}{=} n^{-\frac{1}{2}I}\,\nu^{*n} * \delta_{-\sqrt{n}\,m} \Rightarrow [0,S,0]$$

where m is mean value of ν and the covariance operator S is defined as follows

$$(Sy,z) = \int_{R^d} (y,x-m)(z,x-m)\ (dx)\ , \qquad y,z \in R^d .$$

Moreover, the sequence of probability measures

$$n^{\frac{1}{2}I-1/\lambda\, A}\,\rho_n = \mu_n * \delta_{a'_n}$$

where $a'_n = -a_n - n^{(I-1/\lambda\, A)}(m)$, is conditionally compact and its limits points are of the form $A[0,S,0] = [0,ASA^*,0]$ for some $A \in \mathrm{Inv}([0,D,0])$. In view of (3.5) and (3.6) we get equation

$$A[0,S,0] = [a,D,0] .$$

Hence $a = 0$ i.e. $a'_n \to 0$ and $[0,S,0] = A^{-1}[0,D,0] = [0,D,0]$. Thus $S = D$ which completes the proof of necessity.

<u>The sufficiency</u>. In view of [25] p.194 we have that

$$\rho_n = n^{-\frac{1}{2}I}\,\nu^{*n} * \delta_{-\sqrt{n}\,m} \Rightarrow [0,D,0] ,$$

and by (3.3) and Proposition 2.1 we infer that the sequence of operators $\{n^{-(1/\lambda\, A-\frac{1}{2}I)}\}$ is coditionally compact. Further if we put $a_n = -n^{(I-1/\lambda\, A)}(m)$ we obtain the equality

$$n^{-1/\lambda\, A}\,\nu^{*n} * \delta_{a_n} = n^{-(1/\lambda\, A-\frac{1}{2}I)}\,\rho_n ,$$

where the sequence $\{n^{\frac{1}{2}I-1/\lambda\, A}\rho_n\}$ of probabiliry measures is conditionally compact and all its limit points are equal to $[0,D,0]$. Hence

$$n^{-1/\lambda A}\,\nu^{*n} * \delta_{a_n} \Rightarrow [0,D,0]$$

as $n \to \infty$, which completes the proof of the sufficiency.

Now we procced to investigation of non-Gaussian G-stable measures. By [26], Theorem 5.2. we have that a full measure $[x_0,0,M]$ is G-stable if and only if its spectral measure M is of the form

$$M(E) = \int_S \int_R 1_E(e^{tA}x)e^{-\lambda t}dt\rho(dx) \tag{3.7}$$

where 1_E denotes the indicator of a Borel subset E of $R^d\setminus\{0\}$, S is Borel subset of $R^d\setminus\{0\}$ which intersects each orbit G in $R^d\setminus\{0\}$ in exactly one point, ρ is finite Borel measure on S, λ gives the homomorphism c (see (3.2)) and

$$0 < \lambda < 2\,\mathrm{Re}\,\lambda_k \qquad (k=1,2,\ldots,n)\ , \tag{3.8}$$

and λ_k denotes the eigenvalues of A. In [13] Theorem 1 has been proved the following

THEOREM 3.2. _Let_ $G = \{e^{tA}: t\in R\}$ _be a one parameter group and_ $\mu = [x_0,0,M]$ _where_ M _is of the form_ (3.7), _be a full_ G-_stable measure. Then_ $\nu \in P(R^d)$ _lies in the domain of normal attraction of_ μ _if and only if_

$$\lim_{t\to\infty} e^{\lambda t}\,\nu(e^{sA}: x\in E,\ s\geq t) = \lambda^{-1}\rho(E)$$

for all continuity sets E _of the measure_ ρ.

In the proof of this theorem are used theorems on accompanying laws and convergence of infinitely divisible probability measures on R^d.

Next theorem gives an information about moments of measures in the domain of normal attraction of a full non-Gaussian G-stable measures. Let us put

$$\delta^{-1} = \max_k \mathrm{Re}\,\lambda_k\ , \tag{3.9}$$

where $\lambda_1,\ldots,\lambda_n$ denotes the eigenvalues of A, and let $\lambda>0$ gives the homomorphism c.

THEOREM 3.3. *If* $\nu \in P(R^d)$ *belongs to the domain of normal attraction of a full* G*-stable measure* $\mu = [x_0, 0, M]$ *then* ν *has finite moments of order* r *where* $0 \le r < \lambda\delta$.

The proof of this theorem is given in [13]. Moreover since G-stable measure μ belongs to own domain of normal attraction thus μ has finite moments of order $0 \le r < \lambda\delta$. Converly, by Proposition 2.1 and formula (3.7) we have that μ has infinite moments for $r \ge \lambda\delta$; (for more details see [13]).

COROLLARY 3.1. *Let* $G = \{e^{tA}: t \in R\}$ *and* $\mu = [x_0, 0, M]$ *be a full* G*-stable measure. Then for* $r \ge 0$ $\int_{R^d} |x|^r d\mu(x) < \infty$ *if and only if* $r < \lambda\delta$.

As it has been mentioned above, the class of stable probability measure with respect to the group $\{e^{tI}: t \in R\}$ is equal to the class of stable measures. Now, let us consider stable probability measure with respect to the group $GL(d,R)$ of all nonsingular linear transformations of R^d onto itself. Following K.R.Parthasarathy [23], these measures are called *completely stable*. In [23] was proved the following

THEOREM 3.4. *In* R^d, $d > 1$, *every completely stable probability measure is Gaussian.*

Another more simple proof of Theorem 3.4. is given in [21]. Moreover in [22], are investigated on infinite dimensional Hilbert space H probability neasures stable under $B(H)$ all bounded linear operators. It is very interesting that in such case there exists non-Gaussian (also not infinitely divisible!) $B(H)$-stable probability measures. Also all $B(H)$-stable Gaussian measures are characterized in terms of proper values of their covariance operators.

4. OPERATOR-STABLE MEASURES.

Let $\{X_n\}$ be a sequence of independent identically distributed R^d-valued random vectors. If there exist sequences $\{A_n\}$ and $\{a_n\}$ of continuous linear operators on R^d and vectors of R^d, respectively such that the limit distribution μ of normed sums

$$(4.1) \qquad A_n(X_1 + X_2 + \dots + X_n) + a_n \qquad (n=1,2,\dots)$$

there exists then μ is called <u>operator-stable measure</u>. This concept is due to M.Sharpe who obtained in [28] a characterization of a full operator-stable measures. In particular he proved that: μ is full operator-stable measure if and only if there exists a non-singular operator B such that

$$(4.2) \qquad \forall (t>0)\ \exists (a_t \in R^d) \qquad \mu^{*t} = t^B \mu * \delta_{a_t} .$$

Moreover one can assume that either μ is full Gaussian measure or $\mu = [x_0, 0, M]$ is full and the spectral measure M is of the form

$$(4.3) \qquad M(E) = \int_{S^{d-1}} \int_0^\infty 1_E(t^B u) t^{-2} dt\, m(du)$$

where m is finite Borel measure on the unit sphere S^{d-1} in R^d (see [8]). In [17], by Choquet's Theorem full operator-stable measures are described in terms of characteristic functionals. Another more simple prove is given in [8] too.

We will say that R^d-valued random vector X belongs to <u>the domain of normal attraction</u> of operator-stable measure μ if there exists a sequence $\{a_n\}$ of vectors from R^d such that for any independent copies $X_1, X_2, \dots$ of X we have

$$L(n^{-B}(X_1 + X_2 + \dots + X_n) + a_n) \Rightarrow \mu ,$$

as $n \to \infty$. Of course operator-stable measures belongs to own domain of normal attraction.*)

Let us note that if we put $e^{\lambda u}$ $(\lambda>0,\ u \in R)$ instead of $t>0$ and $\frac{1}{\lambda} A$ instead of B in the formula (4.2) then we can obtain that μ is operator-stable measure if and only if μ is $G = \{e^{uA}: u \in R\}$-stable. Thus as a simple consequence of theorems in section 3 we get the following characterizations of the domains of normal attraction for full operator-stable probability measures.

THEOREM 4.1. <u>The probability measure</u> λ <u>lies in the domain of normal attraction of a full operator-stable measure</u> $\mu = [x_0, D, 0]$ <u>on</u> R^d <u>if and only if</u> λ <u>has second moment and</u>

---

*) The domain of attraction does not depend on the operator **B**; cf. [10].

$$(Dy,y) = \int_{R^d} (x-m,y)^2 \lambda(dx) , \qquad y \in R^d$$

where m *is expectation value of* λ .

THEOREM 4.2. *Let* λ *be a Borel probability measure on* R^d. *Then* λ *lies in the domain of normal attraction of a full operator-stable measure* $\mu = [x_0, 0, M]$ *and the spectral measure* M *is of the form* (4.3) *if and only if*

$$\lim_{t\to\infty} t\lambda(s^B x:\ x\in A,\ s\geq t) = m(A)$$

for all continuity sets A *of the measure* m .

Let a_k $(k=1,2,\ldots,n)$ denotes the eigenvalues of the operator B. Then from (3.8) we have

$$(4.4) \qquad 1 < 2 \operatorname{Re} a_k \qquad (k=1,2,\ldots,n) ,$$

if we consider full non-Gaussian operator-stable measure (see [28], Theorem 3). Further, let

$$\delta^{-1} = \max_{1\leq k\leq n} \operatorname{Re} a_k .$$

THEOREM 4.3. *If* $\lambda \in P(R^d)$ *belongs to the domain of normal attraction of operator-stable measure* $\mu = [x_0, 0, M]$ *then* λ *has finite moments of order* r *where* $0\leq r<\delta$.

COROLLARY 4.1. *For a full operator-stable measure* $\mu = [x_0, 0, M]$ *and* $r\geq 0$ *we have* $\int_{R^d} |x|^r d\mu(x) < \infty$ *if and only if* $r<\delta$.

The above theorems were early proved without using facts from section 3. Namely, Theorem 4.1. in [11] (Theorem 1), Theorems 4.2, 4.3 in [10] (Theorem 1.1, Theorem 3.1) and Corollary 4.1 in [10] (Corollary 3.1) and in [20]. Moreover if we consider in (4.2) the operator B of the form $B = \frac{1}{p} I$ where $p>0$ then μ is stable measure on R^d. Thus the above theorems gives well-known information in theory of stable measures (comp. [2], Chapter VII, § 35).

In our all considerations, the assumption that probability measures are full is very important (comp. Proposition 2.2). J. Kucharczak in

[16] gave characterization of operator-stable measures (i.e. limit distribution of 4.1)) on R^d, not necessarialy full. Namely, he proved the following

THEOREM 4.4. *The measure* $\mu \in P(R^d)$ *is operator-stable if and only if there exist* $\lambda \in P(R^d)$ *and the sequences* $\{B_n\}$, $\{b_n\}$ *of operators and vectors from* R^d *respectively, such that*

$$\mu = B_n \lambda^{*n} * \delta_{b_n}$$

for $n=1,2,\ldots$.

This statement is also true on real separable Banach spaces admitting a homogeneous decomposition (see [18]).

5. OPERATOR-SEMISTABLE MEASURES.

Let $\{k_n\}$ be an increasing sequence of positive integer such that $k_{n+1}/k_n \to \gamma$ for some $1 \le \gamma < \infty$, and $\{X_n\}$ be a sequence of independent identically distributed R^d-valued random vectors. If there exist sequences $\{A_n\}$ and $\{a_n\}$ of linear bounded operators on R^d a vector from R^d respectively, such that the limit distribution μ of the sequence

$$A_n(X_1 + X_2 + \ldots + X_n) + a_n \tag{5.1}$$

there exists then μ is called *operator-semistable measure.* R.Jajte in [6] gave complete characterization of a full operator-semistable measures on R^d analogous to that given by M.Sharpe. Namely, he proved that $\mu \in F(R^d)$ is operator-semistable if and only if there exist a number $0<c<1$ and non-singular operator B on R^d such that the formula

$$\mu^{*c} = B\mu * \delta_b \tag{5.2}$$

holds for some $b \in R^d$. Moreover one can assume that either μ is full Gaussian operator-semistable measure on R^d or $\mu = [x_0, 0, M]$ is full and the spectral measure M is of the form

(5.3) $$M(E) = \sum_{n=-\infty}^{\infty} c^{-n} \int_T 1_E(B^n x)m(dx) ,$$

where m is finite Borel measure on $T = \{x \in R^d: |x| \le 1, |B^{-1}x| > 1\}$, (see [20], Theorem 2.1).

We say that R^d-valued random vector X belongs to *the domain of normal attraction* of operator-semistable measure μ if there exists a sequence $\{a_n\}$ of vectors from R^d such that for any independent copies $X_1, X_2, \dots$ of X we have

$$L(B^n(X_1 + X_2 + \dots + X_{[c^{-n}]} + a_n) \Rightarrow \mu$$

as $n \to \infty$ and by $[\cdot]$ we denote the integral part of real number. As before we use the term "normal" to stress that the norming operators are of the form B^n and the subsequence $\{k_n\}$ is equal $[c^{-n}]$.*) In [12] I proved the following

THEOREM 5.1. *A measure* $\lambda \in P(R^d)$ *belongs to the domain of normal attraction of a full operator-semistable measure* $\mu = [x_0, D, 0]$ *if and only if* λ *has second moment and*

$$(Dy,y) = \int (x-m,y)^2 \lambda(dx)$$

where m *is expection value of* m.

For non-Gaussian operator-semistable probability measures we have the following

THEOREM 5.2. *Let* $\mu = [x_0, 0, M]$ *where* M *is of the form* (5.3), *be a full operator-semistable measure. Then* $\lambda \in P(R^d)$ *belongs to the domain of normal attraction of* μ *if and only if*

$$\lim_{n\to\infty} [c^{-n}]\lambda(B^{-k}x: x \in A, k \ge n) = \frac{m(A)}{1-c}$$

for all continuity sets A *of the measure* m.

For operator-semistable measures we introduce another parameter δ_0 connected with their moments. Namely if $\mu = [x_0, 0, M]$ is a full operator-semistable measure then the spectrum of B is contained in the disc $\{|z|^2 < c\}$. Let $\delta_1 = \min\{|a_k|: a_k$ is eigenvalue of $B\}$ and

*) In the sequel, the number c and the operator B are fixed.

(5.4) $$\delta_0 = \frac{\log c}{\log \delta_1} .$$

THEOREM 5.3. *If* $\lambda \in P(R^d)$ *belongs to the domain of normal attraction of a full operator-semistable measure* $\mu = [x_0, 0, M]$ *then* λ *has finite moments of order* r *for* $0 \le r < \delta_0$ *and* δ_0 *is defined by* (5.4).

We refer the reader to [10] for proofs of the Theorems 5.2. and 5.3. Moreover from Theorem 5.3 and Proposition 2.1. we have the following

COROLLARY 5.1. *If* $\mu = [x_0, 0, M] \in F(R^d)$ *is operator-semistable measure then for* $r \ge 0$ *we have* $\int |x|^r d\mu(x) < \infty$ *if and only if* $r < \delta_0$.

Let us suppose that the operator B in (5.2) is of the form $B = bI$. Then $0 < |b| < 1$ and μ is semi-stable probability measure (see [19]). As a consequence of Theorems 5.1-5.3 and Corollary 5.1 we get characterizations of the domains of normal attraction of semi-stable probability measures on Euclidean spaces.

6. CENTRAL LIMIT THEOREM

In this section we will consider an analoque of Lindeberg-Feller Theorem, when the partial sums of a sequence of independent R^d-valued random vectors are normed by some special operators.

Let $X_1, X_2, \ldots$ be independent R^d-valued random vectors such that

$$E\,X_k = 0 \quad , \qquad E|X_k|^2 < \infty .$$

Further let C_k be a covariance operator of vector X_k and

(6.1) $$B_n = C_1 + \ldots + C_n \quad ; \qquad A_n = (\sqrt{B_n})^{-1}$$

(we assume that the operators B_n are invertible). We introduce the following analoque of Lindeberg condition

(6.2) $$h_n(\varepsilon, t) = \sum_{k=1}^{n} E\{(A_n X_k, t)^2 \, 1_{[|(A_n X_k, t)| > \varepsilon]}\}$$

By the same arguments as in the proof of classical Lindeberg-Feller Theorem we have

THEOREM 6.1. *If for every* $\varepsilon>0$ *and every* $t \in R^d$ $h_n(\varepsilon,t) \to 0$ *as* $n \to \infty$ *then*

$$(6.3) \qquad L(A_n(X_1 + \ldots + X_n)) \Rightarrow \mu = [0,I,0] .$$

Conversly, if $\max_{1\le k\le n} |A_n C_k A_n| \to 0$ *as* $n\to\infty$ *and the formula* (6.3) *holds good then for every* $\varepsilon > 0$ *and every* $t \in R^d$ $h_n(\varepsilon,t) \to 0$, *as* $n \to \infty$.

The proof of this theorem with all details is given in [9].

The central limit problem for independent identically distributed d-dimensional random vectors $X_1, X_2, \ldots$ not necessarialy assuming finite second moments has been proved in [4] and [27]. Following M.G.Hahn and M.J.Klass we will say that random vector X belongs to *the generalized domain of attraction* of a measure μ if there exist a sequence of linear operators $\{A_n\}$ and vectors $\{a_n\}$ from R^d such that for any independent copies $X_1, X_2, \ldots$ of X we have

$$(6.4) \qquad L(A_n(X_1 + \ldots + X_n) + a_n) \Rightarrow \mu .$$

If we suppose that μ is full measure thus by [29] (Theorem 4).we may assume that either μ is a full Gaussian measure on R^d or μ is infinitely divisible measure without Gaussian component. M.G.Hahn and M.J.Klass proved in [4] the following very interesting fact

THEOREM 6.2. *Let* X *be a mean zero, full* R^d*-valued random vector. Then* X *belongs to the generalized domain of attraction of Gaussian measure* $\mu = [0,I,0]$ *if and only if*

$$\lim_{t\to\infty} \sup_{|\theta|=1} \frac{t^2 P\{|(X,\theta)| > t\}}{E(|(X,\theta)|^2 \wedge t^2)} = 0 .$$

Moreover in [4] is given an explicit form of the operators A_n in formula (6.4). References [27] contains an announcement (without proof) another characterization of generalized domain of attraction for Gaussian measure. Namely S.W.Semovskii proved the following

THEOREM 6.3. *Let* X *be non-degenerate* R^d*-valued random vector and*

$$H(t) = \int_{|x|<t} x\,x'\,L(X)(dx) , \qquad t > 0 ,$$

be a matrix where x' denotes the transpose vector. Then X belongs to the generalized domain of attraction of Gaussian measure $\mu = [x_o, S, 0]$ if and only if the function $t \to H(t)$ is slowly varying i.e. for every $k > 0$

$$H(t)[H(kt)]^{-1} \to I$$

as $t \to \infty$.

In view of M.Sharpe paper [28] we see that it remains to give a characterization of generalized domain of atrraction for arbitrary full non--Gaussian operator-stable measure $\mu = [x_o, 0, M]$ on R^d, where M is of the form (4.3). The description of the domains of normal attraction for all operator-stable measures is given in section 4.

7. CONVERGES OF TYPES AND LÉVY'S MEASURES

In this final section we present some theorem on the convergence of types on R^d (see [2], Chapter II, § 10 for d=1). Let us suppose that we have

(7.1) $$L(Y_n) \Longrightarrow L(Y)$$

and

(7.2) $$L(A_n Y_n + a_n) \Longrightarrow L(Y)$$

where $L(Y)$ is full measure on R^d, A_n are linear bounded operators on R^d, and a_n are vectors from R^d. What we can said about the sequence $\alpha_n = \langle A_n, a_n \rangle$ of affine transformations? This and another questions were considered by P.Billingsley in his elegant paper [1], (see also [32]). We quote here the following

THEOREM 7.1. *Let* $L(Y_n) \Rightarrow L(Y)$ *and* $L(Y)$ *is full measure. Then in order that* $L(A_n Y_n + a_n) \Rightarrow L(Y)$, *it is necessary and sufficient that the* A_n, a_n *have the form*

$$A_n = B_n C_n , \qquad a_n = B_n c_n + b_n$$

where $B_n \to I$, $b_n \to 0$, and $L(C_n Y + c_n) = L(Y)$, sequences $\{C_n\}$, $\{c_n\}$ of operators and vectors respectively, are compact.

Next example shows that the convergence types theorems in infinite dimensional spaces does not hold true (comp. [7] were are given unfortunately incorrect propositions).

EXAMPLE. Let H be real separable Hilbert space with an orthonormal complete system $\{e_n\}$. Let Y be H-valued random vector such that $L(Y) = \sum_{n=1}^{\infty} a_n \delta_{e_n}$ $(a_n > 0, \ \Sigma a_n = 1)$. If $H \ni x = \sum_{i=1}^{\infty} x_i e_i$ then by formulae

$$A_n x = \sum_{i \neq n} x_i e_i + n x_n e_n \qquad (n=1,2,\ldots)$$

we define the sequence $\{A_n\}$ of invertible linear bounded operators on H such that $|A_n| = n$. Further, it is easy to see that $L(A_n Y) \Rightarrow L(Y)$, but $|A_n| \to \infty$. In Theorem 7.1 the norm of operators A_n are bounded. Moreover, this example shows that the Propositiom 2.2 is true only on finite dimensional spaces.

Let us note that if instead of (7.1) we assume that Y_n is a partial sums of a sequence $\{X_n\}$ of independent identically distributed R^d-valued random vectors then $L(Y)$ in (7.2) is operator-stable probability measure (comp. section 4). Further if we assume that the probability distributions of X_k are arbitrary and the random vectors $A_n X_k$ $(k=1,2,\ldots,n;\ n=1,2,\ldots)$ are uniformly infinitesimal then following K.Urbanik [29] the limit distribution $L(Y)$ in (7.2) are called Lêvy's measure (operator-selfdecomposable measures). We quote here the following characterization of Lêvy's measures proved in [29] (Theorem 5.1).

THEOREM 7.2. A full probability measure μ on R^d is a Lêvy's measure if and only if there exists semigroup e^{tQ} $(t \geq 0)$, where Q is an operator whose all eigenvalues have negative part, such that

$$\forall t \geq 0 \ \exists \mu_t \in P(R^d) \qquad \mu = e^{tQ} \mu * \mu_t .$$

In the paper [31] K.Urbanik gave complete description of Lêvy's measures on Banach spaces but with some additional condition on norming sequence $\{A_n\}$ in (7.2). Next W.Krakowiak in [15] characterized operator-stable measure on Banach spaces with this same condition on $\{A_n\}$. We quote here this Urbanik's condition because it is rather not expected. Namely the semigroup generated by the operators

$$\{A_m A_n^{-1}: n=1,2,\ldots,m;\ m=1,2,\ldots\}$$

should be compact in the norm topology. With this condition an analoque of Theorem 7.2. holds good in Banach spaces too (see [31], Theorem 4.1).

REFERENCES

[1] P.Billingsley, *Convergence of types in k-spaces*, Z.Wahrscheinlichkeitstheorie verw. Geb. 5(1966), pp.175-179.

[2] B.V.Gnedenko and A.N.Kolmogorov, *Limit distributions for sums of independent random variables*, Moscow 1949 (in Russian).

[3] M.G.Hahn, *The generalized domain of attraction of a Gaussian law on Hilbert spaces*, Lecture Notes in Math. 709 (1979), pp.125-144.

[4] M.G.Hahn and M.J.Klass, *Matrix normalization of sums of i.i.d. random vectors in the domain of attraction of the multivariate normal*, Ann. of Probability (in print).

[5] N.C.Jain, *Central limit theorem in a Banach space*, Lecture Notes in Math. 526 (1975), pp.113-130.

[6] R.Jajte, *Semi-stable probability measures on R^N*, Studia Math. 61 (1977), pp.29-39.

[7] O.Jouandet, *Sur la convergence en type de variables aléatoires â valuers dans des espaces d'Hilbert on de Banach*, C.R. Acad. Sc. Paris 271 (1970), série A, pp.1082-1085.

[8] Z.J.Jurek, *Remarks on operator-stable probability measures*, Coment. Math. XXI (1978).

[9] Z.J.Jurek, *Central limit theorem in Euclidean spaces*, Bull. Acad. Pol. Sci. (in print).

[10] Z.J.Jurek, *Domains of normal attraction of operator-stable measures on Euclidean spaces*, ibidem (in print).

[11] Z.J.Jurek, *Gaussian measure as an operator-stable and operator-semistable distribution on Euclidean space*, Probability and Mathematical Statistics, (in print).

[12] Z.J.Jurek, *On Gaussian measure on R^d*, Procedings of 6th Conference on Probability Theory, Brasov 1979 (in print).

[13] Z.J.Jurek, *Domains of normal attraction for G-stable measures on R^d*, Teor. Verojatnost. i Primenen. (in print).

[14] Z.J.Jurek and J.Smalara, *On integrability with respect to infinitely divisible measures*, Bull. Acad. Pol. Sci. (in print).

[15] W.Krakowiak, *Operator-stable probability measures on Banach spaces*, Colloq. Math. (in print).

[16] J.Kucharczak, *On operator-stable probability measures*, Bull. Acad.Pol.Sci. 23 (1975), pp.571-576.

[17] J.Kucharczak, *Remarks on operator-stable measures*, Colloq. Math. 34 (1976), pp.109-119.

[18] J.Kucharczak and K.Urbanik, *Operator-stable probability measures on some Banach spaces*, Bull. Acad. Pol. Sci. 25 (1977), pp. 585-588.

[19] A.Kumar, *Semi-stable probability measures on Hilbert spaces*, J. Multivar. Anal. 6 (1976), pp.309-318.

[20] A.Łuczak, *Operator-semistable probability measures on R^N*, Thesis, Łódź University (preprint in Polish).

[21] B.Mincer, *Complety stable measures on R^n*, Comentationes Math. (in print).

[22] B.Mincer and K.Urbanik, *Completely stable measures on Hilbert spaces*, Colloq. Math. (in print).

[23] K.R.Parthasarathy, *Every completely stable distribution is normal*, Sankhya 35 (1973), Serie A, pp.35-38.

[24] K.R.Parthasarathy and K.Schmidt, *Stable positive definite functions*, Trans. Amer. Math. Soc. 203 (1975), pp.161-174.

[25] Ju.V.Prohorov, *Convergence of random processes and limit theorems in probability theory*, Teor. Verojatnost. i Primenen. 1 (1956), pp.173-238 (in Russian).

[26] K.Schmidt, *Stable probability measures on R^{ν}*, Z. Wahrscheinlichkeitstheorie verw. Gebiete 33 (1975), pp.19-31.

[27] S,V.Semovskii, *Central limit theorem*, Doklady Akad.Nauk SSSR 245(4), 1979, pp.795-798.

[28] M.Sharpe, *Operator-stable probability measures on vector groups*, Trans. Amer. Math. Soc. 136 (1969), pp.51-65.

[29] K.Urbanik, *Lêvy's probability measures on Euclidean spaces*, Studia Math. 44 (1972), pp.119-148.

[30] K.Urbanik, *Decomposability properties of probability measures*, Sankhya 37 (1975), Serie A, 530-537.

[31] K.Urbanik, *Lêvy's probability measures on Banach spaces*, Studia Math. 63 (1978), pp.283-308.

[32] I.Weismann, *On convergence of types and processes in Euclidean spaces*, Z. Wahrscheinlichkeitstheorie verw. Gebiete 37 (1976), pp. 35-41.

Institute of Mathematics
Wrocław University
Pl. Grunwaldzki 2/4
50-384 Wrocław, Poland.

FOURIER-WIENER TRANSFORM ON BROWNIAN FUNCTIONALS

Hui-Hsiung Kuo*

1. Introduction.

In a series of papers [3;4;5;6;7;8;9], Hida has advocated the study of analysis of Brownian functionals. This study is motivated by Levy's functional analysis, Wiener's theory on nonlinear networks, stochastic evolution equations and Feynman's path integral, among others. Let $\{B(t)\,;\, t \in \mathbb{R}\}$ be a standard Brownian motion, i.e. $\{B(t)\}$ is Gaussian process with $E\,B(t) = 0$ for all t and

$$E(B(t)B(s)) = \frac{1}{2}(|t| + |s| - |t-s|), \ t,\ s \in \mathbb{R} \ .$$

The white noise $\{\dot{B}(t)\,;\, t \in \mathbb{R}\}$ can be regarded as a generalized stochastic process given by $(\dot{B}, \xi) = \int_{\mathbb{R}} \xi(t)dB(t)$ for ξ in the Schwartz space $\mathcal{S}$ of rapidly decreasing real functions on $\mathbb{R}$. Therefore, the probability law μ of the white noise is realized in the space $\mathcal{S}^*$ of tempered distributions and can be shown to exist as follows. Consider the triple $\mathcal{S} \subset L^2(\mathbb{R}) \subset \mathcal{S}^*$ and the characteristic functional

$$C(\xi) = \exp\{-\frac{1}{2}\|\xi\|^2\}, \ \xi \in \mathcal{S},$$

where $\|\cdot\|$ is the $L^2(\mathbb{R})$-norm. By the Bochner-Minlos theorem [1,p.350], there exists a probability measure μ on $\mathcal{S}^*$ such that

$$C(\xi) = \int_{\mathcal{S}^*} \exp\{i(x,\xi)\}\,\mu\,(dx) \ .$$

---

*Research supported by NSF Grant MCS 78-01438.

In this way, an element x of $\mathcal{S}^*$ can be viewed as a sample path $\dot{B}$ of the white noise $\{\dot{B}(t) ; t \in \mathbb{R}\}$. Members of $L^2(\mathcal{S}^*)$ are called Brownian functionals with finite variance.

On the other hand, consider an abstract Wiener space $B^* \subset H \subset B$ (see [2] or [12, Chapter I § 4]). The normal distribution on H with mean 0 and variance 1 extends to a probability measure p_1 on B . The characteristic functional of p_1 is given by

$$C(z) = \exp\{ -\frac{1}{2}\|z\|^2\}, \quad z \in B^* ,$$

where $\|\cdot\|$ is the norm of H . Thus it is not surprising that the triples $\mathcal{S} \subset L^2(\mathbb{R}) \subset \mathcal{S}^*$ and $B^* \subset H \subset B$ have many parallel results. However, the triple $\mathcal{S} \subset L^2(\mathbb{R}) \subset \mathcal{S}^*$ has more structures which lead to many interesting applications. For instance, Ito [10] has used the fact that $\mathcal{S}^*$ is the space of tempered distributions to study a problem arising from statistical mechanics. In the space $\mathcal{S}^*$ we can formally regard $\{\dot{B}(t) ; t \in \mathbb{R}\}$ as coordinate functionals so that we can differentiate Brownian functionals with respect to $\dot{B}(t)$. This leads to what Hida calls causal calculus which has applications in quantum theory. In this paper we will use some of the techniques and results for an abstract Wiener space in [11;12;13;14;15;16] to study Brownian functionals.

2. Integral representation of Brownian functionals.

Let $\mathfrak{F}$ and $\mathfrak{F}_n$ be the reproducing kernel Hilbert spaces of $C(\xi - \eta)$ and $C_n(\xi,\eta) = (n!)^{-1}C(\xi)(\xi,\eta)^n C(\eta)$, respectively. It follows from the Taylor series expansion $C(\xi - \eta) = \sum_{n=0}^{\infty} C_n(\xi,\eta)$ that we have the direct sum decomposition:

$$\mathfrak{F} = \sum_{n=0}^{\infty} \oplus \mathfrak{F}_n .$$

Define a transformation $\mathcal{J}$ from $L^2(\mathscr{S}^*)$ into $\mathfrak{F}$ by

$$(\mathcal{J}\varphi)(\xi) = \int_{\mathscr{S}^*} e^{i(x,\xi)} \varphi(x)\mu(dx) .$$

$\mathcal{J}$ can be shown to be an isomorphism. Let $K_n = \mathcal{J}^{-1}(\mathfrak{F}_n)$. Then we have the Wiener-Ito decomposition of $L^2(\mathscr{S}^*)$:

$$L^2(\mathscr{S}^*) = \sum_{n=0}^{\infty} \oplus K_n .$$

Moreover, each φ in K_n has the following integral representation:

$$(\mathcal{J}\varphi)(\xi) = i^n C(\xi) \int_{\mathbb{R}^n} F(u_1,\ldots,u_n)\xi(u_1)\ldots\xi(u_n)du_1\ldots du_n$$

$$= i^n C(\xi)U(\xi) ,$$

where $F \in \widehat{L^2(\mathbb{R}^n)}$, the symmetric functions in $L^2(\mathbb{R}^n)$.

Example 1. If $\varphi(x) = (x,\xi_0)$, then $\varphi \in K_1$ and

$$U(\xi) = \int_{-\infty}^{\infty} \xi_0(u)\xi(u)du, \quad \xi \in \mathscr{S} .$$

Example 2. If $\varphi(x) = (x,\xi_1)(x,\xi_2)$, $\xi_1 \perp \xi_2$, then $\varphi \in K_2$ and

$$U(\xi) = \int_{-\infty}^{\infty}\int_{-\infty}^{\infty} \{\tfrac{1}{2}[\xi_1(u)\xi_2(v) + \xi_1(v)\xi_2(u)]\}\xi(u)\xi(v)dudv .$$

Example 3. If $\varphi(x) = (x,\xi_0)^2 - \|\xi_0\|^2$, then $\varphi \in K_2$ and

$$U(\xi) = \int_{-\infty}^{\infty}\int_{-\infty}^{\infty} \xi_0(u)\xi_0(v)\xi(u)\xi(v)dudv .$$

Example 4. Let H_n be the Hermite polynomial $H_n(x) = (-1)^n e^{x^2} \frac{d^n}{dx^n} e^{-x^2}$. If $\varphi(x) = \prod_{j=1}^{k} H_{n_j}((x,\xi_j)/\sqrt{2})$, ξ_j's are orthogonal, then $\varphi \in K_n$, $n = n_1 + \dots + n_k$ and

$$U(\xi) = \int_{\mathbb{R}^n} F(u_1,\dots,u_n)\xi(u_1)\dots\xi(u_n)du_1\dots du_n ,$$

where F is the symmetrization of

$$\underbrace{\xi_1 \otimes \dots \otimes \xi_1}_{n_1} \otimes \dots \otimes \underbrace{\xi_k \otimes \dots \otimes \xi_k}_{n_k}$$

As in [10], let $\|\cdot\|_p$, $p \geq 0$, be the norm on $\mathscr{S}$ given by $\|\xi\|_p^2 = \sum_{n=0}^{\infty} (n+\frac{1}{2})^{2p} |\xi_n|^2$, $\xi_n = \int_{-\infty}^{\infty} \xi(t)e_n(t)dt$, where $e_n(t) = \sqrt{g(t)}\, H_n(t/\sqrt{2})(2^n n!\sqrt{\pi})^{-\frac{1}{2}}$ and $g(t) = \frac{1}{\sqrt{2\pi}} e^{-\frac{t^2}{2}}$. Let $\mathscr{S}_p$ be the completion of $\mathscr{S}$ with respect to $\|\cdot\|_p$. Define

$$\|x\|_{-p} = \sup\{|(x,\xi)| ; \|\xi\|_p \leq 1\} .$$

Let $\mathscr{S}_p^* = \{x \in \mathscr{S}^* ; \|x\|_{-p} < \infty\}$. We have

$$\mathscr{S} \downarrow \mathscr{S}_p \subset L^2(\mathbb{R}) \subset \mathscr{S}_p^* \uparrow \mathscr{S}^*, \ p \to \infty .$$

We will call a Brownian functional φ to be k-th differentiable at x in the directions of $\mathscr{S}_p^*$ if the function $f(y) = \varphi(x+y)$, $y \in \mathscr{S}_p^*$, is k-th Fréchet differentiable. If this condition holds for all x in $\mathscr{S}^*$, we say that φ is k-th differentiable in the directions of $\mathscr{S}_p^*$. We use $\varphi^{(j)}(x)$ to denote the j-th Fréchet derivative of f at the origin, $1 \leq j \leq k$. Note that $\varphi'(x) \in \mathscr{S}_p$ and $\varphi''(x) \in \mathscr{L}(\mathscr{S}_p^*, \mathscr{S}_p)$. Moreover, we have the inclusion

$\mathcal{L}(\mathcal{S}_p^*, \mathcal{S}_p) \subset \mathcal{L}(L^2(\mathbb{R}))$ by the restriction so that $\varphi''(x)$ can be regarded as an element in $\mathcal{L}(L^2(\mathbb{R}))$.

Let P_n be the orthogonal projection of $L^2(\mathbb{R})$ onto the span of $\{e_j; j=0,1,\dots,n\}$. By a similar argument as in [15], we can show that if φ is a Brownian functional such that $\|\varphi'\|$ and $\|\varphi''\|_{H-S}$ are in $L^2(\mathcal{S}^*)$, then

$$\varphi_n(x) = -\operatorname{trace} P_n \circ \varphi''(x) + (x, P_n\varphi'(x))$$

is a Cauchy sequence in $L^2(\mathcal{S}^*)$. Define

$$N\varphi = \text{limit of } \varphi_n \text{ in } L^2(\mathcal{S}^*) .$$

<u>Theorem 1</u>. $(\mathcal{J}N\varphi)(\xi) = \|\xi\|^2(\mathcal{J}\varphi)(\xi) + ((\mathcal{J}\varphi)'(\xi), \xi)$.

Proof. We give a formal proof. Let

$$\Theta_n(x) = -\operatorname{trace} P_n \circ \varphi''(x) .$$

Apply the integration of parts formula [13, Theorem 3.1] to obtain the following:

$$\begin{aligned}(\mathcal{J}\Theta_n)(\xi) &= -\sum_{k=0}^{n} \int_{\mathcal{S}^*} e^{i(x,\xi)} (\varphi''(x)e_k, e_k)\mu(dx) \\ &= \sum_{k=0}^{n} \int_{\mathcal{S}^*} e^{i(x,\xi)} (\varphi'(x), e_k)\{i(e_k,\xi) - (x,e_k)\}\mu(dx) \\ &= i\sum_{k=0}^{n} (e_k,\xi) \int_{\mathcal{S}^*} e^{i(x,\xi)} (\varphi'(x), e_k)\mu(dx) \\ &\qquad - \int_{\mathcal{S}^*} e^{i(x,\xi)} (x, P_n\varphi'(x))\mu(dx) .\end{aligned}$$

Therefore, we have

$$(\mathcal{J}\varphi_n)(\xi) = i \sum_{k=o}^{n} (e_k, \xi) \int_{\mathcal{S}^*} e^{i(x,\xi)} (\varphi'(x), e_k)\mu(dx) \ .$$

Apply the integration by parts formula again to get

$$\begin{aligned}(\mathcal{J}\varphi_n)(\xi) &= i \sum_{k=o}^{n} (e_k, \xi) \int_{\mathcal{S}^*} e^{i(x,\xi)} \varphi(x)\{(x, e_k) - i(e_k, \xi)\}\mu(dx) \\ &= \|P_n\xi\|^2 (\mathcal{J}\varphi)(\xi) + i \int_{\mathcal{S}^*} e^{i(x,\xi)} \varphi(x)(x, P_n\xi)\mu(dx) \ .\end{aligned}$$

The theorem follows by letting $n \to \infty$.

Define a differential operator L on $\mathfrak{F}$ by

$$(Lf)(\xi) = \|\xi\|^2 f(\xi) + (f'(\xi), \xi) \ .$$

Then Theorem 1 states that the following diagram is commutative:

$$\begin{array}{ccc} L^2(\mathcal{S}^*) & \xrightarrow{\ \mathcal{J}\ } & \mathfrak{F} \\ {\scriptstyle N}\downarrow & & \downarrow{\scriptstyle L} \\ L^2(\mathcal{S}^*) & \xrightarrow{\ \mathcal{J}\ } & \mathfrak{F} \end{array}$$

The following theorem is well-known. We give a simple proof based on the integral representation.

Theorem 2. $N|_{K_n}$ and $L|_{\mathfrak{F}_n}$ are the multiplication operator by n .

Proof. Let $f \in \mathfrak{F}_n$. Then $f(\xi) = i^n C(\xi)U(\xi)$ and

$$U(\xi) = \int_{\mathbb{R}^n} F(u_1, \ldots, u_n)\xi(u_1)\ldots\xi(u_n)du_1\ldots du_n \ .$$

It follows from a direct computation that for $\xi, \eta \in \mathcal{S}$,

$$(C'(\xi),\eta) = -(\xi,\eta)C(\xi) ,$$

$$(U'(\xi),\eta) = n\int_{\mathbb{R}^n} F(u_1,\ldots,u_n)\eta(u_1)\xi(u_2)\ldots\xi(u_n)du_1\ldots du_n .$$

Therefore, we have $(C'(\xi),\xi) = -\|\xi\|^2 C(\xi)$ and $(U'(\xi),\xi) = nU(\xi)$. Hence $(f'(\xi),\xi) = -\|\xi\|^2 f(\xi) + nf(\xi)$, i.e. $(Lf)(\xi) = nf(\xi)$. Moreover, by the isomorphism $\mathfrak{J}$, we have $N|_{K_n} \approx L|_{\mathfrak{F}_n}$.

3. Fourier-Wiener transform.

Recall that in the previous section, we have $\mathscr{S} \subset \mathscr{S}_p \subset L^2(\mathbb{R}) \subset \mathscr{S}_p^* \subset \mathscr{S}^*$, $p \geq 0$. It is easy to see that the inclusion map $\mathscr{S}_p \to L^2(\mathbb{R})$ is a Hilbert-Schmidt operator when $p > \frac{1}{2}$. Therefore, by a Theorem of Piech [16], the measure μ is supported in $\mathscr{S}_p^*$ when $p > \frac{1}{2}$, i.e., $\mathscr{S}_p \subset L^2(\mathbb{R}) \subset \mathscr{S}_p^*$ is an abstract Wiener space.

For $p > \frac{1}{2}$, we use $\mathcal{A}_p$ to denote the space consisting of all complex-valued functions φ defined on $\mathscr{S}_p^*$ satisfying the following conditions:

(i) φ is infinitely Fréchet differentiable on $\mathscr{S}_p^*$,

(ii) For any $c > 0$, there exist constants K and M such that $\|\varphi^{(n)}(x)\| \leq KM^n$ for all $\|x\|_{-p} \leq c$ and all $n \geq 0$.

We use $\mathcal{A}$ to denote the union of $\mathcal{A}_p$ over $p > \frac{1}{2}$.

The Fourier-Wiener transform $\tau_\alpha\varphi$, $\alpha > 0$, of a Brownian functional φ is defined by

$$(\tau_\alpha\varphi)(y) = \int \varphi(\sqrt{\alpha}\,x + iy)\mu(dx) , \quad y \in \mathscr{S}^* .$$

We will use τ to denote τ_1. $\tau_\alpha\varphi$ can be defined for a Brownian functional φ on the complexication $\mathcal{S}^* + i\mathcal{S}^*$ of $\mathcal{S}^*$ such that $\varphi(\sqrt{\alpha}(\cdot) + iy)$ is μ-integrable for each $y \in \mathcal{S}^*$. It can be shown, by using the Taylor series expansion, that every $\varphi \in \mathcal{A}$ has a unique extension to $\mathcal{S}^* + i\mathcal{S}^*$ and that $\tau_\alpha\varphi$ is defined. Moreover, $\tau_\alpha(\mathcal{A}_p) \subset \mathcal{A}_p$.

<u>Theorem 3</u>. $(\mathfrak{T}\varphi)(\xi) = C(\xi)(\tau\varphi)(\xi)$, $\xi \in \mathcal{S}$.

Proof. The equality holds for $\xi = 0$. Hence let $\xi \neq 0$ be fixed and put $\eta = \xi/\|\xi\|$. Then $x \in \mathcal{S}^*$ can be expressed uniquely as $x = \alpha\eta + z$, $\alpha \in \mathbb{R}$, $z \in \mathcal{S}^*$ and $(z,\eta) = 0$. Moreover, μ can be decomposed into the product measure $(\frac{1}{\sqrt{2\pi}} e^{-\frac{\alpha^2}{2}} d\alpha)\nu(dz)$, where ν is the white noise on $\mathcal{U}^* = \{z \in \mathcal{S}^* ; (z,\eta) = 0\}$. Hence

$$(\mathfrak{T}\varphi)(\xi) = \int_{\mathcal{S}^*} e^{i(x,\xi)} \varphi(x)\mu(dx)$$

$$= \int_{\mathcal{U}^*} \int_{\mathbb{R}} e^{i\alpha\|\xi\|} \varphi(\alpha\eta + z) \frac{1}{\sqrt{2\pi}} e^{-\frac{\alpha^2}{2}} d\alpha\, \nu(dz) .$$

By making a change of variables $\beta = \alpha - i\|\xi\|$, we have

$$(\mathfrak{T}\varphi)(\xi) = C(\xi) \int_{\mathcal{U}^*} \int_{\mathbb{R}} \varphi(\beta\eta + z + i\xi) \frac{1}{\sqrt{2\pi}} e^{-\frac{\beta^2}{2}} d\beta\, \nu(dz)$$

$$= C(\xi) \int_{\mathcal{S}^*} \varphi(x + i\xi)\mu(dx)$$

$$= C(\xi)(\tau\varphi)(\xi) .$$

Define a differential operator acting on Brownian functionals by

$$(\mathcal{L}\varphi)(x) = (x, \varphi'(x)) \ .$$

Here we assume that φ is differentiable in all directions of $\mathcal{S}^*$ so that $\varphi'(x) \in \mathcal{S}$. From Theorem 1 , we have

$$(\mathcal{J}N\varphi)(\xi) = \|\xi\|^2 (\mathcal{J}\varphi)(\xi) + ((\mathcal{J}\varphi)'(\xi), \xi) \ .$$

Hence, by Theorem 3 ,

$$\begin{aligned} C(\xi)(\tau N\varphi)(\xi) &= \|\xi\|^2 C(\xi)(\tau\varphi)(\xi) \\ &\quad + (C'(\xi), \xi)(\tau\varphi)(\xi) + C(\xi)(\tau\varphi)'(\xi), \xi) \\ &= C(\xi)(\mathcal{L}\tau\varphi)(\xi) \ . \end{aligned}$$

Therefore, we have formally proved the following result.

Theorem 4. $\tau N\varphi = \mathcal{L}\tau\varphi$ for all $\varphi \in \mathcal{A}$.

This theorem can be used to solve the following differential equation

$$\begin{cases} \dfrac{\partial}{\partial t} u(t,x) = -N^k u(t,x) \\ u(o,x) = \varphi(x) \ , \end{cases}$$

where k is a positive integer, $t > 0$ and $x \in \mathcal{S}^*$. The special case $k = 1$ has been solved in [11] . Here we adopt the method used in [14] . Assume that $\varphi \in \mathcal{A}$. The idea to solve the above equation is to take Fourier-Wiener transform and then solve the new equation. Hence, let $v(t,y) = \tau u(t,\cdot)(y)$ and $\psi(y) = (\tau\varphi)(y)$.

Then, by Theorem 4, $v(t,y)$ satisfies the following differential equation

$$\begin{cases} \frac{\partial}{\partial t} v(t,y) = -\mathcal{L}^k v(t,y) \\ v(o,y) = \psi(y) . \end{cases}$$

Note that $\psi \in \mathcal{A}$, and so, ψ can be expanded as a Taylor series as follows:

$$\psi(y) = \sum_{n=o}^{\infty} \frac{1}{n!} \psi^{(n)}(0)(y,\ldots,y) .$$

This equality is understood to hold for y in $\mathcal{S}_p^*$ if $\psi \in \mathcal{A}_p$. But it holds also when both sides are regarded as Brownian functionals. Observe that

$$\mathcal{L}\psi^{(n)}(0)(y,\ldots,y) = n\,\psi^{(n)}(0)(y,\ldots y) .$$

This observation suggests that a solution is given by

$$v(t,y) = \sum_{n=o}^{\infty} \frac{1}{n!} e^{-n^k t} \psi^{(n)}(0)(y,\ldots,y) .$$

It can be checked that this series converges absolutely for $t > 0$ and $y \in \mathcal{S}_p^*$ (when $\psi \in \mathcal{A}_p$, i.e., when $\varphi \in \mathcal{A}_p$). Moreover, the convergence is y-uniform on bounded subsets of $\mathcal{S}_p^*$ and t-uniform on $[0,\infty)$, and $v(t,\cdot) \in \mathcal{A}_p$ is a solution. Therefore, a solution of the original equation is given by the inverse Fourier-Wiener transform of $v(t,y)$, i.e.

$$u(t,x) = \tau^{-1} v(t,\cdot)(x) .$$

We mention also that the solution $u(t,x)$ is unique.

4. Causal calculus.

In the space $\mathscr{S}^*$ of tempered distributions we regard the white noise $\{\dot{B}(t) ; t \in \mathbb{R}\}$ as a system of coordinates. This choice of coordinates, rather than using an $L^2(\mathbb{R})$-expansion, is more appropriate for the purpose of time propagation. Therefore, a Brownian functional φ will be thought of as a function of $\{\dot{B}(t) ; t \in \mathbb{R}\}$. We will use $\dot{B}$ to denote the variable of φ. In [4;6;8], Hida defines the differentiation $\frac{d}{d\dot{B}(t)} \varphi(\dot{B})$, t fixed and $\varphi \in K_n$, by employing the integral representation of φ and the functional derivative. Here we define the $\dot{B}(t)$-differentiation in a different way. Recall that in section 2 we define the differentiability of a Brownian functional φ in the directions of $\mathscr{S}_p^*$ and $\varphi'(\dot{B}) \in \mathscr{S}_p$. Assume that φ is differentiable in the directions of $\mathscr{S}_p^*$ for all $p \geq 0$ and that $\varphi'(\dot{B}) \in \mathscr{S}$. Then we define

$$\frac{d}{d\dot{B}(t)} \varphi(\dot{B}) = \varphi'(\dot{B})(t) .$$

Let δ be the Dirac distribution at the origin and $\delta_t = \delta(\cdot - t)$. Then $\delta_t \in \mathscr{S}^*$ and the above definition can be also expressed as follows:

$$\frac{d}{d\dot{B}(t)} \varphi(\dot{B}) = (\delta_t, \varphi'(\dot{B})) .$$

Example 1. For $\varphi(\dot{B}) = \int_{\mathbb{R}} \xi(u)\dot{B}(u)du = (\dot{B}, \xi)$, we have

$$\frac{d}{d\dot{B}(t)} \varphi(\dot{B}) = (\delta_t, \xi) = \int_{\mathbb{R}} \xi(u)\delta_t(u)du = \xi(t) .$$

<u>Example 2</u>. For $\varphi(\dot{B}) = \int_{\mathbb{R}^2} \xi_1(u)\xi_2(v)\dot{B}(u)\dot{B}(v)du\,dv = (\dot{B},\xi_1)(\dot{B},\xi_2)$,

we have

$$\frac{d}{d\dot{B}(t)}\varphi(\dot{B}) = (\delta_t,\xi_1)(\dot{B},\xi_2) + (\dot{B},\xi_1)(\delta_t,\xi_2)$$

$$= \int_{\mathbb{R}^2} \xi_1(u)\xi_2(v)\delta_t(u)\dot{B}(v)du\,dv$$

$$+ \int_{\mathbb{R}^2} \xi_1(u)\xi_2(v)\dot{B}(u)\delta_t(v)du\,dv$$

$$= \xi_1(t)\int_{\mathbb{R}} \xi_2(v)\dot{B}(v)dv + \xi_2(t)\int_{\mathbb{R}} \xi_1(u)\dot{B}(u)du\ .$$

<u>Example 3</u>. For $\varphi(\dot{B}) = H_n(\frac{1}{\sqrt{2}}\int_{\mathbb{R}} \xi(u)\dot{B}(u)du) = H_n(\frac{1}{\sqrt{2}}(\dot{B},\xi))$, where H_n is the Hermite polynomial of degree n , we have

$$\frac{d}{d\dot{B}(t)}\varphi(\dot{B}) = \sqrt{2}\,n\,\xi(t)H_{n-1}(\frac{1}{\sqrt{2}}(\dot{B},\xi))\ .$$

Observe that the first two examples suggest the following symbolic expression:

$$\frac{d}{d\dot{B}(t)}\dot{B}(\cdot) = \delta_t\ .$$

This is what one expects when $\{\dot{B}(t)\ ;\ t\in\mathbb{R}\}$ is taken as a system of coordinates. Moreover, Example 3 suggests that $\frac{d}{d\dot{B}(t)}$ is a map from K_n into K_{n-1} . Indeed, this is so as the following theorem shows.

<u>Theorem 5</u>. If $\varphi\in K_n$, then $\frac{d}{d\dot{B}(t)}\varphi(B)\in K_{n-1}$.

Proof. Suppose ψ is a polynomial Brownian functional of degree $\leq n-2$. Then, by integration by parts formula, we have

$$\int_{\mathscr{S}^*} \psi(\dot{B}) \frac{d}{d\dot{B}(t)} \varphi(\dot{B})\mu(d\dot{B})$$

$$= \int_{\mathscr{S}^*} \psi(\dot{B})(\delta_t, \varphi'(\dot{B}))\mu(d\dot{B})$$

$$= \int_{\mathscr{S}^*} \varphi(\dot{B})\{\psi(\dot{B})(\delta_t, \dot{B}) - (\delta_t, \psi'(\dot{B}))\}\mu(d\dot{B})$$

$$= 0 ,$$

because the quantity inside $\{\ldots.\}$ is a polynomial of degree $\leq n-1$. Therefore, $\frac{d}{d\dot{B}(t)} \varphi(\dot{B}) \in K_{n-1}$.

The following theorem shows that our definition of $\dot{B}(t)$-differentiation agrees with Hida's definition when φ is in K_n.

<u>Theorem 6</u>. Suppose $\varphi \in K_n$ and $(\mathfrak{I}\varphi)(\xi) = i^n C(\xi)U(\xi)$. Then we have

$$\frac{d}{d\dot{B}(t)} \varphi(\dot{B}) = \mathfrak{I}^{-1}\{i^{n-1}C(\xi)U'_\xi(t)\}(\dot{B}) ,$$

where $U'_\xi(t)$ is the functional derivative of U.

Proof. Recall that from Theorem 3 we have

$$\mathfrak{I}(\frac{d}{d\dot{B}(t)} \varphi(\dot{B}))(\xi) = C(\xi)\mathcal{T}(\frac{d}{d\dot{B}(t)} \varphi(\dot{B}))(\xi) .$$

But

$$\mathcal{T}(\frac{d}{d\dot{B}(t)} \varphi(\dot{B}))(\xi) = \int_{\mathscr{S}^*} \varphi'(\dot{B} + i\,\xi)(t)\mu(d\dot{B})$$

$$= \frac{1}{i} (\mathcal{T}\varphi)'_\xi(t)$$

$$= i^{n-1} U'_\xi(t) .$$

Therefore, we have

$$\mathcal{T}(\frac{d}{d\dot{B}(t)} \varphi(\dot{B}))(\xi) = i^{n-1}c(\xi)U_\xi'(t) .$$

Finally, we give a result about the Lévy Laplacian. Suppose φ is a twice differentiable Brownian functional in the directions of $\mathcal{S}_p^*$ for some p. Then $\varphi''(\dot{B})$ can be regarded as a linear operator of $L^2(\mathbb{R})$ such that

$$(\varphi''(\dot{B})f)(t) = \int_{\mathbb{R}} \varphi''(\dot{B})(t,s)f(s)ds, \quad f \in L^2(\mathbb{R}) ,$$

where $\varphi''(\dot{B})(t,s) = \frac{d^2}{d\dot{B}(t)d\dot{B}(s)} \varphi(\dot{B})$. Note that the infinite dimensional analogue of the Laplacian is given symbolically by

$$\int_{-\infty}^{\infty} \frac{d^2}{d\dot{B}(t)^2} \varphi(\dot{B})dt .$$

Theorem 7. Suppose $\varphi''(\dot{B})$ is a Brownian function taking values in the trace class operators of $L^2(\mathbb{R})$. Then

$$\operatorname{trace}_{L^2(\mathbb{R})} \varphi''(\dot{B}) = \int_{\mathbb{R}} \varphi''(\dot{B})(t,t)dt$$

$$= \int_{\mathbb{R}} \frac{d^2}{d\dot{B}(t)^2} \varphi(\dot{B})dt .$$

References.

1. I. M. Gel'fand and N. Ya. Vilenkin, Generalized functions, vol. 4, Applications of harmonic analysis, Academic Press (1964).

2. L. Gross, Abstract Wiener spaces, Proc. 5th Berkeley Sym. Math. Stat. Prob. 2 (1965), 31-42.

3. T. Hida, Quadratic functionals of Brownian motion, J. Multivariate Analysis 1 (1971), 58-69.

4. T. Hida, Analysis of Brownian functionals, Carleton Math. Lecture Notes, no. 13, Carleton Univ. Ottawa (1975).

5. T. Hida, Analysis of Brownian functionals, Mathematical Programming Study 5 (1976), 53-59.

6. T. Hida, White noise and Lévy's functional analysis, Proc. Measure Theory, Oberwolfach (1977), Lecture Notes in Math. vol. 695 (1978), 155-163, Springer-Verlag.

7. T. Hida, Generalized multiple Wiener integrals, Proc. Japan Academy 54 (1978), 55-58.

8. T. Hida, Causal analysis in terms of white noise, (preprint 1979).

9. T. Hida and L. Streit, On quantum theory in terms of white noise, Nagoya Math. J. 68 (1977), 21-34.

10. K. Ito, Stochastic analysis in infinite dimensions, Proc. International Conference on Stochastic Analysis, Evanston, Academic Press (1978), 187-197.

11. H. -H. Kuo, Integration by parts for abstract Wiener measures, Duke Math. J. 41 (1974), 373-379.

12. H. -H. Kuo, Gaussian measures in Banach spaces, Lecture Notes in Math. vol. 463 (1975), Springer-Verlag.

13. H. -H. Kuo, Differential calculus for measures on Banach spaces, Proc. Vector Space Measures and Appl. I, Lecture Notes in Math. vol. 644 (1978), 270-285, Springer-Verlag.

14. Y. J. Lee, Applications of the Fourier-Wiener transform to differential equations on infinite dimensional spaces I, Trans. Amer. Math. Soc. (to appear).

15. M. Ann Piech, The Ornstein-Uhlenbeck semigroup in an infinite dimensional L^2 setting, J. Functional Analysis 18 (1975), 271-285.

16. M. Ann Piech, Support properties of Gaussian Processes over Schwartz space, Proc. Amer. Math. Soc. 53 (1975), 460-462.

Department of Mathematics
Louisiana State University
Baton Rouge, La. 70803
U. S. A.

ON UNCONDITIONAL CONVERGENCE OF RANDOM SERIES IN BANACH SPACES

V.V. Kvaratskhelia

Let X be a Banach space. We say that a series $\sum_k a_k$, $a_k \in X$, $k = 1,2,\ldots$ converges unconditionally in X, if its every permutation is convergent. In finite dimensional Banach spaces, as it is well-known, the notions of unconditionally convergence and and absolute convergence are the same. In 1950 Dvoretzky and Rogers have proved that in any infinite dimensional Banach space these two notions are not equivalent.

Now we turn into to the case of random series. How the unconditional convergence for random series can be defined? In the function theory the usual definition of unconditional convergence of measurable functions is as follows: a series $\sum_k \xi_k(\omega)$ is called to be convergent a.s. unconditionally, if its every permutation is convergent a.s. But this definition does not reflect always the entity of unconditional convergence. We illustrate this by giving an example. Let ε_k, $k = 1,2,\ldots$ be independent Bernoulli random variables. At it is well-known the series $\sum_k \frac{1}{k} \varepsilon_k$ is convergent a.s. since $\sum_k \frac{1}{k^2} < \infty$, moreover it is clear that each permutation of this series is a.s. convergent, but this series does not converge unconditionally, since $\sum \left| \frac{1}{k} \varepsilon_k \right| = \sum \frac{1}{k} = \infty$.

Now we give an explicit definition. Let $(\Omega, \mathcal{A}, P)$ be a probability space and let $\{\xi_k\}, \xi_k : \Omega \longrightarrow X$, $k = 1,2,\ldots$ be X - valued random variables. We say that the series $\sum \xi_k$ is convergent a.s. unconditionally if there exist a set $\Omega_0 \in \mathcal{A}$ of a full probability $(P(\Omega_0) = 1)$ such that for all $\omega \in \Omega_0$ the series $\sum_k \xi_k(\omega)$ is convergent unconditionally in X.

Let γ_k, $k = 1,2,\ldots$ be the standard Gaussian random variables and let the sequence $\{\gamma_k\}$ be Gaussian.

<u>Theorem.</u> For a Banach space X the following conditions are equivalent:

1° X does not contain l^n_∞ uniformly.

2° A series $\sum_k a_k$ is unconditionally convergent in X iff there exists $p \geq 2$, a linear bounded operator $B: l_p \longrightarrow X$ and a sequence of numbers $\{\alpha_k\} \in l_p$ such that $a_k = \alpha_k B e_k$, $k = 1,2,\dots$, $\{e_k\}$ being the natural basis in l_p.

3° A series $\sum_k a_k \gamma_k$ converges unconditionally a.s. iff the series $\sum_k a_k$ is unconditionally convergent.

<u>Remark.</u> 1. For L_r - spaces ($1 \leq r \leq 2$) the statement which is similar to 2° has been proved by P. Ørno [1].

2. This Theorem is contained in [2] without proof.

The proof of this Theorem uses the following

<u>Lemma.</u> Let a Banach space X does not contain l^n_∞ uniformly and let the series $\sum a_k$ be unconditionally convergent in X. Then there exists a sequence of positive numbers $\{\beta_k\}$ with $\sum_k \beta_k \leq 1$, such that for some q, $1 < q \leq 2$ we have $\sum \frac{|x^*(a_k)|^q}{\beta_k^{q-1}} < \infty$ for all $x^* \in X^*$, where X^* is the dual space.

<u>Proof of the Lemma.</u> Let $\mu = \{\mu_k\}$ be a sequence of positive numbers with $\sum_k \mu_k = 1$. Let us define the operator mapping X^* into $L_1(\mu)$: $T x^* = \{ \frac{x^*(a_k)}{\mu_k} \}$ for all $x^* \in X^*$.
Here $L_1(\mu)$ is the Banach space of number sequences $\{x_k\}$, which are absolutely convergent with weights $\mu = \{\mu_k\}$. The norm of $L_1(\mu)$ is defined naturally: $\|x\|_{L_1(\mu)} = \sum |x_k| \mu_k$.

It is clear that T is a linear operator and that the boundedness of T is an easy consequence of unconditionally convergence of the series $\sum a_k$:

$$\|T x^*\|_{L_1(\mu)} = \sum_k |x^*(a_k)| \leq \sup_{y^* \in S^*} \sum_k |y^*(a_k)| \, \|x^*\| ,$$

for all $x^* \in X^*$. Here S^* is the unit ball of X^*.

It is easy to see that the conjugate operator T^* maps $L_1^*(\mu) = L_\infty(\mu)$ into X. Combining results of papers [3] and [4'] ([3], theorem 92; [4'], proposition 3.) we get, that the operator

$T^* : L_\infty(\mu) \longrightarrow X$ is p-absolutely summing for some $p \geq 2$. Then using the result of [4] we have that there exists a sequence of positive numbers φ_k, $k = 1,2,\ldots$, such that $\sum_k \varphi_k \mu_k \leq 1$ and

$$\sum_k \frac{|x^*(a_k)|^q}{\varphi_k^{q-1} \mu_k^{q-1}} < \infty \quad \text{for all } x^* \in X^*, \text{ where } 1/p + 1/q = 1.$$

If we denote $\beta_k = \varphi_k \mu_k$, $k = 1,2,\ldots$ we get the required statement.

The Lemma is proved.

<u>Corollary</u>. If X is a space of cotype 2 then in Lemma we can take $q = 2$.

<u>Proof of the Theorem</u>. $1^\circ \Longrightarrow 2^\circ$. Let X does not contain l_∞^n uniformly. It is clear that if $a_k = \alpha_k Be_k$, $k = 1,2,\ldots$, where $\sum_k |\alpha_k|^p < \infty$, $\{e_k\}$ is the natural basis in l_p and $B: l_p \longrightarrow X$ is a linear bounded operator, then $\sum a_k$ is unconditionally convergent. Let us proof the inverse assertion. Let $\sum a_k$ be unconditionally convergent in X. Then by Lemma there exists a sequence of positive numbers $\{\beta_k\}$ with $\sum_k \beta_k \leq 1$, such that $\sum_k \frac{|x^*(a_k)|^q}{\beta_k^{q-1}} < \infty$ for all $x^* \in X^*$ and some $1 < q \leq 2$. Let us define an operator $B: l_p \longrightarrow X$ put $Be_k = \frac{a_k}{\beta_k^{1/p}}$ where $1/p + 1/q = 1$. and $\{e_k\}$ is the natural basis in l_p.

B is a linear and bounded operator, since applying Hölder's inequality we have

$$\|Bx\| = \| \sum_k x_k Be_k \| = \| \sum_k \frac{a_k}{\beta_k^{1/p}} x_k \| \leq$$

$$\leq \sup_{x^* \in S^*} \left(\sum_k \frac{|x^*(a_k)|^q}{\beta_k^{q-1}} \right)^{1/q} \|x\|, \quad x \in l_p.$$

Thus we have $a_k = \alpha_k Be_k$, $k = 1,2,\ldots$, where B is a linear and bounded operator and $\alpha_k = \beta_k^{1/p}$, $k = 1,2,\ldots$

This proves $1^\circ \Longrightarrow 2^\circ$

$2^\circ \Longrightarrow 3^\circ$. If $\sum_k a_k \gamma_k$ is a.s. unconditionally convergent, then $\sum_k a_k$ is unconditionally convergent for each Banach space X [2]. Now let us assume that $\sum_k a_k$ is unconditionally convergent. Then there exists a linear bounded operator $B: l_p \longrightarrow X$ and a sequence of positive numbers (α_k) with $\sum \alpha_k^p < \infty$, where $p \geq 2$, such that $a_k = \alpha_k Be_k$, where (e_k) is the natural basis in l_p. Let us consider the series $\sum_k \alpha_k e_k \gamma_k$. Since $\sum_k \alpha_k^p < \infty$ and (e_k) is the natural basis in l_p, then this series is a.s. unconditionally convergent in l_p. Consequently, it is clear, that $\sum_k a_k \gamma_k$ is a.s. unconditionally convergent (since $\sum_k a_k \gamma_k = B\sum_k \alpha_k e_k \gamma_k$, where B is the linear bounded operator).

$2^o \Longrightarrow 3^o$ is proved.

$3^o \Longrightarrow 1^o$. Suppose that $\sum_k a_k \gamma_k$ is a.s. unconditionally convergent in X iff $\sum_k a_k$ is unconditionally convergent. Let us prove that X does not contain l_∞^n uniformly. The proof will be done ab contrario. Let us assume that X contains l_∞^n uniformly. Then there exists an unconditionally convergent series $\sum_k a_k$ such that $\| a_k \| = \frac{1}{\sqrt{\ln(k+1)}}$ k = 1,2,..., (see [3], [5]).
But the series $\sum_k a_k \gamma_k$ does not converge a.s. since the sequence $(\| a_k \| \gamma_k)$ is not convergent a.s. to zero (see [6], p.72).

Thus we get the contradiction and our proof is finished.

References

[1] Ørno, P., A note on unconditional converging series in L_p, Proc.Amer.Math.Soc., 59, 2, 1976.

[2] Kvaratskhelia, V.V., On unconditional convergence in Banach spaces, (in Russian), Bull.Acad. Sci. Georg.SSR, 90, 1978, 533-536.

[3] Maurey, B., Thèorémes de factorization pour les operateurs linéaires à valeurs dans les espaces L_p, Astérisque, 11,1974

[4] Rosenthal, H.P., Some applications of p-suming operators to Banach space theory,Studia Math., 58,1976, 21-43.

[4'] Rosenthal, H.P., On subspaces of Lp, Ann.Math.37, 1973, 344-373.

[5] Rakov, S.A., (in Russian), Mat.Zametki, 14, 1, 1973.

[6] Vakhania, N.N., Probability distributions on linear space, (in Russian), Mecniereba, Tbilisi, 1971.

Academy of Sciences of the Georgian SSR
Cumputing Center, Tbilisi, 380093 USSR

p-STABLE MEASURES AND p-ABSOLUTELY SUMMING OPERATORS

by

W. LINDE

Sektion Mathematik
Fridrich-Schiller University, Jena, DDR

V. MANDREKAR

Department of Statistics and Probability
Michigan State University,E.Lansing, USA

and

A. WERON

Institute of Mathematics
Wrocław Technical University, Wrocław, Poland

We investigate operators T from some separable Banach space $L_{p'}$, $1 < p \leq 2$, $1/p + 1/p' = 1$, into a Banch space E for which $\exp(-\|T'a\|^p)$ is the charaoteristic function of a Radon measure on E. It turns out that the set $\Sigma_p(L_{p'},E)$ of all those operators becomes a Banach space under the equivalent norms

$$\sigma_{pr}(T) = \{ \int_E \|x\|^r d\mu \}^{1/r} , \qquad 1 \leq r < p \leq 2 ,$$

where μ is the Radon measure with characteristic functional $\exp(-\|T'a\|^p)$. In case $p = 2$ we get the set of so-called γ-Radonifying operators which are very useful in the investigation of Gaussian measures on Banach spaces, cf. [9] and references therein. For $p \neq 2$ similar class of operators was studied in [14]. If Π_p denotes the ideal of p-absolutely summing operators, then for $1 < p < 2$ in general, neither $\Sigma_p(L_{p'},E) \subseteq \Pi_p(L_{p'},E)$ nor the converse inclusion hold. We investigate Banach spaces E where one of the inclusions is valid. Moreover, we characterize spaces E where $T \in \Sigma_p(L_{p'},E)$ iff T' is p-absolutely summing. In case $p = 2$, these are exactly spaces of type 2, see [4]. For $p < 2$ the condition is much stronger because these are spaces of stable type p isomorphic to a subspace of some L_p.

The paper is motivated from the works [8, 13 and 21]. Some of the results presented here are taken from [24].

1. Notations and Definitions.

Let E be a real Banach space. For a real number p $(1 < p \leq 2)$, L_p be a separable Banach space of measurable functions having p-integrable absolute value. As in [7], we say that E is of S_p-type (resp. SQ_p-type) if E is isomorphic to a subspace (resp. to a subspace of a quotient) of some space L_p. If μ is measure, or more generally, a cylindrical measure on E then $\hat{\mu}(a) = \int_E \exp(i\langle x,a\rangle)d\mu(x)$ denotes the characteristic function of μ and is a mapping from the dual E' into the field of complex numbers. Let $\xi_1, \xi_2, \ldots$ be a sequence of independent identically distributed random variables with the characteristic function $\exp(-|t|^p)$ for their distribution. Then we say that E is of stable type p if there exists a constant $c > 0$ such that for all $x_1, \ldots, x_n \in E$,

$$\{\mathbb{E} \, \|\sum_i^n x_i \xi_i\|^r\}^{1/r} \leq c \{\sum_i^n \|x_i\|^p\}^{1/p}$$

for some (each) r with $0 < r < p$.

As usual, we denote p' the conjugate of p, i.e., $1/p + 1/p' = 1$ and note that $\exp(-\|g\|^p)$ $(g \in L_p)$ is the characteristic function of a cylinder measure θ_p on $L_{p'}$. It is known that there exists an operator X from L_p into $L_r(\Omega, P)$ such that for each $f \in L_{p'}$, $X(f)$ has distribution $f(\theta_p)$ and $\|X(f)\|_r = c_{rp}\|f\|_p$ where $c_{rp} = \{\mathbb{E}|\xi_1|^r\}^{1/r}$. Finally, Π_p will denote the ideal of p-absolutely summing operators. For the definition of p-absolutely summing operators and their properties we refer the reader to [17] or [18].

2. θ_p-Radonifying Operators.

An operator T from $L_{p'}$ into E is said to be θ_p-Radonifying if $T(\theta_p)$ extends to a Radon measure on E. In this case the Radon extension is a p-stable symmetric measure on E. $\Sigma_p(L_{p'},E)$ denotes the set of all θ_p-Radonifying operators from $L_{p'}$ into E.

Let us recall that an operator S from E into $L_r(\Omega,P)$ is decomposable if there is a strongly measurable E-valued mapping ϕ from Ω such that $Sa = \langle\phi(\cdot),a\rangle$ for all $a \in E'$.

PROPOSITION 1. The following are equivalent:

(1) $T \in \Sigma_p(L_{p'},E)$

(2) $\exp(-\|T'a\|^p)$ is the characteristic function of a Radon measure on E.

(3) XT' is decomposable, where X denotes the operator from L_p into $L_r(\Omega,P)$ generating θ_p.

Proof: The equivalence of (1) and (2) follows from the fact that the characteristic function of $T(\theta_p)$ is equal to $\exp(-\|T'a\|^p)$. The equivalence of (2) and (3) may be established by using the Badrikian--Schwartz theorem (cf. [18], Prop. 25.3.7).

For $T \in \Sigma_p(L_{p'},E)$ we put

$$\sigma_{pr}(T) = \{\int_E \|x\|^r d\mu\}^{1/r} ,$$

where $\hat{\mu}(a) = \exp(-\|T'a\|^p)$, $a \in E'$ and $1 \le r < p$. This integral is finite because of the result of [1].

PROPOSITION 2. $\Sigma_p(L_{p'},E)$ is a Banach space under the norms σ_{pr}.

Proof: Let's remark that by the equivalence of (1) and (2) in Prop. 1. Σ_p is closed under scalar multiplication and by the equivalence of (1) and (3) in Prop. 1. Σ_p is closed under addition. Using again Prop. 1. and the observation

$$\sigma_{pr}(T) = \{\int_\Omega \|\phi(\omega)\|^r dP\}^{1/r} ,$$

where ϕ is the mapping which decomposes XT', we have that σ_{pr} is a norm on Σ_p. Moreover, the completness of Σ_p w.r.t. σ_{pr} carries over by the completness of corresponding $L^r(\Omega,P;E)$.

In some cases it is possible to describe θ_p-Radonifying operators. The following fact follows from the equivalence of (1) and (2) in Prop. 1. and by Ito-Nisio's theorem (see [23] p.274).

COROLLARY 1. An operator T from $l_{p'}$ into E is θ_p-Radonifying if and only if the sum

$$\sum_1^{\infty} Te_i\xi_i$$

exists a.e., where $e_1,e_2,\ldots$ are unit vectors in $l_{p'}$.

The following result shows why θ_p-Radonifying operators are useful in the study of p-stable symmetric measures on Banach spaces.

PROPOSITION 3. Let μ be a p-stable symmetric Radon measure on a Banach space E. Then there exists an operator $T \in \Sigma_p(L_{p'}[0,1],E)$ with $\hat{\mu}(a) = \exp(-\|T'a\|^p)$.

Proof: In view of the result of [22] we know that for p-stable symmetric measure μ on E, $\hat{\mu}(a) = \exp(-\int_U |\langle x,a\rangle|^p dm$ where m is a finite measure on the unit ball U of E. Let $T_1\colon L_{p'}(U,m) \to E$ given by the formula

$$T_1 f = \int_U f(x)x\, dm(x) .$$

Then we define $T\colon L_{p'}[0,1] \to E$ by $T = T_1J'$, where J is the isometric imbedding of $L_p(U,m)$ into $L_p[0,1]$ (cf. [11]). Hence

$$\|T'a\|^p = \|JT_1'a\|^p = \|T_1'a\|^p = \int_U |\langle x,a\rangle|^p\, dm(x)$$

and consequently $\hat{\mu}(a) = \exp(-\|T'a\|^p)$.

THEOREM 1. Given real numbers r and q with $0 \le r \le q < p$, then there exists a constant $d > 0$ such that for all Banach spaces E and all p-stable symmetric measures μ the following estimation holds:

$$\left(\int_E \|x\|^q d\mu\right)^{1/q} \le d\left(\int_E \|x\|^r d\mu\right)^{1/r} .$$

Proof: Let μ be a p-stable symmetric measure on E. By Prop. 3. there exists an operator $T \in \Sigma_p(L_{p'}, E)$ with $\hat{\mu}(a) = \exp(-\|T a\|^p)$. By Prop. 2. $\Sigma_p(L_{p'}, E)$ is a Banach space w.r.t. the norms $\sigma_{pq}(T)$ and $\sigma_{pr}(T)$, where $r \le q < p$. Consequently, there exists a constant $c > 0$ such that $\sigma_{pr}(T) \le c\sigma_{pq}(T)$, and the identity operator $I:(\Sigma_p, \sigma_{pq}) \to (\Sigma_p, \sigma_{pr})$ is continuous. By Banach's theorem the inverse I^{-1} is also continuous and hence there exists a constant $d(E)$ such that for each p-stable symmetric measure μ on E we have:

$$\left(\int_E \|x\|^q d\mu\right)^{1/q} \le d(E) \left(\int_E \|x\|^r d\mu\right)^{1/r}.$$

Without loss of generality we may assume that E is separable, so one may consider E as a close subspace of $C[0,1]$ and in the above inequality, put $d(E) = d(C[0,1]) = d$.

COROLLARY 2. The all norms σ_{pr}, $1 \le r < p$, on the space $\Sigma_p(L_{p'}, E)$ are equivalent.

3. θ_p-Radonifying and p-Absolutely Summing Operators.

In this section we want to generalize the result of [4]: E is of cotype 2 (cf. [9] and [14] for the definition) iff $\Sigma_2(L_2, E) = \Pi_2(L_2, E)$.

The following result is a consequence of the fact that for $r > 1$ the ideals of r-Radonifying and r-absolutely summing operators coincide (see [18], 25.4.8) and the theorem of [1] which says that each p-stable measure has strong r-order for $1 < r < p$.

PROPOSITION 4. If $1 < r < p$ then $\Pi_r(L_{p'}, E) \subseteq \Sigma_p(L_{p'}, E)$.

In the next theorem we characterize those Banach spaces E for which such inclusion holds when $r = p$.

THEOREM 2. Let $1 < p < 2$, then the following are equivalent:

(1) E is of stable type p.

(2) For each (one infinite dimensional) space L_p we have $\Pi_p(L_{p'}, E) \subseteq \Sigma_p(L_{p'}, E)$.

Proof: Let us assume that E is of stable type p and let T be p-absolutely summing operator from $L_{p'}$ into E. Since we only investigate separable spaces L_p there exists an isometric imbedding J from L_p into $L_p[0,1]$ (cf. [11]). Then TJ' is p-absolutely summing. Using the results of [6] there exists a strongly measurable function ϕ from [0,1] into E with $\mathbb{E}\|\phi\|^p < \infty$;

$$\|T'a\|^p = \|JT'a\|^p = \int_0^1 |\langle\phi(t),a\rangle|^p dt .$$

By [2] or [16] $\exp(-\|T'a\|^p)$ is the characteristic function of a Radon measure.

To prove the converse we assume that the inclusion of (2) holds and that E is not of stable type p . Then by ([5] and [15] cf. also [23] p.371) there are, for each natural number n , elements $x_1,\ldots,x_n \in E$ such that for all real numbers $t_1,\ldots,t_n$, the inequalities

$$\{\sum_{i=1}^n |t_i|^p\}^{1/p} \le \|\sum_{i=1}^n t_i x_i\| \le 2\{\sum_{i=1}^n |t_i|^p\}^{1/p}$$

hold. If we define an operator T from $L_{p'}$ into E with

$$\|T'a\|^p = \sum_{i=1}^n |\langle x_i,a\rangle|^p$$

we get

$$\sigma_{pr}(T) = \{\mathbb{E}\|\sum_{i=1}^n x_i\xi_i\|^r\}^{1/r} \ge \{\mathbb{E}(\sum_{i=1}^n |\xi_i|^p)^{r/p}\}^{1/r}$$

$$\ge c(n \log n)^{1/p} \quad \text{for } n \ge n_o \quad \text{(cf. [3],[23] p.277).}$$

On the other side, if for instance

$$T(g) = \sum_{i=1}^n \langle g,1_{A_i}\rangle\lambda(A_i)^{-1/p} x_i$$

where 1_{A_i} denotes indicator function of the measurable set A_i, $A_1,\ldots,A_n$ are disjoint and $\lambda(A_i)$ denotes the measure of A_i, then by the Pietsch inequality

$$\|T(g)\| \le 2\{\sum_{i=1}^n |\langle g,1_{A_i}/\lambda(A_i)^{1/p}\rangle|^p\}^{1/p}$$

we get $\pi_p(T) \le 2n^{1/p}$.

Thus, the estimation

$$\sigma_{pr}(T) \le c\pi_p(T)$$

cannot hold for any $c > 0$. This contradicts (2) proving the theorem.

Next we want to investigate Banach spaces E for which always the inclusion

$$\Sigma_p(L_{p'},E) \subseteq \Pi_p(L_{p'},E)$$

is true. Unfortunately we have in this case not complete characterization of spaces with this property.

PROPOSITION 5. Assume E has the following property: There are real numbers r and q with $0 < r \le q \le p$ such that each E-valued operator T with $T' \in \Pi_r$ is q-absolutely summing. Then

$$\Sigma_p(L_{p'},E) \subseteq \Pi_p(L_{p'}, E) .$$

Proof: Choose a θ_p-Radonifying operator T from $L_{p'}$ into E. By Prop. 6. below T' is r-absolutely summing. This implies $T \in \Pi_q(L_{p'},E) \subseteq \Pi_p(L_{p'},E)$ and proves the theorem.

COROLLARY 3. If E is a subspace of a Banach lattice which is r-convex and q-concave for some $1 \le r \le q \le p$ then each E-valued θ_p-Radonifying operator is p-absolutely summing (cf. [19]). If. moreover, E is of stable type p, for instance if E is contained in L_s, $p < s \le 2$, then the two classes of operators coincide. If $1 \le s \le 2$ and E is isomorphic to a subspace of L_s then

$$\Sigma_p(L_{p'},E) \subseteq \Pi_p(L_{p'},E) .$$

THEOREM 3. $\Sigma_p(l_{p'},E) = \Pi_p(l_{p'},E)$ if and only if E is of type SQ_p and of stable type p.

Proof: Let us first assume that the identity holds. Then E is of stable type p because of Th. 2. Now, we choose $x_1,\dots,x_n \in E$ and a matrix $A = (\alpha_{ij})$ $1 \le i,j \le n$, of real numbers. Then we define T from $l_{p'}$ into E by

$Te_i = x_i$ for $1 \le i \le n$ and $Te_i = 0$ for $n < i < \infty$, where e_i is the i-th unit vector of $l_{p'}$. If A denotes the mapping from l_p^n into l_p generated by the matrix above we get

$$\{\sum_{i=1}^{n}\|\sum_{j=1}^{n}\alpha_{ij}x_j\|^p\}^{1/p}\le c\sigma_{pr}(TA')\le c_1\pi_p(TA')\le c_1\|A\|\,\pi_p(T)$$

$$\le c_2\|A\|\sigma_{pr}(T)\le c_3\|A\|\,\{\sum_{i=1}^{n}\|x_i\|^p\}^{1/p}\ .$$

Using [7] it follows that E is of type SQ_p.

To prove the converse it suffices to show (in view of Th. 2.) $\Sigma_p(l_{p'},E)\subseteq\Pi_p(l_{p'},E)$.

Take $T\in\Sigma_p(l_{p'},E)$ and let A be a mapping from l_p into l_p. Then since E is of type SQ_p by ([7])

$$\{\sum_{i=1}^{\infty}\|TA'e_i\|^p\}^{1/p}\le c\|A\|\{\sum_{i=1}^{\infty}\|Te_i\|^p\}^{1/p}\le c_1\|A\|\sigma_{pr}(T)$$

which proves $T\in\Pi_p(l_{p'},E)$.

<u>COROLLARY 4</u>. If $2<q<\infty$ then there are θ_p-Radonifying operators from $l_{p'}$ into L_q which are not p-absolutely summing.

4. <u>Duals of θ_p-Radonifying Operators</u>.

In this section we want to generalize the well known result of [4] about spaces of (stable) type 2: the following two properties are equivalent:

(1) E is of (stable) type 2,

(2) an operator T from L_2 into E belongs to $\Sigma_2(L_2,E)$ if and only if T' is 2-absolutely summing.

To do so we need

<u>PROPOSITION 6</u>. If $T\in\Sigma_p(L_{p'},E)$ and $0<r\le\infty$, then T' is r-absolutely summing.

<u>Proof</u>: Because of the inclusion properties of the ideals of r-absolutely summing operators (see [17]) we only need to treat case $0<r<p$. Let X be the operator from L_p into $L_r(\Omega,P)$ generating θ_p and choose $a_1,\dots,a_n\in E'$. Then we get

$$\{\sum_{i=1}^{n} \|T'a_i\|^r\}^{1/r} = c_{rp}^{-1}\{\sum_{i=1}^{n} \|XT'a_i\|^r\}^{1/r}$$

$$= c_{rp}^{-1}\{\sum_{i=1}^{n} \int_\Omega |\langle\phi(\omega),a_i\rangle|^r \, dP\}^{1/r}$$

$$\leq c_{rp}^{-1} \sup_{\|x\|\leq 1} \{\sum_{i=1}^{n} |\langle x,a_i\rangle|^r\}^{1/r} \{\int_\Omega \|\phi\|^r \, dP\}^{1/r} .$$

This proves $T' \in \Pi_r(E',L_p)$ and

$$\pi_r(T') \leq c_{rp}^{-1}\{\int_\Omega \|\phi\|^r \, dP\}^{1/r} = c_{rp}^{-1}\sigma_{pr}(T) .$$

THEOREM 4. Suppose $1 < p < 2$. Then the following are equivalent:

(1) $T \in \Sigma_p(L_{p'},E)$ if and only if $T' \in \Pi_p(E',L_p)$ for each (one infinite dimensional) space L_p.

(2) E is of type S_p and of stable type p.

Remark: (2) is equivalent to

(2') E is of type S_p and does not contain an isomorphic copy of l_p (cf. [20],[23])

Proof: (1) implies the existence of a constant $c > 0$ such that $\sigma_{pr}(T) \leq c\pi_p(T')$ for all finite rank operators T from $L_{p'}$ into E. If $x_1,\ldots,x_n \in E$ we define an operator T from $L_{p'}$ into E with

$$\|T'a\| = \{\sum_{i=1}^{n} |\langle x_i,a\rangle|^p\}^{1/p} , \qquad a \in E' .$$

Then from

$$\{\mathbb{E} \|\sum_{i=1}^{n} x_i\xi_i\|^r\}^{1/r} = \sigma_{pr}(T) \leq c\pi_p(T') \leq c\{\sum_{i=1}^{n} \|x_i\|^p\}^{1/p}$$

it follows that E is of stable type p.

To prove that E is of type S_p we choose $x_1,\ldots,x_n , y_1,\ldots,y_n \in E$ with

$$\sum_{i=1}^{n} |\langle x_i,a\rangle|^p \leq \sum_{i=1}^{n} |\langle y_i,a\rangle|^p \qquad \text{for all } a \in E' .$$

Now we define operators T and S from $L_{p'}$ into E with

$$\|T'a\|^p = \sum_{i=1}^{n} |\langle x_i, a\rangle|^p \quad \text{and} \quad \|S'a\|^p = \sum_{i=1}^{n} |\langle y_i, a\rangle|^p$$

for all $a \in E'$.
Because of $\|T'a\| \le \|S'a\|$ for all $a \in E'$ we get

$$\pi_p(T') \le \pi_p(S').$$

Consequently,

$$\{\sum_{i=1}^{n} \|x_i\|^p\}^{1/p} < c_1 \{ \mathbb{E} \|\sum_{i=1}^{n} x_i \xi_i\|^r\}^{1/r} = c_1 \sigma_{pr}(T)$$

$$\le c_1 c_2 \pi_p(T') \le c_1 c_2 \pi_p(S') \le c_1 c_2 \{\sum_{i=1}^{n} \|y_i\|^p\}^{1/p}.$$

From [10], Th. 7.3. it follows that E is of type S_p.

To prove the converse take an operator T from $L_{p'}$ into E such that T' is p-absolutely summing. Because E is of type S_p this implies $T \in \Pi_p(L_{p'}, E)$ (see [7] Cor. 4 or [18] 19.5.3). Since E is of stable type p then by Th. 2. $T \in \Sigma_p(L_{p'}, E)$. Now if we apply Prop. 6. then we see that the condition (1) of the theorem is satisfied, which completes the proof.

COROLLARY 5. Suppose $p < q \le 2$. Then T belongs to $\Sigma_p(L_{p'}, L_q)$ if and only if T' is p-absolutely summing.

5. Problems:

(1). Suppose E is of type SQ_p. Does this imply

$$\Sigma_p(L_{p'}, E) \subseteq \Pi_p(L_{p'}, E)$$

for all spaces L_p (not only for l_p, as in Th. 2.)?

(2). If $T \in \Sigma_p(L_{p'}, E)$ then there is θ_p-Radonifying operator S such that S' is p-decomposable and

$$\|T'a\| = \|S'a\| \quad \text{for all } a \in E'.$$

In which cases all θ_p-Radonifying operators have a p-decomposable adjoint?

(3). What is the connection between Σ_p and the ideal of p-integral operators? More precisely, when does $T \in \Sigma_p(L_{p'},E)$ imply that T' is p-integral? When does the converse implication hold? For the definition of p-integral operators and their properties we refer the reader to [18].

References:

[1] de Acosta, A., *Asymptotic behavior of stable measures.* Ann. Probability 5 (1977), 494-499.

[2] de Acosta, A., *Banach spaces of stable type and the generation of stable measures.* Preprint 1975.

[3] Baumbach, G., Linde, W., *Asymptotic behavior of p-absolutely summing norms of identity operators.* Math. Nachr. 78 (1977), 193-196.

[4] Chobanjan, S.A., Tarieladze, V.I., *Gaussian characterization of Banach spaces.* J. Multivariate Anal. 7 (1977), 183-203.

[5] Krivine, J.L., *Sous-espaces de dimension finite des espaces de Banach reticules.* Ann. of Math. 104 (1976), 1-29

[6] Kwapień, S., *On a theorem of L.Schwartz and its application to absolutely summing operators.* Studia Math. 38 (1970), 193-201.

[7] Kwapień, S., *On operators factorizable through L_p space.* Bull. Soc. Math. France, Mem. 31-32 (1972), 215-225.

[8] Linde, W., *Estimation of integral of stable measures on Banach spaces.* (to appear).

[9] Linde, W., Pietsch, W., *Mappings of Gaussian measures of cylindrical sets in Banach spaces.* Teor. Verojatnost. i Primenen. 19(1974), 472-487.

[10] Lindenstrauss, J., Pełczyński, A., *Absolutely summing operators in L_p-spaces and their applications.* Studia Math. 29 (1968), 275-326.

[11] Lindenstrauss, J., Tzafriri, L., *Classical Banach spaces,* Springer -Verlag Lecture Notes in Math., vol. 338, Berlin-Heidelberg-New York 1973.

[12] Lindenstrauss, J., Tzafriri, L., *Classical Banach apaces II, Function spaces.* Springer-Verlag Berlin-Heidelberg-New York 1979.

[13] Mandrekar, V., Weron, A., *α-stable characterization of Banach Spaces ($1 < \alpha < 2$).* Preprint 1979.

[14] Maurey, B., *Espaces de cotype p, $0 < p \leq 2$.* Seminairé Maurey--Schwartz 1972-1973, Exposé VII.

[15] Maurey, B., Pisier, G., *Series de variables aleatoires vectorielles independantes et proprietes geometriques des espaces de Banach.* Studia Math., 58 (1976), 45-90.

[16] Mouchtari, D., *Sur l'existence d'une topologie de type Sazanov sur une espace de Banach.* Seminairé Maurey-Schwartz 1975-1976, Exposé XVII.

[17] Pietsch, A., *Absolut p-summierende Abbildungen in normierten Raumen.* Studia Math. 28 (1966/67), 333-353.

[18] Pietsch. A., *Operator ideals.* VEB Deutscher Verlag der Wissenschaften, Berlin 1978.

[19] Pisier, G., *Some results on Banach spaces without local unconditional structure.* Theses, Chapitre 9, Universite Paris VII, 1977.

[20] Rosenthal, H.P., *On subspaces of L_p.* Ann. Math. 97 (1973), 344-373.

[21] Tien, N.Z., Weron, A., *Banach spaces related to α-stable measures.* Preprint 1979.

[22] Tortrat, A., *Sur les lois e(λ) dans les espaces vectoriels,* Applications aux lois stables. Preprint 1975.

[23] Woyczyński, W.A., *Geometry and martingales in Banach spaces.* Part II, in Advances in Probability and Related Topics, vol. 4 Ed. J.Kuelbs, Marcel Dekker, Inc., New York 1978, 267-517.

[24] Linde, W., Mandrekar, V., Weron, A., *Radonifying operators related to p-stable measures on Banach spaces,* (to appear).

SUPPORT AND SEMINORM INTEGRABILITY THEOREMS FOR r-SEMISTABLE PROBABILITY MEASURES ON LCTVS

by

Donald Louie
Department of Mathematics
The University of Tennessee, Knoxville, USA ,

and

Balram S. Rajput*
Department of Mathematics
The University of Tennessee, Knoxville, USA,

and

Indian Statistical Institute, New Delhi, India.

ABSTRACT

Let μ be an r-semistable K-regular probability measure of index $\alpha \in (0, 2]$ on a complete locally convex topological vector space E. It is shown that the topological support S_μ of μ is a translated convex cone if $\alpha \in (0, 1)$, and a translated truncated cone if $\alpha \in (1, 2]$. Further, if $\alpha = 1$ and μ is symmetric, then it is shown that S_μ is a vector subspace of E. These results subsume all earlier known results regarding the support of stable measures. A result regarding the support of infinitely divisible probability measure on E is also obtained. A seminorm integrability theorem is obtained for K-regular r-semistable probability measures μ on E. The result of de Acosta (Ann. of Probability, 3(1975), 865 - 875)and Kanter (Trans. Seventh Prague Conf., (1974), 317 - 323) is included in this theorem as long as the measures are defined on LCTVS and seminorm is continuous.

The research of this author was partially supported by the Office of Naval Research under contract No. N00014-78-C-0468.

1. INTRODUCTION

Let E be a complete locally convex topological vector space (LCTVS) and let μ be a stable probability measure (p.m.) of index $\alpha \in (0, 2]$; then it is shown by Tortrat [15] that for $\alpha \neq 1$, S_μ, the support of μ, is a certain cone (if μ is symmetric, then it is shown by Rajput [13, 14] that S_μ is a subspace for all α; this result for $1 < \alpha \leq 2$ is also obtained by de Acosta [1]). Furthermore, if p is a continuous seminorm (in fact measurability is enough) on E, then it is shown by de Acosta [1] and Kantor [8] that

$$\int_E p^\delta(x)\ \mu(dx) < \infty\ , \text{ for all } 0 \leq \delta < \alpha\ .$$

A natural and nontrivial generalization of stable measures is the class of r-semistable measures, which was first introduced and studied on the real line R by Paul Lévy [12]. Later, Kruglov, in an interesting paper [9], obtained a quite explicit form of the characteristic function of r-semistable p. measures on R and showed that this class has properties similar to those of stable p. measures (similar situation is true in Hilbert space is shown by Kruglov [10] and by Kumar [11]). Partialy motivated from these papers, we raised and completely answered, in this paper, the question of whether r-semistable p. measures have properties similar to those of stable p. measures mentioned above. Explicitly, we obtain the following results: Let μ be a K-regular r-semistable p. measure (see Definition 2.1) of index $\alpha \in (0, 2]$ on a complete LCTVS E, then S_μ, the support of μ, is a translated convex cone or a translated truncated cone according as whether $0 < \alpha < 1$ or $1 < \alpha \leq 2$; further, if $\alpha = 1$ and μ is symmetric, we prove that S_μ is a subspace (Theorem 3.2). This result subsumes all earlier known results regarding the support of stable measures [1, 4, 13, 14, 15]. (A general theorem which gives a formula for the support of K-regular infinitely divisible (i.d.) p. measures on E and which includes some results for the supports of i.d. measures derived in [4, 14, 15] is also obtained). Let μ and E be as above and p a continuous seminorm on E ; then $\int_E p^\delta(x)\,\mu(dx) < \infty$, if $0 \leq \delta < \alpha$. This result includes the seminorm integrability theorem for stable measures in [1, 8], as long as the measures are defined on LCTVS and p is continuous.

Our proof of the support theorem for i.d. measures uses similar ideas to those of Brockett [4], who proved part of our result in Hilbert spaces, and Tortrat [15, 16], who proved similar results under different hypotheses in certain LC spaces. Our techniques of proof of the support theorem for r-semistable measures, however, seem new and quite interesting. Our proof of the seminorm integrability result is classical and has the drawback in that it uses a strong central limit theorem in Banach spaces [2].

2. PRELIMINARIES

Unless otherwise stated, the following conventions and notation will remain fixed in this paper:

All vector spaces considered are over the real field R and all topological spaces are assumed Handsdorff. If μ and ν are two finite K-regular p. measures on the Borel σ-algebra $\mathcal{B}$ of a topological vector space E , then μ^{*n} and $\mu * \nu$ will denote, respectively, μ convoluted n-times and the convolution of μ and ν . If $a \neq 0$, then T_a will denote the map on E defined by $T_a(x) = ax$, $x \in E$; further $T_a\mu$ will denote the measure $\mu \circ T_a^{-1}$. For any $x \in E$, δ_x will denote the degenerate measure at x . E and E^* will, respectively, denote a complete LCTVS and its topological dual, and $M_K(E)$ will denote the class of all K-regular p. measures on E . If A is a subset of a topological space, then $\bar{A}$ will denote its closure; finally, θ will denote the zero element of E .

We will now give the definition of r-semistable p. measures and some of their properties pertinent to this paper. This definition and results are taken from Chung, Rajput and Tortrat [5], which may be referred to for other properties of r-semistable p. measures. The first result below dealing with i.d. p.m. is taken from [6, 7] .

Definition 2.1: Let E be a LCTVS , $\mu \in M_K(E)$ and $0 < r \leq 1$. Then μ is said to be r-semistable if there exists a K-regular p. measure ν , sequences $\{a_n\} \subseteq R$, $a_n > 0$, and $\{x_n\} \subseteq E$, and an increasing sequence of positive integers $\{k_n\}$ such that

$$\frac{k_n}{k_{n+1}} \longrightarrow r$$

and

$$T_{a_n} \nu^{*k_n} * \delta_{x_n} \xrightarrow{w} \mu ,$$

as $n \longrightarrow \infty$ (the symbol ' $\xrightarrow{w}$ ' will always denote the weak convergence).

(i) Let $\mu \in M_K(E)$ be i.d. then there exists a measure F (called the Lévy measure), a quadratic form Q on E^* , an $x_0 \in E$, and a compact convex circled subset K of E with $F(K^c) < \infty$ such that, for every $f \in E^*$, the characteristic function $\hat{\mu}$ of μ has the representation

$$\hat{\mu}(f) = \exp\{if(x_0) - \tfrac{1}{2}Q(f) + \int_E \psi(f, x)dF(x)\} ,$$

where $\psi(f, x) = e^{if(x)} - 1 - if(x)\, I_K(x)$ (I_K is the indicator of K) ; further, Q and F are unique and x_0 depends on the choice of K . For the sake of simplicity of notation we will use the notation $[x_0, Q, K, F]$ to denote the above representation for μ .

(ii) Let μ be as above with the representation $[x_0, Q, K, F]$, then there exists a unique continuous (in weak topology) semigroup $\{\mu^s : s > 0\}$ of K-regular i.d. p. measures with $\mu = \mu^1$ (μ^s is referred to as the sth root of μ and has the representation $[s\,x_0, s\,Q, K, sF]$), and

$$(\mu^s)^t = \mu^{st}. \tag{2.1}$$

(iii) Let $\mu \in M_K(E)$ and $r \in (0, 1)$, then μ is r-semistable if and only if μ is i.d. and there exist a unique $\alpha \in (0, 2)$ and $x(r_n) \in E$ such that

$$\mu^{r^n} = T_{r^{n/\alpha}} \mu * \delta_{x(n)} , \tag{2.2}$$

for all $n = 1, 2, \ldots$. The number α is referred to as the index of μ ($\alpha = 2$ corresponds to the Gaussian case).

(iv) Let $\mu \in M_K(E)$ then μ is 1-semistable $\Leftrightarrow \mu$ is r-semistable for every $r \in (0, 1) \Leftrightarrow \mu$ is stable.

(v) The class of stable K-regular p. measures are properly contained in the class of r-semistable p. measures for every fixed $r \in (0, 1)$.

3. SUPPORT THEOREMS FOR I.D. AND r-SEMISTABLE PROBABILITY MEASURES

We recall that the support of a finite Borel measure μ on a topological space is, by definition, the smallest closed set (if it exists) with full μ-measure. If μ is K-regular (or even τ-regular) the support of μ always exists. The main purpose of this section is to prove the following two theorems.

Theorem 3.1: Let μ be an i.d. K-regular p.m. on E with representation $[\theta, 0, K, F]$:

(i) Let $\mathcal{U}$ be the class of all convex circled Borel nbds. of θ directed by reverse set inclusion; set $F_0 = F/K^c$, $F_U = F/K \cap U^c$, $a_U = \int_E x dF_U(x)$, $\nu_0 = e(F_0)$, and $\nu_U = e(F_U)$ (note $a_U \in E$, see [7]), then

$$S_\mu = [\bigcap_V \{\bigcup_{U \geq V} (S_{\nu_U} + a_U)\}^- + S_{\nu_0}]^- . \tag{3.1}$$

In addition if $\{\delta_{a_U}\}$ is tight and δ_a is any limit pt. of $\{\delta_{a_U}\}$, then

---

K^c and U^c , respectively, denote the complements of K and U .

$$S_\mu = a + \overline{G(F)} \ ,$$

where $G(F)$ is the semigroup with zero element which is generated by S_F, the support of F ($S_F = \{x \in E: F(V_x) > 0$, for every open nbd. V_x of $x\}$).

(ii) (Tortrat) If $\int_K p_K(x)\, dF(x) < \infty$, where p_K is the Minkowski functional of K which is assumed to take the value $+\infty$ off the set $\bigcup_{n=1}^{\infty} nK$, then $\{\delta_{a_U}\}$ is tight (a_U is as in (i)); hence $S_\mu = a + \overline{G(F)}$, where δ_a is any limit point of $\{\delta_{a_U}\}$.

(iii) If $\int_K p_K^2(x)\, dF(x) < \infty$, then $S_\mu = \overline{G(F) + A}$, where A is a closed set.

Theorem 3.2: Let μ be a K-regular r-semistable p.m., $r \in (0, 1)$, of index $\alpha \in (0, 2]$ on E.

(i) If $\alpha \in (1, 2]$, then S_μ is a translate of a truncated cone; further, if μ is strictly r-semistable (i.e. $x(r) = \theta$ in (2.2)), then S_μ is a truncated cone.

(ii) If $\alpha \in (0, 1)$, then S_μ is a translate of a convex cone; further, if μ is strictly r-semistable, then S_μ is a convex cone.

(iii) If $\alpha = 1$ and μ is symmetric, then S_μ is a subspace.

Remark 3.3: As hinted in Section 1, part (iii) of Theorem 3.1 and the fact that $S_\mu = a + \overline{G(F)}$ under a hypothesis similar to $\int_K p_K(x) dF(x) < \infty$, was obtained, in the Hilbert space setting, by Brockett [4] and the last statement, under certain other hypotheses, was obtained, in LCTV setting, by Tortrat [15, 16]. Our proof of Theorem 3.1 uses similar ideas as those of [4]; however, because of the weaker structure available in arbitrary LCTV spaces, modifications of techniques are required. Since clearly, from Definition 2.1, every stable measure is r-semistable for all r, Theorem 3.2 includes the support results regarding stable measures obtained in [1, 4, 13, 14, 15] ; and, in view of Section 2, the above theorem also provides the corresponding results for 1-semistable measures.

For the proof of Theorems 3.1 and 3.2, we will need the following lemmas. The proof of Lemma 3.4 is elementary and Lemma 3.5 is well known. Lemma 3.6 was first conceived in [17] in the locally compact group setting; the proof presented here is similar to the one in [17], but certain details need to be verified. The last Lemma is taken from [5].

Lemma 3.4: Let $r \in (0, 1)$ and $\alpha \geq 1$. Set $A = \{r^{m/\alpha}k: k = 1, 2, \ldots, [1/r^m], m = 1, 2, \ldots\}$, where $[x]$ denotes the integral part of the number x. Then A is dense in $[0, \infty)$ if $\alpha > 1$, and A is dense in $[0, 1]$ if $\alpha = 1$.

Lemma 3.5: Let μ and ν be two K-regular p. measures on a LCTVS E and $a \in R$, $a \neq 0$. Then

$$S_{T_a\mu} = aS_\mu \quad \text{and} \quad S_{\mu*\nu} = \overline{[S_\mu + S_\nu]} .$$

Lemma 3.6: Let $\{\nu_n\}$ and $\{\lambda_n\}$ be two nets of K-regular p. measures on a LCTVS E and let ν be a K-regular p.m. on E. Assume $\nu = \nu_n * \lambda_n$, for each n, $\{\nu_n\}$ is tight, and $\nu_n \xrightarrow{w} \nu$. Then $\lambda_n \longrightarrow \delta_\theta$ and $S_\nu = \bigcap_m \overline{[\bigcup_{n\geq m} S_{\nu_n}]}$. Further, if $S_{\nu_n} \uparrow$ with n, then $S_\nu = \overline{[\bigcup_n S_{\nu_n}]}$.

Proof: From [6], $\{\lambda_n\}$ is tight; hence it has a subnet which converges to a K-regular p.m. λ. This implies $\nu = \nu * \lambda$. Hence (using characteristic functions) $\lambda = \delta_\theta$. Now, by repeating the above argument replacing $\{\lambda_n\}$ by any subnet of it, we have that each subnet of $\{\lambda_n\}$ in turn has a subnet converging to δ_θ. This shows $\lambda_n \xrightarrow{w} \delta_\theta$.

Now we prove the second part. For each fixed m, let $U_m = E \setminus \overline{\bigcup_{n>m} S_{\nu_n}}$. Then $\nu_n(U_m) = 0$ (by the definition of the support), for all $n \geq m$. But, since $\nu_n \xrightarrow{w} \nu$, $\liminf_n \nu_n(U_m) \geq \nu(U_m)$. This implies $\nu(U_m) = 0$, for every m. So $S_\nu \subseteq \bigcap_m \overline{[\bigcup_{n\geq m} S_{\nu_n}]}$. To prove the reverse inclusion, let $x \in \bigcap_m \overline{[\bigcup_{n\geq m} S_{\nu_n}]}$ and U be an arbitrary open nbd. of θ. It follows that there exists a subnet $\{m_k\}$ of $\{m\}$ such that $(x + W) \cap S_{\nu_{m_k}} \neq \emptyset$, where W is a closed nbd. of θ such that $W + W \subseteq U$. Then $W \subseteq U - y$, for every $y \in W$. From this and $\nu = \nu_{m_k} * \lambda_{m_k}$, we have

$$\nu(x + U) = \int_E \nu_{m_k} (U - y + x)\, \lambda_{m_k}(dy) \geq \nu_{m_k}(W + x)\, \lambda_{m_k}(W),$$

for all k. Taking k large and noting that $\lambda_{m_k} \xrightarrow{W} \delta_\theta$ and $\nu_{m_k}(W + x) > 0$, for all k (as shown above), we have $\nu(x + U) > 0$. This shows $x \in S_\nu$, which completes the proof of the second part. The proof of the last part is now obvious.

Note that in the above the hypothesis of tightness on $\{\nu_n\}$ is needed only to conclude $\lambda_n \xrightarrow{W} \delta_\theta$. Thus if $\lambda_n \xrightarrow{W} \delta_\theta$ were already in the hypothesis of the lemma, then the conclusions would hold without the tightness hypothesis on $\{\nu_n\}$. This observation will be used in the proofs of Theorems 3.1 and 3.2.

Lemma 3.7: Let μ be a K-regular strictly r-semistable p.m. of index $\alpha \in (0, 1)$ on E. Then $\hat{\mu}(f) = \exp\{\int_E (e^{if(x)} - 1)dF(x)\}$ and $\int_K p_K(x)\, dF(x) < \infty$, where F is the Lévy measure of μ and K is the compact convex circuled set appearing in the Lévy representation of μ (note μ is i.d.).

We are now ready to prove Theorems 3.1 and 3.2.

Proof of Theorem 3.1 (i): It is shown in [7] that $\nu_U * \delta_{a_U} \xrightarrow{W} \mu_0$, with $\hat{\mu}_0(f) = \exp\{\int_K (e^{if(x)} - 1 - i f(x))dF(x)\}$, that $\mu_0 = \nu_U * \delta_{a_U} * \lambda_U$, with λ_U i.d. and K-regular, for every $U \in \mathcal{U}$, and that $\lambda_U \xrightarrow{W} \delta_\theta$ (note $\mu = [\theta, 0, K, F]$). Lemma 3.6 applies and we get $S_{\mu_0} = \bigcap_V [\bigcup_{U \geq V} (S_{\nu_U} + a_U)]^{-}$. Then, since $\mu = \mu_0 * \nu_0$, we get (3.1). To prove the second part denote by δ_a the limit of a subnet of $\{\delta_{a_U}\}$ and use the same notation for the subnet. Then $\mu_0 * \delta_{-a} = \nu_U * \lambda_U * \delta_{a_U - a}$, $\lambda_U * \delta_{a_U - a} \xrightarrow{W} \delta_\theta$ and $\nu_U \xrightarrow{W} \mu_0 * \delta_{-a}$. Thus, since $S_{\nu_U} \uparrow$ with U, we have, from Lemma 3.6, $S_{\mu_0} = \overline{\bigcup S_{\nu_U} + a}$. Therefore, $S_\mu = a + \overline{\bigcup S_{\nu_U} + S_{\nu_0}}$. But (see, for example, [14]), $\overline{\bigcup S_{\nu_u} + S_{\nu_0}} = \overline{G(F)}$, we have $S_\mu = a + \overline{G(F)}$.

<u>Proof of Theorem 3.1 (ii)</u>: It is shown in [7] that $\{\nu_U * \delta_{a_U}\}$ is shift tight; hence $\{\delta_{a_U}\}$ is shift tight [7]. Now, for any $f \in E^*$,

$$|e^{if(a_U)} - 1| \le |f(a_U)| = |\int_{K\cap U^c} f(x)\, dF(x)| \le p^+_{K^0}(f) \int_K p_K(x) dF(x),$$

implying that $\{e^{if(a_U)} : U \in \mathscr{U}\}$ is equicontinuous at θ in E^* , for the topology of uniform convergence on compact convex circled subnets of E . Hence, from [6], $\{\delta_{a_U}\}$ is tight. This completes the proof of (ii) .

<u>Proof of Theorem 3.1 (iii)</u>: Denote by M the measure which is equal to F on K and 0 off K and recall that $\mu = \mu_0 * \nu_0$ (see the proof of (i)) . The condition $\int_K p_K^2(x)\, dF(x) < \infty$ implies

$$\mu_0(f) = e^{if(a_0)} \exp\{\int_E (e^{if(x)} - 1 - \frac{if(x)}{1 + p_K^2(x)})\, dM(x)\}, \text{ for some } a_0 \in E,$$

(see [6]). Now define, for every $U \in \mathscr{U}$ ($\mathscr{U}$ is as in (i)) , $M_U = M$ on $(K \cap U)^c$ and $M_U(B) = \int_B p_K(x)\, dM(x)$, if B is a Borel subset of $K \cap U$. Clearly M_U is equivalent to M and, since $M_U \le M$, M_U is a Lévy measure [6]. Denote by α_U the K-regular i.d. p.m. with ch. function

$$\hat{\alpha}_U(f) = \exp\{\int_E (e^{if(x)} - 1 - \frac{if(x)}{1 + p_K^2(x)})dM_U(x)\} ;$$

it follows that $\mu_0 * \delta_{-a_0} = \alpha_U * \beta_U$, for some K-regular i.d. p.m. β_U , for every $U \in \mathscr{U}$. Now, since for $f \in E^*$,

$$|\int_E [\frac{f(x)}{1 + p_K^2(x)}]\, dM_U(x)|$$

$$\le |\int_{(K\cap U)^c} [\frac{f(x)}{1 + p_K^2(x)}]dM(x)| + |\int_{K\cap U} [\frac{p_K(x)\, f(x)}{1 + p_K^2(x)}]\, dF(x)|$$

+ p_{K^0} denotes the Minkowski functional of K^0 , the polar of K .

$$= \int_{K\cap U^c} \frac{f(x)}{1 + p_k^2(x)} \; dF(x)| + |\int_{K\cap U} \frac{p_K(x)f(x)}{1 + p_k^2(x)} \; dF(x)|$$

$$\leq p_{K^\circ}(f) \; [F(U^c) + \int_K p_k^2(x) \; dF(x)] \; ;$$

it follows that $b_U = \int_E [\frac{x}{1 + p_K^2(x)}] dM_U(x)$ belongs to E and $\hat{\alpha}_U(f) = e^{if(b_U)} \exp\{\int_E (e^{if(x)} - 1) \; dM_U(x)\}$. Therefore, since $\int_{K\cap U} p_K(x) \; dM_U(x) = \int_K p_K^2(x) \; dF(x) < \infty$, using what we have proved in (ii) and replacing K by $K \cap U$ (with U a closed nbd. of θ), we have, for some $b'_U \in E$, $S_{\alpha_U} = b'_U + \overline{G(M_U)} = b'_U + \overline{G(M)}$, since M is equivalent to M_U .

Hence

$$S_\mu = \overline{[S_{\mu_0} + S_{\nu_0}]} = a_0 + \overline{[S_{\alpha_U} + S_{\beta_U} + S_{\nu_0}]}$$

(for a fixed closed nbd. U of θ) ,

$$= a_0 + b'_U + \overline{[G(M) + S_{\nu_0} + S_{\beta_U}]}$$

$$= \overline{G(F) + A} \; ,$$

where $A = S_{\beta_U} + a_0 + b'_U$ (note $\overline{G(F)} = \overline{[G(M) + S_{\nu_0}]}$) .

This completes the proof of Theorem 3.1.

<u>Proof of Theorem 3.2 (i)</u>: According to [5], μ can be centered, i.e., there exists an $x_0 \in E$ and a strictly r-semistable p.m. ν with the same index such that $\mu = \nu * \delta_{x_0}$. Thus, to complete the proof of (i), we need to show that S_ν is a truncated cone. We first show that $sS_\nu \subseteq S_\nu$, for any $s \geq 1$. Let $s \geq 1$ and set $t = s - 1 \geq 0$. Using Lemma 3.4, we

choose a sequence $\{k_n\}$ of positive integers such that $1 \leq k_n \leq [1/r^n]$ and $t_n \equiv r^{n/\alpha} k_n \to t$, as $n \to \infty$. Then, since $r^{n(1-1/\alpha)} \to 0$ (note $1 < \alpha$), as $n \to \infty$, and $r^{n(1-1/\alpha)} r^{n/\alpha} k_n = r^n k_n$, we have $r^n k_n \to 0$, as $n \to \infty$. Therefore, by semigroup and continuity property of $\{\mu^p : p > 0\}$ (see Section 2), we have

$$\mu^{r^n k_n} * \mu^{1-r^n k_n} = \mu$$

and

$$\mu_n \equiv \mu^{1-r^n k_n} \xrightarrow{w} \mu ,$$

as $n \to \infty$. Therefore, using the fact that $\mu^{r^n k_n} = (\mu^{r^n})^{*k_n} = T_{r^{n/\alpha}} \mu^{*k_n}$, it follows, from Lemmas 3.5 and 3.6 (note that $\{\mu^p : 0 < p \leq p_0\}$ is tight (see Section 2)), that

$$S_\mu = [r^{n/\alpha} S_\mu^{(k_n)} + S_{\mu_n}]^{-} , \tag{3.2}$$

for each $n = 1, 2, \ldots$, and

$$S_\mu = \bigcap_{j=1}^{\infty} [\bigcup_{n \geq j} S_{\mu_n}]^{-} , \tag{3.3}$$

where $S_\mu^{(k_n)}$ denotes the k_n-fold sum of S_μ. Now let $x \in S_\mu$. Then, by (3.3), for each $j = 1, 2, \ldots$,

$$x \in [\bigcup_{n \geq j} S_{\mu_n}]^{-} . \tag{3.4}$$

Let $\mathcal{D}$ be the set of pairs (W, n), where W is an open nbd. of x and n is a positive integer such that $W \cap S_{\mu_n} \neq \phi$. Define the relation

$\leq$ on $\mathscr{D}$ by $(W_1, n_1) \leq (W_2, n_2)$ if and only if $W_2 \subseteq W_1$ and $n_1 \leq n_2$. Using (3.4), we can easily verify that $(\mathscr{D}, \leq)$ is a directed set. Let $x_{(W, n)}$ be any element in $W \cap S_{\mu_n}$ and let $t_{(W, n)} = t_n$. Then $\{t_{(W, n)}\}$ is a subnet of $\{t_n\}$ and $x_{(W, n)} \to x$. Now, by (3.2), $t_{(W, n)}x + x_{(W, n)} \in S_\mu$; and, clearly, $t_{(W, n)}x + x_{(W, n)} \to tx + x = sx \in S_\mu$, since S_μ is closed. We will now show that S_μ is a semigroup. Let $x, y \in S_\mu$. Choose, as before, k_n's such that $t_n \equiv r^{n/\alpha} k_n \to 1$. Since $y \in [\overline{\bigcup_{n \geq j} S_{\mu_n}}]$, for each $j = 1, 2, \ldots$, (from (3.3)), we can define, as above, a net $\{y_{(W, n)}\}$ such that $y_{(W, n)} \in W \cap S_{\mu_n}$ and $y_{(W, n)} \to y$. Also, if $t_{(W, n)} = t_n$, then, as before, $\{t_{(W, n)}\}$ is a subnet of $\{t_n\}$. Now $t_{(W, n)}x + y_{(W, n)} \in S_\mu$ (by (3.2)); hence, since $t_{(W, n)}x + y_{(W, n)} \to x + y$ and S_μ is closed, $x + y \in S_\mu$.

Proof of Theorem 3.2 (ii): Again we write $\mu = \mu_0 * \delta_{x_0}$ with μ_0 strictly r-semistable p.m. of index $\alpha \in (0, 1)$ [5], and show that S_{μ_0} is a convex cone. First we show that S_{μ_0} is a semigroup. Let B be a Banach space and g a continuous linear map from E to B. Let $\lambda = \mu_0 \circ g^{-1}$, then we assert that λ is strictly r-semistable with the same index α. To see this one first notes that λ is K-regular i.d. and that for any rational $s > 0$, $\lambda^s = \mu_0^s \circ g^{-1}$ (this uses the fact that the factor measure appearing in the definition of a K-regular i.d. measure on a LCTVS is unique). Then using continuity of the semigroup, one obtains that $\lambda^s = \mu_0^s \circ g^{-1}$, for all reals $s > 0$. Hence

$\lambda^{r^n} = \mu_0^{r^n} \circ g^{-1} = T_{r^{n/\alpha}} \mu_0 \circ g^{-1} = T_{r^{n/\alpha}}\lambda$, showing λ is strictly r-semistable of index α. Now using the fact that S_{μ_0} is the projective limit of supports of measures of the type $\mu_0 \circ g^{-1}$ (see [13]), it will follow that S_{μ_0} is a semigroup, if we can show that S_λ is a semigroup.

From Lemma 3.7, $\hat{\lambda}(f) = \exp\{\int_B (e^{if(x)} - 1)dF_\lambda(x)\}$, $f \in B^*$, where B^* is the topological dual of B . Let $\nu = \lambda * \delta_a$, where $a = \int_K x\, dF(x)$ (note that since, by Lemma 3.7, $\int_K p_K\, dF_\lambda < \infty$, $a \in B$; here K and p_K are as in Theorem 3.1). Let U_n denote the closed unit disc around θ in B of radius $1/n$, $n = 1, 2, \ldots$; we will show $\delta_{a_n} \equiv \delta_{a_{U_n}} \overset{w}{\to} \delta_a$, where $\delta_{a_{U_n}}$ is as defined in Theorem 3.1(i) . Since we already know that $\{\delta_{a_{U_n}}\}$ is tight (Theorem 3.1(ii)) , to prove $\delta_{a_n} \xrightarrow{w} \delta_a$, it is sufficient to prove that $\delta_{a_n}(f) \longrightarrow \delta_a(f)$, for every $f \in B^*$. But this follows from $|e^{if(a_n)} - e^{if(a)}| \leq |\int_{K \cap U_n} f(x)\, dF_\lambda(x)| \leq p_{K^0}(f) \int_{K \cap U_n} p_K\, dF_\lambda$, for every $f \in B^*$ and the dominated convergence theorem. Thus, since $S_\lambda = \overline{G(F_\lambda)} + a$ (Theorem 3.1(ii)) $= S_\lambda + a$, we have $S_\lambda = \overline{G(F_\lambda)}$. Showing S_λ is a semigroup, and hence S_{μ_0} is a semigroup. Now we will show that $S_{\mu_0^t} = S_{\mu_0}$, for $t > 0$. Let F be the Lévy measure of μ_0 ; then, by Lemma 3.7, $\mu_0(f) = \exp\{\int_E (e^{if(x)} - 1)dF(x)\}$. Therefore, letting g as above,

$$\hat{\lambda}(f) = \exp\{\int_E (e^{if(g(x))} - 1)dF(x)\} = \exp\{\int_{\{g \neq 0\}} (e^{if(g(x))} - 1)dF(x)\}$$

$$= \exp\{\int_{B\setminus\{\theta\}} (e^{if(x)} - 1)Fog^{-1}(dx)\} = \exp\{\int_B (e^{if(x)} - 1)dG(x)\},$$

for $f \in B^*$, where $G = Fog^{-1}/B\setminus\{\theta\}$. This, the fact that G is Levy (this can be proved directly by just using the definition of a Lévy measure), and the uniqueness of Lévy measure, imply that $G = F_\lambda$. Thus $\lambda^t = \exp\{\int_B (e^{if(x)} - 1)tF_\lambda\}$ (see Section 2(ii)) ; therefore $S_{\lambda^t} = \overline{G(tF_\lambda)} = \overline{G(F_\lambda)}$. Hence, since $S_{\mu_0^t}$ is the projective limit of supports of measures of the type λ^t [13], we have $S_{\mu_0^t} = S_{\mu_0}$. To finish the proof we need only show that $sS_{\mu_0} \subseteq S_{\mu_0}$, for $0 < s < 1$. This we do in the following:

For $s \in (0, 1)$, choose by Lemma 3.4, $k_n \in \{1,\dots, [\frac{1}{r^n}]\}$ such that $r^{n/\alpha} k_n \longrightarrow s$, as $n \longrightarrow \infty$. Now by using the facts $\mu_0^{r^n k_n} = T_{r^{n/\alpha}} \mu_0^{\star k_n}$ and $S_{\mu_0^t} = S_{\mu_0}$, $t > 0$, we get

$$S_{\mu_0} = r^{n/\alpha} \overline{[S_{\mu_0}^{(k_n)}]},$$

where $S_{\mu_0}^{(k_n)}$ is the k_n-fold sum of S_{μ_0}. Hence for $x \in S_{\mu_0}$, $r^{n/\alpha} k_n x \in S_{\mu_0}$, so $sx \in S_\mu$, since $r^{n/\alpha} k_n x \longrightarrow sx$, as $n \longrightarrow \infty$.

Proof of Theorem 3.2(iii): Since μ is symmetric and i.d., S_μ is a subgroup, by Theorem 3.1. Now, $\mu^{r^n} \star \mu^{1-r^n} = \mu$ and the fact that μ^t is symmetric i.d. imply that

$$\overline{[r^n S_\mu + S_{\mu^{1-r^n}}]} = S_\mu ,$$

and $\theta \in S_{1-r^n}$. Consequently, $r^n S_\mu \subseteq S_\mu$, for all $n = 1, 2, \dots$, and hence S_μ is a subspace.

Remark 3.8: The fact that S_{μ_0} is a subgroup and that $S_{\mu_0^t} = S_{\mu_0}$ shown above in the proof of part (ii) can also be recovered from [16]. But in order to keep the paper self contained we relied on our result rather than using [16].

4. SEMINORM INTEGRABILITY THEOREM FOR r-SEMISTABLE MEASURES

As we noted in the introduction, the proof of the result of this section is classical (see, for example, [3]); therefore, we will only give an outline of the proof and refer the reader to [11] for details, where a similar result is obtained in Hilbert spaces.

Theorem 4.1: Let μ be a K-regular r-semistable p.m. of index $\alpha \in (0, 2)$ on E and let p be a continuous seminorm on E. Then

$$\int_E p^\delta(x)\, \mu(dx) < \infty , \tag{4.1}$$

if $\delta < \alpha$.

Outline of the Proof: Let $\nu = \mu * \bar{\mu}$, $(\bar{\mu} \equiv T_{-1}\mu)$, the symmetrization of μ. By Fubini's theorem, it is sufficient to prove (4.1) for ν. Using some arguments of the proof of Theorem 3.2(ii), we note that ν is (K-regular symmetric) r-semistable of the same index α. Let N be the quotient space $E/p^{-1}(\theta)$; if $\mathring{x} = x + p^{-1}(\theta)$, set $\|\mathring{x}\| = p(x)$, then $(N, \|\cdot\|)$ is a normed space, and $\lambda \equiv \nu \circ T^{-1}$ is a symmetric K-regular r-semitstable p.m. of index α (here T is the usual quotient map). Since a K-regular p.m. on a metric space has a separable support, we can assume that there exists a separable Banach subspace B of the completion $\tilde{B}$ of $(N, \|\cdot\|)$ such that $\lambda^s(B) = 1$, for all $s > 0$, (one such B is the closure in $\tilde{B}$ of the supports of λ^s, s positive rationals). Since

$$\int_E p^\delta d(\mu * \bar{\mu}) = \int_B \|\mathring{x}\|^\delta\, d\lambda ,$$

by the change of variable, we need to prove (4.1) for a symmetric r-semistable p.m. of index α defined on a separable Banach space B. This is outlined below:

According to [5], we have

$$T_{r^{n/\alpha}}\, \lambda^{*k_n} \xrightarrow{W} \lambda ,$$

where $k_n = [\frac{1}{r^n}]$. This and Theorem 10 of [2] implies that

$$k_n \cdot T_{r^{n/\alpha}} \lambda \xrightarrow{w} F ,$$

on complements of nbds of θ in B , where F is the Lévy measure of λ . Now repeating the proof of Theorem 3.4 of [11], for given $\varepsilon > 0$ and positive integer m, one can choose t_0 such that if $t \geq t_0$, then

$$\frac{b^{m\alpha}}{a} (1 + \varepsilon)^{-1} \leq \frac{Q_\lambda(t)}{Q_\lambda(b^m t)} \leq a\ b^{m\alpha}(1 + \varepsilon) , \tag{4.2}$$

where $a = 1/r$, $b = r^{1/\alpha}$ and $Q_\lambda(t) = \lambda\{\dot{x} \in B: \|\dot{x}\| \geq t\}$. Now using (4.2) and following the proof of Theorem 3.5 of [11], one obtains $\int_B \|\dot{x}\|^\delta \, d\lambda < \infty$; which completes the proof.

Remark 4.2: It is worth noting that this theorem also provides a third proof of the seminorm integrability result for stable p. measures, which is different from the first two (obtained by de Acosta [1] and Kanter [8]), as long as the measures are defined on LCTV spaces and p is a continuous seminorm.

REFERENCES

[1] A. de Acosta, Stable measures and seminorms, Ann. of Probability, 3(1975), 365 - 375.

[2] A. de Acosta, A. Araujo, and E. Giné, On Poisson measures, Gaussian measures, and the central limit theorem in Banach spaces, Advances in Probability, 5(1978), 1 - 68.

[3] L. Breiman, Probability, Addison-Wesley, Reading, Massachusetts, 1968.

[4] P.L. Brockett, Supports of infinitely divisible measures on Hilbert spaces, Ann. of Probability, 5(1977), 1012 - 1017.

[5] D.M. Chung, B.S. Rajput, and A. Tortrat, Semistable laws on TVS, 1979 (to appear).

[6] E. Dettweiler, Grenzwertsätze für Wahrscheinlichkeitsmasse auf Badrikianschen Räumen, Dissertation, Eberhard-Karls-Universitat zu Tübingen, 1974.

[7] E. Dettweiller, Grenzwertsätze für Warhscheinlichkeitmasse auf Badrikianschen Räumen, Wahr. Verw. Gebetie, 34(1976), 285 - 311.

[8] M. Kanter, On the boundedness of stable processes, Trans. Seventh Prague Conference on Inf. Theory and Statist. Dec. Theory, Prague, Vol. A, (1974), 317 - 323.

[9] V.M. Kruglov, on the extension of the class of stable distributions, Theory Prob. Applications, 17(1972), 685 - 694.

[10] V.M. Kruglov, on a class of limit laws in a Hilbert space, Lit. Mat. Sbornik, 12(1972), 85 - 88 (in Russian).

[11] A. Kumar, Semistable probability measures on Hilbert spaces, J. Multivariate Anal., 6(1976), 309 - 318.

[12] P. Lévy, <u>Théorie de l'Addition Variables Aléatoires</u>, Bautier-Villars, Paris, France, 1937.

[13] B.S. Rajput, On the support of certain symmetric stable probability measures on TVS, Proc. Amer. Math. Soc., 63(1977), 306 - 312.

[14] B.S. Rajput, On the support of symmetric infinitely divisible and stable probability measures on LCTVS, Proc. Amer. Math. Soc., 66(1977), 331 - 334.

[15] A. Tortrat, Sur le support des lois indéfinitment divisibles dans les espaces vectoriels localment convexes, Ann. Inst. Henri Poincaré, 13(1977), 27 - 43.

[16] A. Tortrat, Second complément sur le support des lois indéfiniment divisibles, Ann. Inst. Henri Poincaré, 14(1978), 349 - 354.

[17] J. Yaun and T. Liang, On the supports and absolute continuity of infinitely divisible probability measures, Semigroup Forum, 12(1976), 34 - 44.

REMARK ON THE EXTRAPOLATION OF BANACH SPACE VALUED STATIONARY PROCESSES

by

A. Makagon

Institute of Mathematics
Wrocław Technical University
Wybrzeże Wyspiańskiego 27
Wrocław, 50-370 Poland

In previous paper [3] we have constructed the Hellinger square integral with respect to an $L(B,B^*)$ -valued measure, where B is a Banach space. Also some of its applications for the interpolation of Banach space valued stationary processes were given there.

The purpose of this paper is to look at the extrapolation problem in the language of the Hellinger integral theory. As we will see, a simple application of this technics allows us to get a slight generalization of Rosanov's theorem. It seems to be interesting to note that the space L used in original papers of Rosanov ([6],[7]; see also [5]) is simply the space of all *-weak densities of the measures which are Hellinger square integrable with respect to the spectral measure of the process.

In Introduction we repeat some basic points of the Hellinger integral theory. Although the proofs are included, for more details we refer to [3] .

1. Introduction

For any two Banach spaces X and Y let $L(X,Y)$ denote the space of all linear and continuous operators from X into Y. If X is a Hilbert space and $A \in L(X,Y)$ then the generalized inverse operator $A^{\#}$ is defined to be a mapping from $AX \subset Y$ into X such that $A^{\#} y = x$, provided $Ax=y$ and x is orthogonal to the null space of A.

Now let B be a complex Banach space and let B^* be the space of all antilinear, continuous functionals on B. Suppose that (S, Σ) is a measurable space in the sense that Σ is a σ-algebra. Throughout this paper the letter F stands for a semispectral measure defined to be a mapping from Σ into $L(B,B^*)$ such that for every $x \in B$ the function $(F(\cdot)x)(x)$ is a non-negative measure on Σ . If F is a semispectral measure then there exists a Hilbert space H, an operator $R \in L(B,H)$ and a spectral measure E in H such that $H = \overline{sp}\,\{E(\Delta)Rx:\ x \in B,\ \Delta \in \Sigma\}$ and for every $\Delta \in \Sigma$ $F(\Delta) = R^*E(\Delta)R$ ([2],[8]). The triple (H,E.R) is said to be a minimal dilation of F.

1.1 Definition. Suppose that F is a semispectral measure defined on a measurable space (S,Σ). For every fixed set $\Delta \in \Sigma$ let $\| \ \|_\Delta$ denote the norm in $F(\Delta)B$ defined by equality

$$\| F(\Delta)x \|_\Delta = \sqrt{(F(\Delta)x)(x)} \qquad x \in B$$

Since $\| F(\Delta)x\|_{B^*} \leq \mathrm{const}(\Delta) \cdot \| F(\Delta)x \|_\Delta$, $x \in B$, the completion $S(\Delta)$ of $F(\Delta)B$ in $\| \ \|_\Delta$-norm is a subset of B^*.

A B-valued measure m on Σ is said to be Hellinger square integrable with respect to F(we will write $m \in H_{2,F}$) if

(i) $m(\Delta) \in S(\Delta)$ for every $\Delta \in \Sigma$

(ii) $$\|m\|_F^2 = \sup_{\sigma \in \mathfrak{F}} \sum_{\Delta \in \sigma} \| m(\Delta)\|_\Delta^2 < \infty$$

where the supremum is taken over the family $\mathfrak{F}$ of all finite measurable partitions σ of S.

1.2 Remark. Since for every fixed $\Delta \in \Sigma$, $F(\Delta)$ is a non-negative operator (i.e., $(F(\Delta)x)(x) \geq 0$, $x \in B$), there exists a Hilbert space H_Δ and an operator $W(\Delta) \in L(B,H_\Delta)$ such that $F(\Delta) = W(\Delta)^*W(\Delta)$. It is clear that the operator $W(\Delta)^{*\#}$ is an isometry from $(S(\Delta), \| \ \|_\Delta)$ onto $\overline{W(\Delta)B}$. Thus $m \in H_{2,F}$ if and only if

$m(\Delta) \in W(\Delta)^* H_\Delta$ for each $\Delta \in \Sigma$ and

$$\sup_{\mathfrak{F}} \sum_{\Delta\in\sigma} \|W(\Delta)^{*\#} m(\Delta)\|^2 < \infty$$

If $B=H$ is a Hilbert space, then $m \in H_{2,F}$ if and only if $m(\Delta) \in F(\Delta)^{1/2} H$ and $\sup \sum_{\Delta\in\sigma} \| F(\Delta)^{1/2\#} m(\Delta)\|^2 < \infty$.
If B is a one-dimensional Banach space then we obtain the usual definition of the Hellinger square integral.

1.3. Theorem. With the above notations

(1) $(H_{2,F}, \| \ \|_F)$ is a Hilbert space.

(2) The measure F has the following minimal dilation:

$$\begin{array}{ccc} B & \xrightarrow{F(\Delta)} & B^* \\ \downarrow T & & \uparrow T^* \\ H_{2,F} & \xrightarrow{P(\Delta)} & H_{2,F} \end{array}$$

where $(Tx)(\Delta) = F(\Delta)x$ $x \in B$, $\Delta \in \Sigma$ and $P(\Delta)$ is the operator of restriction of a measure to the set Δ, i.e., $(P(\Delta)m)(\Delta') = m(\Delta \cap \Delta')$.

Proof. Let (H,E,R) be a minimal dilation of F. For every $h \in H$ we define $(Vh)(\Delta) = R^*E(\Delta)h$, $\Delta \in \Sigma$. We shall prove that V maps H onto $H_{2,F}$ and that for each $h \in H$ $\|Vh\|_F = \|h\|$. Since V is linear, it follows that $(H_{2,F} \| \ \|_F)$ is a Hilbert space and V is isometry from H onto $H_{2,F}$.

Let $h \in H$. Since

$$\sum_{\Delta\in\sigma} \|(E(\Delta)R)^{*\#}(Vh)(\Delta)\|^2 = \sum_{\Delta\in\sigma} \| P_{\overline{E(\Delta)RB}} h \|^2 =$$

$$= \|P_{\bigoplus_\Delta \overline{E(\Delta)RB}} h\|^2 \leq \|h\|^2, \qquad \sigma \in \mathfrak{F},$$

by Remark 1.2 we have that $Vh \in H_{2,F}$ and $\|Vh\|_F \leq \|h\|$.

Now suppose that $m \in H_{2,F}$. We shall prove that $m = Vh$ for some $h \in H$. For each $\sigma \in \mathfrak{F}$ let us define

$$f_\sigma = \sum_{\Delta\in\sigma} (E(\Delta)R)^{*\#} m(\Delta).$$

The set $\mathfrak{F}$ with relation $\sigma_1 > \sigma_2$ if σ_1 is finer than σ_2, is a partially ordered directed set. Note that $\|f_\sigma\|^2 \leq \|m\|_F^2$ and that for every x B, $\Delta \in \Sigma$ and $\sigma > \{\Delta, S-\Delta\}$ we have

$$(f_\sigma, E(\Delta)Rx) = (\sum_{\Delta'\epsilon\sigma} (E(\Delta)R)^* (E(\Delta')R)^{*\#} m(\Delta'))(x) =$$

$$= \sum_{\substack{\Delta'\epsilon\sigma \\ \Delta'\subset\Delta}} ((E(\Delta')R)^* (E(\Delta')R)^{*\#} m(\Delta'))x) = (m(\Delta))(x)$$

Thus, from the weak compactness of the unit ball in a Hilbert space, it follows that there exists $h \epsilon H$ such that $h = \lim_{\mathcal{F}} f_\sigma$. Since $(E(\Delta)R)^*$ is continuous as a mapping from $(H,\sigma(H,H))$ into $(B^*,\sigma(B^*,B))$, we have

$$((Vh)(\Delta))(x) = ((E(\Delta)R)^*h)(x) = \lim ((E(\Delta)R)^* f_\sigma)(x) =$$

$$= (m(\Delta))(x) \qquad x \epsilon B ,$$

To complete the proof we shall show that $\|Vh\|_F \geq \|h\|$. Let $\epsilon > 0$ and $h \in H$. Since $H = \overline{sp}\{E(\Delta)Rx : x \epsilon B, \Delta \epsilon \Sigma\}$, there exists $\sigma = \{\Delta_1,\dots,\Delta_n\}$ and $x_1,x_2,\dots,x_n \epsilon B$ such that

$$\| h - \sum_{i=1}^{n} E(\Delta_i)Rx_i \| < \epsilon .$$

We have $\|h\|^2 - \|Vh\|_F^2 \leq \|h\| - \sum_{i=1}^{n} \|(E(\Delta_i)R)^{*\#}(Vh)(\Delta_i)\|^2 =$

$$= \|h - P_{\oplus \overline{E(\Delta_i)RB}} h\|^2 \leq \| h - \sum_{i=1}^{n} E(\Delta_i)Rx_i \|^2 < \epsilon^2 .$$

Thus $\|Vh\|_F = \|h\|$ and V is a unitary operator. Consequently $(VH = H_{2,F}, VE(\Delta), VR)$ is a minimal dilation of F. But for each $x \epsilon B$ and $\Delta \in \Sigma$

$$(VRx)(\Delta) = (E(\Delta)R)^* Rx = F(\Delta)x$$

$$(VE(\Delta)Rx)(\Delta') = (E(\Delta')R)^* E(\Delta)Rx = F(\Delta\cap\Delta')x = (P(\Delta)Rx)(\Delta').$$

<u>1.4 Remark</u>. To see the connection between the dilation constructed in the original proof of Sz.-Nagy (cf. [2]) and the one obove, we observe that if

$$m(\Delta) = \sum_{i=1}^{k} F(\Delta\cap\Delta_i)x_i \quad \text{and} \quad n(\Delta) = \sum_{j=1}^{l} F(\Delta\cap\Delta_j')x_j'$$

then, from the proof of Th.1.3., we have

$$m(\Delta) = \sum_{i=1}^{k} R^*E(\Delta)E(\Delta_i)Rx_i = (V(\sum_1^k E(\Delta_i)Rx_i))(\Delta) \quad \text{and}$$

$$(m,n)_F = (\sum_i E(\Delta_i)Rx_i, \sum_j E(\Delta_j)Rx_j') = \sum_i \sum_j (F(\Delta_i\cap\Delta_j')x_i)(x_j').$$

Let us suppose that there exists a σ-finite, non-negative scalar measure μ on Σ such that F is absolutely continuous with respect to μ. Then there exists a Hilbert space K and a linear, bounded operator Q from B into $L^2(S,\Sigma,\mu; K)$, the space of all K-valued μ-square (Bochner) inregrable functions on S, such that for every $x,y \in B$

$$\frac{d(F(.)x)(y)}{d\mu}(s) = ((Qx)(s), \ (Qy)(s))_K \qquad \mu - a.e,$$

where $(.,.)_K$ denotes the inner product in K([4]). The operator Q is said to be a quasi square root of the density of F. If B is separable and there exists a function $\frac{dF}{d\mu} : S \longrightarrow L(B,B^*)$ such that

$$\int_\Delta \left(\frac{dF}{d\mu}(s)\ x\right)(y)\mu(ds) = (F(\Delta)x)(y)$$

$x,y \in B$, $\Delta \in \Sigma$, then the operator Q has the form $(Qx)(s) = \hat{Q}(s)x$, $x \in B$, where $\hat{Q}$ is a strongly measurable operator function on S with values in $L(B,K)$ ([5]). We will refer to $\frac{dF}{d\mu}$ as the operator density of F. The function $\hat{Q}(.)$ is called a square root of the density $\frac{dF}{d\mu}$.

1.5. Theorem. Suppose that F is absolutely continuous w.r. to a non-negative σ-finite scalar measure μ and that $Q \in L(B,L^2(\mu,K))$ is a quasi square root of its density. The measure $m \in H_{2,F}$ if and only if there exists a function $f \in L^2(\mu,K)$ such that for each $x \in B$ and $\Delta \in \Sigma$

$$(*) \qquad (m(\Delta))(x) = \int_\Delta (f(s), \ (Qx)(s))\mu(ds)$$

Moreover, if B is separable and F has the operator density with respect to μ, then $m \in H_{2,F}$ if and only if it has the form

$$(m(\Delta))(x) = \int_\Delta (Q^*(s)f(s))(x)\mu(ds) \qquad x \in B$$

with $f \in L^2(S,\Sigma,\mu; K)$.

Proof. Let 1_Δ denote the operator of multiplication by the indicator of a set Δ in $L^2(\mu, K)$, $\Delta \in \Sigma$. First we note that the triple $(L^2(\mu, K), 1_\Delta, Q)$ is a dilation of F (not minimal in general). By [2] Th.2, there exists a unitary operator U from $M(Q) \overset{df}{=} \overline{sp}\{1_\Delta Qx : x \in B, \ \Delta \in \Sigma\}$ onto $H_{2,F}$ such that $UQ = T$ and $U1_\Delta = P(\Delta)U$, $\Delta \in \Sigma$ (P and T as in Th. 1.3). If $f = 1_\Delta Qx$, then $(Uf)(\Delta') = (P(\Delta)Tx)(\Delta') = F(\Delta \cap \Delta')x = Q^*1_{\Delta'}f$. It follows that for every $f \in M(Q)$, $(Uf)(\Delta') = Q^*1_{\Delta'} f$, i.e.,

$$((Uf)(\Delta'))(x) = (Q^*1_{\Delta'}f)(x) = (f, 1_{\Delta'}Qx)_{L^2(\mu,K)} =$$
$$= \int_{\Delta'}(f(s),(Qx)(s))\mu(ds) .$$

Since U is a unitary operator, every measure in $H_{2,F}$ has the form (*) with $f \in M(Q) \subset L^2(S, \Sigma, \mu; K)$.

Now suppose that $m(\Delta) = Q^* 1_\Delta f$ for some $f \in L^2(\mu, K)$, Setting $g = P_{M(Q)} f$ we have $g \in M(Q)$ and

$$m(\Delta) = Q^* 1_\Delta f = (Q^* 1_\Delta P_{M(Q)})f = Q^* 1_\Delta g = (Ug)(\Delta) .$$

Thus $m \in H_{2,F}$.

The second part of the theorem follows immediately from the first one.

2. Extrapolation of stationary processes

In this section G will stand for the group of real numbers or for the group of integers. Γ will denote the dual group of G. Let B be a complex Banach space and let H be a Hilbert space. By a B-valued stationary process (SP) we will mean a function $X = (X_g)_{g \in G}$ from G into $L(B,H)$ such that its correlation $X_h^* X_g = K(g-h)$ depends on $g-h$ and is weakly continuous. The time domain $M(X)$ of the process is defined to be the closed subspace of H generated by all $X_g x$, $g \in G$, $x \in B$.

Let $(U_g)_{g \in G}$ be the group of shifts of the process X. By Stone theorem there exists a spectral measure E in $M(X)$ such that $U_g = \int_\Gamma \exp(igt) E(dt)$. The measure $F(\Delta) \overset{df}{=} X_o^* E(\Delta) X_o$, $\Delta \in \mathcal{B}(\Gamma)$, is said to be the spectral measure of the process. By definition the measure F has the following minimal dilation

$$\begin{array}{ccc} B & \xrightarrow{P(\Delta)} & B^* \\ {\scriptstyle X_o}\downarrow & & \uparrow{\scriptstyle X_o^*} \\ M(X) & \xrightarrow[E(\Delta)]{} & M(X) \end{array}$$

From the proof of Th.1.3 it follows that the operator V defined by $(Vv)(\Delta) = (E(\Delta) X_o)^* v$, $v \in M(X)$, is a linear isometry from $M(X)$ onto $H_{2,F}$. By Stone theorem we obtain that for each $x \in B$, $g \in G$ and $\Delta \in \mathcal{B}(\Gamma)$

$$(VX_g x)(\Delta) = \int_\Delta \exp(igt) \ F(dt)x$$

Thus we have the following

<u>2.1. Theorem.</u> Let $X = (X_g)_{g \in G}$ be a B-valued SP with the spectral measure F.

(1) The operator V defined by

$$(VX_g x)(\Delta) = \int_\Delta \exp(igt)F(dt)x$$

$g \in G$, $x \in B$, $\Delta \in \mathcal{B}(\Gamma)$, maps isometrically $M(X)$ onto $H_{2,F}$.

(2) If we write $N_o(X) = M(X) \ominus \overline{sp}\{X_g x: \ x \in B, \ g \leq 0\}$, then

$$V(N_o(X)) = \{m \in H_{2,F} : \int_\Gamma \exp(-igt)m(dt) = \hat{m}(g) = 0 \ \text{ for}$$

every $g \leq 0$ } (here all integrals are meant in a *-weak sense).

Proof. It remains to prove the last statement. We observe that if $m \in H_{2,F}$, then $\hat{m}(g) = X_g^* V^{-1}m \in B^*$ and so $\hat{m}$ is well defined. Since $v \in N_o(X)$ if and only if $(v, X_g x) = 0$ for every $x \in B$ and $g \leq 0$, if follows that

$$((\widehat{Vv})(g)(x) = \int_\Gamma \exp(-igt)(Vv)dt)(x) = \int_\Gamma \exp(-igt)(v, E(dt)X_o x) =$$

$$= (v, X_g x) = 0 \ ,$$

provided $v \in N_o(X)$, $g \leq 0$ and $x \in B$. The proof is completed.

<u>2.2 Definition</u> A B-valued stationary process $X = (X_g)_{g \in G}$ is said to be

(1) regular, if $\bigcap_{g \in G} \overline{sp}\{X_h x : h \leq g, \ x \in B\} = \{0\}$

(2) singular, if $N_o(X) = \{0\}$.

In the following theorem a characterization of the regularity and singularity in terms of the $H_{2,F}$ space is given. Latter we will see that the Rosanov's theorems [6],[7] and their Banach space versions given by Miamee and Salehi [5] arise from the one below.

<u>2.3. Theorem.</u> Let $X = (X_g)_{g \in G}$ be a B-valued SP with the spectral measure F.

(1) The process X is not singular if and only if there exists a non-zero measure $m \in H_{2,F}$ such that $\hat{m}(g) = 0$ for each $g \leq 0$.

(2) The process X is regular if and only if there exists a subset $\mathcal{M} \subset H_{2,F}$ such that

(i) $\hat{m}(g) = 0$ for every $m \in \mathcal{M}$ and $g \leq 0$, and

(ii) $\overline{sp}\,\{P(\Delta)m : \Delta \in \mathfrak{B}(\Gamma),\ \ m \in \mathcal{M}\} = H_{2,F}$

Proof. The first part of the theorem is obvious. To prove (2) it suffices to observe that X is regular if and only if $\overline{sp}\,\{E(\Delta)v :\ v \in N_0(X), \Delta \in \mathfrak{B}(\Gamma)\} = M(X)$ and to apply Theorem 2.1.

Now suppose that $F(\Gamma)B$ is a separable subset of B^* and that F is absolutely continuous w.r. to a non-negative, σ-finite scalar measure μ. Let $Q \in L(B, L^2(\mu, K))$ be a quasi square root of the density of F. Since $F(\Gamma)B = Q^*QB$ is separable, the space QB in separable, too ([1] , Lemma 2.5). We choose a complete system of functions $\{ g_j : j = 1,2,\dots \}$ in QB and define

$$\overline{\overline{QB}}(t) = \overline{sp}\,\{g_j(t):\ j = 1,2,\dots,\} \subset K , \qquad t \in \Gamma$$

The following lemma, as well as Lemma 2.7, seems to be well known. For the clarity we sketch the proof.

<u>2.4. Lemma</u> (cf. [7], p.87)

(1) The function $\overline{\overline{QB}}(\cdot)$ does not depend on the choice of a system $\{g_j: j = 1,2,\dots,\}$ in the sense of μ-almost everywhere equality.

(2) For every $h \in K$ a function $P_{\overline{\overline{QB}}(\cdot)}h$ is Bochner measurable.

(3) $M(Q) \overset{df}{=} \overline{sp}\,\{ 1_\Delta\ Qx:\ x \in B , \Delta \in \Sigma \} =$

$$= \{f \in L^2(\mu,K):\ f(t) \in \overline{\overline{QB}}(t)\ \ \mu\text{-a.e}\}$$

Proof. The first statement follows immediately from the fact,that every sequence convergent in $L^2(\mu,K)$ contains a subsequence that is convergent μ-almost surely.

The same fact makes clear that $M(Q) \subset \{f \in L^2(\mu,K):\ f(t) \in \overline{\overline{QB}}(t)\ \mu\text{-a.e}\}$. Conversely, if $g \in \{f \in L^2(\mu,K):\ f(t) \in \overline{\overline{QB}}(t)\ \mu\text{-a.e}\} \ominus M(Q)$, then for every $\Delta \in \mathfrak{B}(\Gamma)$ and $j = 1,2,\dots$ we have

$$\int_\Delta (g(s), g_j(s))\mu(ds) = (g,\ 1_\Delta\, g_j) = 0$$

Therefore there exists a set $\Delta, \mu(\Delta) = 0$, such that

$$g(s) \perp \overline{sp}\,\{g_j(s) :\ j=1,2,\dots\} = \overline{\overline{QB}}(s)$$

provided $s \notin \Delta$. Hence $g = 0$.

To prove 2 we define (by recurrence) a sequence of functions $\{f_j:\ j = 1,2,\dots,\} \subset M(Q)$ such that $\overline{sp}\,\{ 1_\Delta\, f_j : j = 1,2,\dots, \Delta \in \mathfrak{B}(\Gamma)\} = M(Q)$ and $f_i \perp 1_\Delta\, f_j$ for each $\Delta \in \mathfrak{B}(\Gamma)$ and $j \neq i$.

The last property implies that there exist a set $\Delta \in \mathcal{B}(\Gamma)$ such that $\mu(\Delta) = 0$ and for every $s \notin \Delta$ and $k \neq j$ $(f_j(s), f_k(s)) = 0$. Observe that $\overline{QB}(s) = \overline{sp}\{f_j(s): j = 1,2,\dots\}$ μ-a.e. . Setting

$$a_k(s) = \begin{cases} \|f_k(s)\|^{-1} & , \text{ if } \|f_k(s)\| \neq 0 \\ 0 & , \text{ if } \|f_k(s)\| = 0 \end{cases} \qquad k = 1,2,\dots$$

we get that for every $h \in K$

$$P_{\overline{QB}(s)}h = \sum_k (h, a_k(s) f_k(s))\, a_k(s) f_k(s)$$

is μ-measurable.

We note that, since every Bochner measurable function is μ-almost surely separably valued, one can assume that the space K is separable.

<u>2.5. Theorem</u>. Let $X = (X_g)_{g \in G}$ be a B-valued SP with the spectral measure F. Suppose that $F(\Gamma)B$ is separable and that F is absolutely continuous w.r. to a σ-finite non-negative measure μ. Let Q be a quasi square root of the density of F.

(1) The process X is not singular if and only if there exists a function $f \in L^2(\mu, K)$ such that for each $x \in B$

$$\int_\Gamma \exp(-igt)(f(t), (Qx)(t))\mu(dt) = 0$$

for every $g \leq 0$ and it is not identically zero.

(2) The process X is regular if and only if there exists a sequence $\{f_j: j = 1,2,\dots\} \subset L^2(\mu, K)$ such that

(i) $\int_\Gamma \exp(-igt)(f_j(t), (Qx)(t))\mu(dt) = 0$

for every $x \in B$, $j = 1,2,\dots$ and $g \leq 0$, and

(ii) $\overline{sp}\{f_j(t) : j = 1,2,\dots\} = \overline{\overline{QB}}(t)$ μ-a.e. .

<u>Remark.</u> Since every non-zero scalar measure with the Fourier transform vanishing for $g \leq 0$ is equivalent to the Lebesque measure l, we have $l \ll F$, provided X is not singular. If X is regular, then from (2) (ii) it follows that $F \ll l$ (see [3], 3.9), so in the second part of the theorem we may take $\mu = l$.

<u>Proof</u>. The first part of the theorem follows from Th.2.3 and Th. 1.5. To prove (2) it suffices to observe that X is regular

if and only if there exists a family $\{f_\alpha : \alpha \in \mathcal{M}\} \subset M(Q)$ such that

(i) $(\widehat{Uf_\alpha})(g) = 0$ for every $\alpha \in \mathcal{M}$ and $g \leq 0$, and

(ii) $\overline{sp}\ \{1_\Delta f_\alpha : \alpha \in \mathcal{M}, \Delta \in \mathcal{B}(\Gamma)\} = M(Q)$

where U is defined in the proof of Th. 1.5.
By separability of $M(Q)$ one can assume that $\mathcal{M}$ is a countable set and so, by Lemma 2.4, the condition (ii) is equivalent to

$$(ii)' \quad \overline{sp}\ \{f_\alpha(s) : \alpha = 1,2,\ldots,\} = \overline{QB}(s) \quad \mu\text{-a.e.}.$$

At last we consider the case when B is a separable Banach space and F has the operator density $\frac{dF}{d\mu}$. The following theorem is proved in [6],[7] for a Hilbert space and in [5] for a Banach space case. We state it in some different form.

2.6. Definition. For every $t \in \Gamma$ let $\|\ \|_t$ denote the norm in $\frac{dF}{d\mu}(t)B$ defined by

$$\left\| \frac{dF}{d\mu}(t)x \right\|_t = \sqrt{\left(\frac{dF}{d\mu}(t)x\right)(x)} \qquad x \in B,$$

(we assume that for every $t \in \Gamma$ $\frac{dF}{d\mu}(t)$ is a non-negative, linear and continuous operator). The completion of $\frac{dF}{d\mu}(t)B$ in $\|\ \|_t$ -norm will be denoted by S_t. S_t is a subspace (not necessarity closed) of B^*.

We note that the above definition does not depend on the choise of a measure μ and of a version of the density $\frac{dF}{d\mu}$; i.e., if μ' is another σ-finite non-negative measure such that $F \ll \mu'$ and if $\frac{dF}{d\mu'}$ is the operator density of F w.r. to μ', then there exists a Borel set Δ such that $F(\Delta') = 0$ for each Borel $\Delta' \subset \Delta$ and $S'(t) = S(t)$ and $\|\ \|'_t = \|\ \|_t$, provided $t \notin \Delta$.

2.7. Lemma. Under the above assumption, let $Q : \Gamma \to L(B,K)$ be a square root of the density $\frac{dF}{d\mu}$. Then

(1) For μ-a.e. $s \in \Gamma$ the operator $Q(s)^{*\#}$ maps isometrically $(S(s), \|\ \|_s)$ onto $\overline{Q(s)B}$.

(2) For every *-weak measurable function $\varphi : \Gamma \to B^*$ such that $\varphi(s) \in Q(s)^*K$ μ-a.e., the function

$$f(s) = Q(s)^{*\#}\varphi(s)$$

is Bochner measurable.

Proof. (1) follows immediately from the observation that for μ-a.e. $s \varepsilon \Gamma$ $\| \frac{dF}{d\mu}(s)x \|_s = \|\hat{Q}(s)^{*\#}(\frac{dF}{d\mu}(s)x)\|$, $x \varepsilon B$.

(2). By separability of B one can assume that K is a separable Hilbert space (see remark following Lemma 2.4). Hence it suffices to prove the weak measurability of f. Let $k \varepsilon K$. Then by Lemma 2.4(2), $P_{\overline{\hat{Q}(s)B}}\, k$ is a μ-measurable function and we have

$$(f(s),k) = (\hat{Q}(s)^{*\#}\varphi(s),\, k) = (\hat{Q}(s)^{*\#}\varphi(s),\, P_{\overline{\hat{Q}(s)B}}k) \quad s \varepsilon \Gamma$$

The proof will be completed if we show that the function

$$\tilde{g}(s) = (\hat{Q}(s)^{*\#}\varphi(s),\, g(s))$$

is μ-measurable, provided g is bounded, measurable and $g(t) \varepsilon \overline{\hat{Q}(t)B}$ μ-a.e., .

It is obviously true if $g(\cdot) = \hat{Q}(\cdot)x$ for some $x \varepsilon B$. Hence $\tilde{g}$ is measurable for every $g \varepsilon \overline{sp}\,\{1_\Delta(\cdot)\hat{Q}(\cdot)x: \ x \varepsilon B,\ \Delta \varepsilon \mathcal{B}(\Gamma)\} = \{f \in L^2(\mu,K): \ f(s) \varepsilon \overline{\hat{Q}(s)B}\ \ \mu\text{-a.e.}\}$.
Let now g be bounded, measurable and $g(t) \varepsilon \overline{\hat{Q}(t)B}$ μ-a.e. .
From the above it follows that the function

$$(\hat{Q}(\cdot)^{*\#}\varphi(\cdot)\ ,\ 1_\Delta(\cdot)\, g(\cdot))$$

is measurable, provided $\mu(\Delta) < \infty$. Since μ is a σ-finite measure, the proof is finished.

<u>2.8. Theorem</u> (cf. [5], [6], [7] p.85). Let B be a separable Banach space and let $X = (X_g)_{g\varepsilon G}$ be a B-valued SP with the spectral measure F. Suppose that F has the operator density with respect to some σ-finite, non-negative measure μ. Then the following statements are equivalent

(1) The process X is regular.

(2) The measure F is equivalent to the Lebesgue measure and there exists a sequence of *-weakly measurable functions $\varphi_j : \Gamma \longrightarrow B^*$, $j = 1,2,\ldots$ such that

(i) $(\varphi_k(\cdot))(x) \varepsilon L^1(1)$ and

$$\int_\Gamma \exp(-igt)(\varphi_k(t)(x)\, dt = 0$$

for every $x \varepsilon B$, $k = 1,2,\ldots$ and $g \leqslant 0$,

(ii) $\varphi_k(t) \varepsilon S(t)$ 1-a.e. and $\int_\Gamma \| \varphi_k(t) \|_t^2\, dt < \infty$, $k=1,2,\ldots$,

(iii) $\overline{sp}\,\{\varphi_k(t) : \ k = 1,2,\ldots\} = S(t)$ 1-a.e , where the closure is taken in the $\| \ \|_t$ -norm.

Proof. If X is regular, then the functions $\varphi_k(t) = \hat{Q}(t)^* f_k(t)$, $k = 1,2,\ldots$, where $\hat{Q}(.)$ is a square root of $\frac{dF}{dI}$ and f_k, $k=1,2,\ldots$ are taken from Th.2.5, satisfy the conditions (i)-(iii).
Conversely, put $f_k(t) = \hat{Q}(t)^{*\#}\varphi_k(t)$.
By Lemma 2.7, f_k, $k= 1,2,\ldots$ are well defined and belong to $L^2(1,K)$. From Th.2.5 it follows that X is regular.

As it was shown in [7], pp. 90-97, the above theorem is closely related to the "logarithm condition" of regularity.

References

[1] Chobanjan, S.A., Weron, A., Banach space valued stationary processes and their linear prediction, Dissertationes Math. 125 (1975), 1-45.

[2] Górniak, J., Weron, A., An analogue of Sz.-Nagy's dilation theorem, Bull.Acad.Polon.Sci.,Ser.Math., Astronom.,Phys. XXIV, No 10, (1976), 867-872,

[3] Makagon, A., On the Hellinger square integral with respect to an operator valued measure and stationary processes, to appear.

[4] Makagon, A., Schmidt, F., The decomposition theorem for densities of positive operator valued measures, to appear.

[5] Miamee, A.G., Salehi, H., Necessary and sufficient conditions for factorability of non-negative operator valued functions on a Banach space, Proc.Amer.Math.Soc. 46 1(1974), 43-50.

[6] Rosanov, Yu.A., Some approximation problems in the theory of stationary processes, J.Mult.Anal. 2 (1972), 135-144.

[7] Rosanov, Yu.A., Theory of innovation processes, Moscow 1975 (in Russian).

[8] Weron, A., Remarks on positive definite operator valued functions in Banach spaces, Bull.Acad.Polon.Sci.,Ser.Math., Astronom.,Phys., XXIV No. 10 (1976), 873-876.

DILATIONS WITH OPERATOR MULTIPLIERS

W. Mlak and F.H.Szafraniec

The present paper deals with dilations related to positive definite operator kernels. The dilations involved here are projective-like representations. We refer to [3] for basic ideas of dilation theory.

1. Suppose we are given two complex Hilbert spaces H and H_1. We denote by $L(H,H_1)$ the space of all linear bounded operators on H with values in H_1. We write $L(H) = L(H,H)$.

Let S be a set. We say that the operator valued function (called a kernel) $K: S \times S \to L(H)$ is positive definite if for every n, every $s_1,\dots,s_n \in S$, $f_1,\dots,f_n \in H$ the inequality

$$\sum_{i,k=1}^{n} (K(s_i,s_k)f_k,f_i) \geq 0$$

holds true. We write then that K is P.D.

The Kolmogorov-Aronszajn factorization theorem [5] says that if K is P.D. then there is a Hilbert space H_K and an operator valued function $X: S \to L(H,H_K)$, such that $K(s,t) = X(s)^*X(t)$. The minimality requirement $H_K = [X(S)H]$ (= the closure of the linear span of $X(S)H$) determines H_K and X up to unitary equivalence.

Let G be a to semigroup of actions on S. We will always require they map S onto itself. Denote by $g(S)$ the action of $g \in G$ on $s \in S$. We have

(1.0) $\quad (gh)(s) = g(h(s)), \quad g,h \in G, \quad s \in S.$

If G has a unit e then

(1.1) $\quad e(s) = s, \quad s \in S.$

Given an operator valued function $A: G \times S \to L(H)$. The kernel K is said to be A-projectively invariant (shortly A-p.i.)if

(1.2) $\quad K(g(s),g(t))= A(g,s)^*K(s,t)A(g,t): \; g \in G, \quad s,t \in S.$

THEOREM 1. Suppose the positive definite kernel $K: S\times S \to L(H)$ is A-projectively invariant. Let H_K be the minimal space for Kolmogorov-Aronszajn factorization $K(s,t)=X(s)^*X(t)$. Then there are operator functions $V: G\times G \to L(H_K)$, $T: G \to L(H_K)$ such that $V(g,h)$ is a partial isometry for $g,h \in G$ and $T(g)$ a coisometry for $g \in G$, and operators $R_s \in L(H,H_K)$ where $s \in S$, such that the following conditions hold true :

(1.3) $K(g(s),h(t)) = A(g,s)^* R_s^* T(g)^* T(h) R_t A(h,t)$
for $s,t \in S$, $g,h \in G$;

(1.4) $T(g)T(h)V(g,h) = T(g,h)$ for $g,h \in G$;

(1.5) $V(g,h)R_s A(g\,h,s) = R_s A(h,s)A(g,h(s))$
for $g,h \in G$ and $s \in S$.

<u>Proof.</u> Let us fix $g,h \in G$ and let H' be the subspace of H_K spanned by vectors $X(s)A(g\,h,s)f$, $s \in S$, $f \in H$. For $s_1,\dots,s_n \in S$, $f_1,\dots,f_n \in H$ we have by (1.2)

$$\Delta' = \|\sum_{j=1}^{n} X(s_j)A(g\,h,s_j)f_j\|^2 =$$

$$= \sum_{i,k} A(g\,h,s_k)^* X(s_k)^* X(s_i)A(gh,s_i)f_i,f_k) =$$

$$= \sum_{i,k} (A(gh,s_k)^* K(s_k,s_i)A(gh,s_i)f_i,f_k) =$$

$$= \sum_{i,k} (K(gh)(s_k),(gh)(s_i))f_i,f_k).$$

Since $K((gh)(s_k),(g\,h)(s_i)) = K(g(h(s_k)),g(h(s_i))) =$

$$= A(g,h(s_k))^* K(h(s_k),h(s_i))A(g,h(s_i)) =$$

$$= A(g,h(s_k))^* A(h,s_k)^* K(s_k,s_i)A(h,s_i)A(g,h(s_i)),$$

then

$$\Delta' = \sum_{i,k} (X(s_i)A(h,x_i)A(g,h(s_i))f_i, X(s_k)A(h,s_k)A(g,h(s_k)f_k) =$$

$$= \|\sum_{j=1} X(s_j)A(h,s_j)A(g,h(s_j)f_j\|^2.$$

It follows that there is an isometry $V(g,h)$ from H' into H_K such that

$$V(g,h)\sum_{j=1}^{n} X(s_j)A(g\,h,s_j)f_j = \sum_{j=1}^{n} X(s_j)A(h,s_j)A(g,h(s_j)f_j.$$

Defining $V(g,h) = 0$ on $H_K \ominus H'$ we get the desired partial isometry.

In order to construct $T(g)$ we define for fixed g the space H'' spanned by vectors $X(s)A(g,s)f$, $s \in S$, $f \in H$.

We have

$$\Delta'' = \|\sum_{j=1} X(s_j)A(g,s_j)f_j\|^2 =$$

$$= \sum_{i,k} (A(g,s_k)^* X(s_k)^* X(s_i) A(g,s_i) f_i, f_k) =$$

$$= \sum_{i,k} (A(g,s_k)^* K(s_k,s_i) A(g,s_i) f_i, f_k) =$$

$$= \sum_{i,k} ((K(g(s_k), g(s_i)) f_i, f_k) = \| \sum_{j=1}^{n} X(g(s_j)) f_j \|^2$$

by A-projective invariance of K. It follows that there is the unique isometry $T(g)$ from H" into H such that

$$T(g) \sum_j X(s_j) A(g,s_j) f_j = \sum_j X(g(s_j)) f_j .$$

Let $t \in S$ and take $s \in g^{-1}(\{t\})$. Then $T(g)X(s)A(g,s) = X(t)f$ which implies that the range of $T(g)$ is the whole space H_K. It is plain now that $T(g)$ is a coisometry, if we extend it taking $T(g) = 0$ on $H_K \ominus H"$.

Define now $R_s = X(s)$ and for $f_1, f_2 \in H$ and $g,h \in G$ we just have that for $s,t \in S$

$$(A(g,s)^* R_s^* T(g)^* T(h) R_t A(h,t) f_1, f_2) =$$

$$= (T(h) R_t A(h,t) f_1, T(g) R_s A(g,s) f_2) =$$

$$= (X(h(t)) f_1, X(g(s)) f_2)$$

shows that

$$K(g(s), h(t)) = A(g,s)^* R_s^* T(g)^* T(h) R_t A(h,t)$$

which completes the proof of (1.3).

Next, by definitions of V and T we have for $g,h \in G$, $s \in S$ and $f \in H$

$$T(g)T(h)V(g,h)X(s)A(gh,s)f =$$

$$= T(g)T(h)X(s)A(h,s)A(g,h(s))f =$$

$$= T(g)X(h(s))A(g,h(s))f = X(g(h(s)))f =$$

$$= T(g,h)X(s)A(gh,s)f$$

which proves (1.4) because the initial spaces of $V(g,h)$ and of $T(g,h)$ coincide. This concludes the proof of (1.4).

Suppose $f \in H$. Then $V(g,h)R_s A(gh,s)f = V(g,h)X(s)A(gh,s)f =$ $= X(s)A(h,s)A(g,h(s))f = R_s A(h,s)A(g,h(s))f$ and (1.5) is proved.

<u>Corollary 1.</u> If $S = G$, G has a unite and $g(s) = gs$ then taking in (1.3) $s = e$ and s,t in place of g,h respectively and defining $R = R_e$ we get an "orthodox" dilation formula

(1.3′) $\quad K(s,t) = A(s,e)^*R^*T(s)^*T(t)RA(t,e).$

Formula (1.5) reduces then to

(1,5′) $\quad V(s,t)RA(st,e) = RA(s,e)A(s,t).$

<u>Remark 1.</u> In (1.4) the explicit form of R_s does not appear. In the context of Cor. 1 no explicit use of R_s is needed either.

<u>Remark 2.</u> If G is a topological semi-group and S an topological space and K is weakly continuous on $S \times S$ and the map $(g,s) \to g(s)$ is continuous as well the map $(g,s) \to A(g,s)$ is strongly continuous then $V(\cdot,\cdot)$ and $T(\cdot)$ are strongly continuous.

<u>Remark 3.</u> If $A(e,s) = I_H$ in Th.1 then $T(e) = I_{H_K}$

<u>Remark 4.</u> If $A(g,s)H = H$ then every $T(g)$ of Th.1 is unitary. Consequently, by (1.4), every $V(g,h)$ is unitary.

2. We will discuss now Theorem 1 in case when $A(g,s) = \alpha(g,s)I$ where α is a scalar complex valued function. We will assume in all what follows that G is a group and

(2.0) $\quad \alpha(g,s) \neq 0$ for $g \in G$, $s \in S$.

The above condition holds true if G acts on S transitively and $K \not\equiv 0$. Indeed, suppose $\alpha(g_o,s_o) = 0$ for some $g_o \in G$, $s_o \in S$. Then, since K is A-p.i. ($A(g,s) = \alpha(g,s)I$) we get that

$$K(g_o(s_o),g_o(t)) = 0 \quad \text{for every} \quad t \in S$$

Since g_o maps S onto S we conclude that $K(g_o(s_o),z) = 0$ for each $z \in S$. Hence

$$K(g(g_o(s_o)),g(z)) = \overline{\alpha(g,g_o(s_o))}K(g_o(s_o),z)\,\alpha(g_o,z) = 0.$$

Let $u_o \in S$. There is a $g \in G$ such that $g(g_o(s_o)) = u_o$. Hence $K(u_o,g(z)) = 0$ for $z \in S$. Since $g(S) = S$ we conclude that $K(u_o,s) = 0$ for $s \in S$. Since u_o is arbitraty we get that $K \equiv 0$. the desired contradiction. Our arguments simplify essentially the proof of [4], Lem. 2.2, p.6, because, as seen above, there is no need to discuss the case when $\alpha(g,s) = 0$ for some g,s.

Going back to condition (2.0), we see that, if it is satisfied, the operators $T(g)$, $V(g,h)$ of Th.1 are unitary. Since in this case

$$V(g,h)X(s)A(gh,s)f = V(g,h)\alpha(gh,s)X(s)f =$$

$$= \alpha(g,s)\alpha(g,h(s))X(s)f$$

for $g,h \in G$, $s \in S$, $f \in H$, we conclude that

$$V(g,h) = \frac{\alpha(g,s)\alpha(g,h(s))}{\alpha(gh,s)} I_H$$

It follows that the ratio appearing on the right hand side of the last equality does not depend on s and has a constant modulus equal to one. Let

$$(2.1) \qquad \sigma(g,h) = \frac{\alpha(gh,s)}{\alpha(g,s)\alpha(g,h(s))}$$

Theorem 1 , namely (1.4) imply the following theorem:

<u>Theorem</u> 2. Suppose G is a group and the P.D. kernel K satisfies (1.2) where $A(g,s) = \alpha(g,s)I$, $\alpha(g,s) \in \mathbf{C}$ and $\alpha(g,s) \neq 0$ for $g \in G$, $s \in S$. Then $T(g)$ of Th. 1 is a projective unitary representation of G namely $T(g)$ are unitary operators and

$$T(g)T(h) = \sigma(g,h)T(g,h)$$

for $g,h \in G$, where σ is defined by (2.1).

The above theorem gives an extension of Th.2.7 of [4] to operator valued kernels. It is obvious that if $K \neq 0$, then σ of the above theorem is a second order cocycle with values in the multiplicative group of complex numbers of modulus one.

3. Suppose in the sequel G is a group. Furthermore suppose the function V is <u>normalized</u>, i.e. $V(g^{-1},g) = I$ for all $g \in G$. Then, by (1.4), $T(g^{-1})T(g) = T(g^{-1}g) = T(e)$. Provided $T(e) = I_{H_K}$ we have $T(g^{-1})T(g) = T(g)T(g^{-1}) = I$ which simply means that all $T(g)$'s are unitary.

This enables us to state the following

<u>Theorem 3.</u> Suppose G is a group and

$$(3.1) \qquad A(e,g) = I_H \text{ and } A(g^{-1},s)A(g^{-1},g(s)) = I_H, \; g \in G.$$

Then $T(g)$ as well as $V(g,h)$ are unitary operators for all $g,h \in G$ and the formula (1.3) takes the form

$$(3.2) \qquad K(g(s),h(t)) = A(g,s)^*R_s^*T(g^{-1}h)R_tA(h,t)$$

while, in case $S = G$, the formula (1,3′) takes the form

$$K(s,t) = A(s,e)^*R^*T(s^{-1}t)RA(t,e)$$

and, when $A(g,e) = I$, this reduces to

$$K(s,t) = R^*T(s^{-1}t)R.$$

Proof. The only thing we have to prove is that V is normalized. This follows immediately from the definition of V(g,h) (cf. the proof of Th. 1) because of (3.1).

Remark 5. Kunze [2] has considered the case of an A-p.i. operator kernel with A satisfying

$$A(gh,s) = A(h,s)A(g,h(s)) \text{ and } A(e,s) = I_H.$$

This case fits in with what we have proposed here and his kernel is of the form (3.2) with $g \longrightarrow T(g)$ being an "ordinary" unitary group representation (V(g,h) = I).

4. Our approach gives authomatically a generalization of a theorem of Evans [1], p.370. This theorem is an operator version of the well known Segal's construction of generating functionals connected with Weyl's commutation relation.
Suppose namely that K and A are as in Theorem 3 (for $A(g,s) = \exp\{\frac{1}{2}\, i\, \mathrm{Im} < g,s > \}$ where $g,s \in G = S =$ a complex Hilbert space, the condition (3.1) holds trivialy). Then T satisfies the generalized Weyl's relation of the form (1.4) with V satisfying (1.5'); as easily seen, if $A(g,s) = \exp\{\frac{1}{2}\, i\, \mathrm{Im} < g,s > \}$ we get precisely Evan's result.

5. Recall [6] that for scalar type multipliers V some their properties holds automatically. For example, V must necessarily satisfy the following cocycle relation

(5.1) $V(h,k)V(g,hk) = V(g,h)V(gh,k)$.

If multipliers are allowed to be operator valued, some new phenomena appear. In this general case a multiplier V (i.e. V satisfying (1.4) satisfies (5.1) if and only if each V(g,h) commutes with each T(k), provided T(k)'s are unitary operators.

References

[1] D.E. Evans, J.T.Lewis, Some semigroups of completely positive maps on the CCR algebra, J.Functional Analysis, 24 (1977), 369-377.

[2] R.A.Kunze, Positive definite operator valued kernels and unitary representations, Proceedings of the Conference hold at UC, Irvine; ed.B.R.Geldbaum, Academic Press, London, 1967, 235-247.

[3] W.Mlak, Dilations of Hilbert space operators (general theory) Dissertationes Math., 153 (1978), pp. 65.

[4] K.R.Parthasaratny, K.Schmidt, Positive definite kernels, continuous tensor products, and central limit theorems in Probability Theory, Lecture Notes in Math., vol. 278, Springer Verlag, New York, 1972.

[5] G.B. Pedrick, Theory of reproducing kernels for Hilbert spaces of vector valued functions, University of Kansas, Technical Report 19 (1957) (unpublished).

[6] V.S.Varadarajan, Geometry of quantum theory, vol. II, Van Nostrand, New York, 1970.

Instytut Matematyczny PAN
ul. Solskiego 30
31-027 Kraków

Instytut Matematyki UJ
ul. Reymonta 4
30-059 Kraków

ON THE CONSTRUCTION OF WOLD-CRAMÉR DECOMPOSITION FOR BIVARIATE STATIONARY PROCESSES

Hannu Niemi
University of Helsinki
Department of Mathematics
SF-00100 Helsinki 10, Finland

Introduction

A method to construct the Wold-Cramér decomposition for a q-variate stationary stochastic process $\underline{\underline{x}}_k = (x_k^1,\ldots,x_k^q)$, $k \in Z$, is considered. The method is based on the forming of orthogonal decompositions for $\underline{\underline{x}}_k$, $k \in Z$, by applying orthogonal projections of $\underline{\underline{x}}_k$, $k \in Z$, onto its component processes x_k^j, $k \in Z$; $j = 1,\ldots,q$; and then use criteria given by Robertson [8] and Jang Ze-Pei [2] to decide whether the result is the desired Wold-Cramér decomposition. In this paper we present an improvement, in the special case $q = 2$, of our general result obtained in [7]. Unfortunately our method does not give a complete solution to the problem how to construct the Wold-Cramér decomposition (cf. Example 4).

1. Dominated decompositions

Let H be a (fixed) complex Hilbert space; one can choose e.g. $H = L^2(\Omega,A,P)$, where (Ω,A,P) is a probability space. Let $\underline{\underline{x}}_k = (x_k^1,\ldots,x_k^q)$, $k \in Z$, be a q-variate second order stochastic process, i.e., $x_k^j \in H$, $j = 1,\ldots,q$; $k \in Z$. By $\overline{sp}\{\underline{\underline{x}}\}$ we denote the closed linear subspace in H spanned by the set $\{x_k^j \mid j = 1,\ldots,q;\ k \in Z\}$. For $n \in Z$, by $\overline{sp}\{\underline{\underline{x}};n\}$ we denote the closed linear subspace in H spanned by the set $\{x_k^j \mid j = 1,\ldots,q;\ k \le n\}$. Furthermore, we use the notation

$$\overline{sp}\{\underline{\underline{x}};-\infty\} = \bigcap_{k\in Z} \overline{sp}\{\underline{\underline{x}};k\}.$$

Recall that the sequence $\underline{\underline{x}}_k$, $k \in Z$, is called <u>regular</u> (or purely <u>non-deterministic</u>), if $\overline{sp}\{\underline{\underline{x}};-\infty\} = \{0\}$; it is called <u>singular</u> (or <u>deterministic</u>), if $\overline{sp}\{\underline{\underline{x}};-\infty\} = \overline{sp}\{\underline{\underline{x}}\}$.

In what follows, by P_M we denote the orthogonal projection of H onto a given closed linear subspace M in H.

Let $\underline{\underline{x}}_k$, $k \in Z$, be a q-variate stochastic process. A decomposition

$$\underline{\underline{x}}_k = \underline{\underline{y}}_k + \underline{\underline{z}}_k,\quad k \in Z,$$

for $\underline{\underline{x}}_k$, $k \in Z$, is called orthogonal, if $\overline{sp}\{\underline{y}\} \perp \overline{sp}\{\underline{z}\}$; it is called dominated, if $\overline{sp}\{\underline{y},n\} \subset \overline{sp}\{\underline{x};n\}$ and $\overline{sp}\{\underline{z};n\} \subset \overline{sp}\{\underline{x};n\}$ for all $n \in Z$. Recall that the decomposition $\underline{\underline{x}}_k = \underline{\underline{y}}_k + \underline{\underline{z}}_k$, $k \in Z$,

$$\underline{\underline{y}}_k = (P_{\overline{sp}\{\underline{x};-\infty\}}\, x_k^1, \ldots, P_{\overline{sp}\{\underline{x};-\infty\}}\, x_k^q), \quad k \in Z,$$

the Wold-Cramér decomposition for $\underline{\underline{x}}_k$, $k \in Z$, is the only dominated orthogonal decomposition for $\underline{\underline{x}}_k$, $k \in Z$, with the property that $\underline{\underline{y}}_k$, $k \in Z$, is a singular and $\underline{\underline{z}}_k$, $k \in Z$, is a regular q-variate stochastic process (cf. Cramér [1]).

We assume that the reader is familiar with the basic covariance and spectral properties of q-variate stationary stochastic processes $\underline{\underline{x}}_k$, $k \in Z$ (cf. Masani [3]). We use the same terminology as used in [3].

Remark. Let $\underline{\underline{x}}_k = (x_k^1, \ldots, x_k^q)$, $k \in Z$, be a q-variate stationary stochastic process. The closed linear subspaces

$$(1) \qquad M = \overline{sp}\{\underline{x};-\infty\}, \quad M = \bigvee_{j \in K} \overline{sp}\{x^j;-\infty\} \quad \text{and} \quad M = \bigvee_{j \in K} \overline{sp}\{x^j\}, \quad K \subset \{1,\ldots,q\},$$

in H are invariant under the shift operator group of $\underline{\underline{x}}_k$, $k \in Z$. Thus, the decomposition $\underline{\underline{x}}_k = \underline{\underline{y}}_k + \underline{\underline{z}}_k$, $k \in Z$, for $\underline{\underline{x}}_k$, $k \in Z$, where

$$\underline{y} = (P_M x_k^1, \ldots, P_M x_k^q), \quad k \in Z,$$

with M of the form (1), has the property that both of the sequences $\underline{\underline{y}}_k$, $k \in Z$, and $\underline{\underline{z}}_k$, $k \in Z$, are stationary.

Theorem 1. *A q-variate stationary stochastic process $\underline{\underline{x}}_k = (x_k^1, \ldots, x_k^q)$, $k \in Z$, is singular, if and only if for any and, a fortiori, for all $j = 1, \ldots, q$, the orthogonal decomposition*

$$(2) \qquad \begin{cases} \underline{\underline{x}}_k = \underline{\underline{y}}_k + \underline{\underline{z}}_k, & k \in Z, \\ \underline{\underline{y}}_k = (P_{\overline{sp}\{x^j\}} x_k^1, \ldots, P_{\overline{sp}\{x^j\}} x_k^q), & k \in Z, \end{cases}$$

has the properties:

(i) *the decomposition $\underline{\underline{x}}_k = \underline{\underline{y}}_k + \underline{\underline{z}}_k$, $k \in Z$, is dominated;*

(ii) *both of the q-variate stationary stochastic processes $\underline{\underline{y}}_k$, $k \in Z$, and $\underline{\underline{z}}_k$, $k \in Z$, are singular.*

Proof. (a): Suppose $\underline{\underline{x}}_k$, $k \in Z$, is singular. Then, for any fixed $j = 1, \ldots, q$,

$$\overline{sp}\{x^j\} \subset \overline{sp}\{\underline{x};-\infty\} = \overline{sp}\{\underline{x}\}.$$

Thus, the orthogonal decomposition $\underline{\underline{x}}_k = \underline{\underline{y}}_k + \underline{\underline{z}}_k$, $k \in Z$, is dominated. Furthermore, it is clear that both of the sequences $\underline{\underline{y}}_k$, $k \in Z$, and $\underline{\underline{z}}_k$, $k \in Z$, are singular, since they are obtained by applying bounded linear transformations to the singular stochastic process $\underline{\underline{x}}_k$, $k \in Z$.

(b): The second part of the theorem follows immediately from a result by Robertson [8; Corollary 2.9].

In what follows, by $S(\underline{\underline{x}})$ and $R(\underline{\underline{x}})$ we denote the singular and regular parts, respectively, of a q-variate stationary stochastic process $\underline{\underline{x}}_k$, $k \in Z$; and by $r(\underline{\underline{x}})$ we denote the rank of $\underline{\underline{x}}_k$, $k \in Z$.

Theorem 2. *Let $\underline{\underline{x}}_k = (x_k^1, \dots, x_k^q)$, $k \in Z$, be a q-variate stationary stochastic process with $r(\underline{\underline{x}}) \le 1$. Suppose there exists an integer $1 \le j \le q$ such that the q-variate stationary stochastic process $\underline{\underline{y}}_k$, $k \in Z$, (resp. $\underline{\underline{z}}_k$, $k \in Z$) in the decomposition (2) is regular. Then $r(\underline{\underline{x}}) = 1$, the corresponding q-variate stationary stochastic process $\underline{\underline{z}}_k$, $k \in Z$, (resp. $\underline{\underline{y}}_k$, $k \in Z$) is singular and*

$$S(\underline{\underline{x}}) = \underline{\underline{z}}, \quad R(\underline{\underline{x}}) = \underline{\underline{y}}$$

(resp.

$$S(\underline{\underline{x}}) = \underline{\underline{y}}, \quad R(\underline{\underline{x}}) = \underline{\underline{z}}).$$

Proof. Suppose e.g. $\underline{\underline{y}}_k$, $k \in Z$, is regular. Since

$$r(\underline{\underline{x}}) \ge r(\underline{\underline{u}}) + r(\underline{\underline{v}})$$

for all q-variate stationary stochastic processes $\underline{\underline{u}}_k$, $k \in Z$; $\underline{\underline{v}}_k$, $k \in Z$, such that $\underline{\underline{x}}_k = \underline{\underline{u}}_k + \underline{\underline{v}}_k$, $k \in Z$; $\overline{sp}\{\underline{\underline{u}}\} \perp \overline{sp}\{\underline{\underline{v}}\}$ and $\overline{sp}\{\underline{\underline{u}}\} \subset \overline{sp}\{\underline{\underline{x}}\}$, $\overline{sp}\{\underline{\underline{v}}\} \subset \overline{sp}\{\underline{\underline{x}}\}$ (cf. [8; Theorem 2.6]), it follows that

$$r(\underline{\underline{x}}) = r(\underline{\underline{y}}) + r(\underline{\underline{z}}). \tag{3}$$

Thus, $r(\underline{\underline{x}}) = r(\underline{\underline{y}}) = 1$, $r(\underline{\underline{z}}) = 0$; and a fortiori $\underline{\underline{z}}_k$, $k \in Z$, must be singular. Finally, it follows from the equation (3) that even

$$R(\underline{\underline{x}}) = \underline{\underline{y}}, \quad S(\underline{\underline{x}}) = \underline{\underline{z}}$$

(cf. [8; Corollary 2.7, Theorem 3.1]).

The other part of the theorem can be proved in a similar way.

Remark. (i) Let $\underline{\underline{x}}_k$, $k \in Z$, be a q-variate stationary stochastic process such that its spectral measure is absolutely continuous with respect to m, the Lebesgue measure of $[0,2\pi)$. The hypothesis $r(\underline{\underline{x}}) \leq 1$ is then satisfied e.g. in the following two cases:

(a) the spectral density $\underline{\underline{f}}$ of $\underline{\underline{x}}_k$, $k \in Z$, satisfies the condition

$$\text{essinf rank}(\underline{\underline{f}}) \leq 1 \quad (m)$$

(cf. [3; § 11]).

(b) $q = 2$ and $\underline{\underline{x}}_k$, $k \in Z$, is not regular (cf. [3; § 11]).

(ii) There exist q-variate stationary stochastic processes $\underline{\underline{x}}_k$, $k \in Z$, with an m-continuous spectral measure having the properties:

(a') $r(\underline{\underline{x}}) = 1$;

(b') the q-variate stationary stochastic processes $\underline{\underline{y}}_k$, $k \in Z$, and $\underline{\underline{z}}_k$, $k \in Z$, in the decomposition (2) are singular for all $j = 1,\ldots,q$.

One can choose even $q = 2$ (cf. Example 4).

Example 3. Let $\underline{\underline{x}}_k = (x_k^1,\ldots,x_k^q)$, $k \in Z$, be a q-variate stationary stochastic process. Suppose $\underline{\underline{\mu}} = (\mu_1,\ldots,\mu_q)$ is the spectral measure of $\underline{\underline{x}}_k$, $k \in Z$, i.e.,

$$x_k^j = \int e^{ik\lambda} d\mu_j(\lambda), \quad j = 1,\ldots,q;\ k \in Z.$$

For $j = 1,\ldots,q$, by $\overline{sp}\{\mu_j\}$ we denote the closed linear subspace in H spanned by the set $\{\mu_j(A) \mid A \subset [0,2\pi) \text{ is a Borel set}\}$. Furthermore, by ν_{jk} we denote the uniquely determined bounded complex-valued measure on $[0,2\pi)$ for which

$$(4) \qquad (\int u\, d\mu_j \mid \int v\, d\mu_k) = \int u\,\bar{v}\, d\nu_{jk} \quad \text{for all } u \in L^1(\mu_j),\ v \in L^{\hat{}}(\mu_k);$$

$j,k = 1,\ldots,q$ (cf. [7; § 1]). Recall that

(i) $\overline{sp}\{\mu_j\} = \overline{sp}\{x^j\}$, $j = 1,\ldots,q$;

(ii) ν_{jk} is absolutely continuous w.r.t. ν_{jj}; $j,k = 1,\ldots,q$;

(iii) for all $h = 1,\ldots,q$

$$P_{\overline{sp}\{x^h\}} x_k^j = \int e^{ik\lambda} \frac{d\nu_{jh}}{d\nu_{hh}}(\lambda)\, d\mu_h(\lambda), \quad j = 1,\ldots,q;\ k \in Z$$

(cf. [7; Theorem 2]).

2. On the construction of the Wold-Cramér decomposition

The results stated in Theorems 1 and 2 can be used to improve our method [7; § 3] to construct the Wold-Cramér decomposition for q-variate stationary sequences.

We point out the improvements in the case $q = 2$.

Algorithm. Let $\underline{\underline{x}}_k = (x_k^1, x_k^2)$, $k \in Z$, be a bivariate stationary stochastic processes and let $\underline{\underline{\mu}} = (\mu_1, \mu_2)$ be the spectral measure of $\underline{\underline{x}}_k$, $k \in Z$.

Step 1. Let $\underline{\underline{\mu}}_s = (\mu_{1,s}, \mu_{2,s})$ and $\underline{\underline{\mu}}_c = (\mu_{1,c}, \mu_{2,c})$ be the m-singular and m-continuous parts of $\underline{\underline{\mu}}$, respectively (cf. [7; Example 3 and the references given there]). Put

$$u_k^j = \int e^{ik\lambda} d\mu_{j,s}(\lambda), \quad v_k^j = \int e^{ik\lambda} d\mu_{j,c}(\lambda), \quad j = 1,2; \quad k \in Z.$$

Step 2. If $R(\underline{\underline{v}}) = 0$, i.e., if $\underline{\underline{v}}_k$, $k \in Z$, satisfies the singularity conditions stated in Theorem 1 or equivalently the sigularity conditions given by Matveev [5] or by Jang Ze-Pei [2; Part I, Theorem 15 and Part II], then

$$S(\underline{\underline{x}}) = \underline{\underline{x}}, \quad R(\underline{\underline{x}}) = 0$$

(cf. [7; § 3]).

If $R(\underline{\underline{v}}) \neq 0$, go to Step 3.

Step 3. If $S(\underline{\underline{v}}) = 0$, i.e., if $\underline{\underline{v}}_k$, $k \in Z$, satisfies the regularity conditions given by Wiener and Masani [9], [10] (cf. Matveev [4], [6]), then

$$S(\underline{\underline{x}}) = \underline{\underline{u}}, \quad R(\underline{\underline{x}}) = \underline{\underline{v}}$$

(cf. [7; § 3]).

If $S(\underline{\underline{v}}) \neq 0$, go to Step 4.

Step 4. If there does not exist any (non-zero) singular component v_k^j, $k \in Z$, of $\underline{\underline{v}}_k$, $k \in Z$, go to Step 5.

If there exists a (non-zero) singular component, say v_k^1, $k \in Z$, of $\underline{\underline{v}}_k$, $k \in Z$, put

$$(5) \qquad w_k^j = \int e^{ik\lambda} \frac{d\tilde{\nu}_{j1}}{d\tilde{\nu}_{11}}(\lambda)\, d\tilde{\mu}_1(\lambda), \quad j = 1,2; \quad k \in Z,$$

where $\tilde{\underline{\underline{\mu}}} = (\tilde{\mu}_1, \tilde{\mu}_2)$ is the spectral measure of $\underline{\underline{v}}_k$, $k \in Z$, and $\tilde{\nu}_{jk}$; $j,k = 1,2$, are the corresponding measures satisfying (4). Then the bivariate stationary sequence $\underline{\underline{w}}'_k = \underline{\underline{v}}_k - \underline{\underline{w}}_k = (0, v_k^2 - w_k^2)$, $k \in Z$, is either singular or regular; or equivalently the stationary sequence $w'_k = v_k^2 - w_k^2$, $k \in Z$, is either singular or regular. If w'_k, $k \in Z$, is singular, then

$$S(\underline{\underline{x}}) = \underline{\underline{x}}, \quad R(\underline{\underline{x}}) = 0.$$

If w'_k, $k \in Z$, is regular, then

$$S(\underline{\underline{x}}) = \underline{\underline{u}} + \underline{\underline{w}}, \quad R(\underline{\underline{x}}) = \underline{\underline{w}}' = \underline{\underline{v}} - \underline{\underline{w}}$$

(cf. [7; § 3]).

Step 5. Suppose there exists a (non-zero) component, say v_k^1, $k \in Z$, of $\underline{\underline{v}}_k$, $k \in Z$, such that the stationary stochastic processes $\underline{\underline{w}}_k$, $k \in Z$, defined in (5), and $\underline{\underline{w}}_k' = \underline{\underline{v}}_k - \underline{\underline{w}}_k = (0, v_k^2 - w_k^2)$, $k \in Z$, satisfy one of the following two conditions:

(R.1) $\underline{\underline{w}}_k$, $k \in Z$, is a regular bivariate stationary stochastic process (necessarily $r(\underline{\underline{w}}) = 1$), i.e., it satisfies the regularity condition given in [10];

(R.2) $\underline{\underline{w}}_k'$, $k \in Z$, or equivalently $w_k' = v_k^2 - w_k^2$, $k \in Z$, is regular.

If $\underline{\underline{w}}_k$, $k \in Z$, is regular, then

$$S(\underline{\underline{x}}) = \underline{\underline{u}} + \underline{\underline{w}}', \quad R(\underline{\underline{x}}) = \underline{\underline{w}}$$

(cf. Theorem 2, [7; § 3]). If w_k', $k \in Z$, is regular, then

$$S(\underline{\underline{x}}) = \underline{\underline{u}} + \underline{\underline{w}}, \quad R(\underline{\underline{x}}) = \underline{\underline{w}}'$$

(cf. Theorem 2, [7; § 3]).

If no components v_k^j, $k \in Z$, of $\underline{\underline{v}}_k$, $k \in Z$, satisfying the condition (R.1) or (R.2) can be found, go to Step 6.

Step 6. $r(\underline{\underline{x}}) = 1$ and

$$S(\underline{\underline{x}}) = \underline{\underline{u}} + S(\underline{\underline{v}}), \quad R(\underline{\underline{x}}) = R(\underline{\underline{v}}).$$

The Wold-Cramér decomposition $\underline{\underline{v}} = S(\underline{\underline{v}}) + R(\underline{\underline{v}})$ cannot be formed by applying the algorithm (cf. Theorem 1, Example 4).

Remark. (i) The conditions given in [2], [5] and Theorem 1 for a q-variate stationary stochastic process to be singular seem to be rather unpractical. The algorithm can be applied also without applying Step 2. However, in this case the statement "$r(\underline{\underline{x}}) = 1$" in Step 6 must be replaced with the statement "$r(\underline{\underline{x}}) = r(\underline{\underline{v}}) \leq 1$".

(ii) Jang Ze-Pei [2] has presented the spectral representation for the covariance kernels of the singular and regular components, respectively, of a q-variate stationary stochastic process.

Example 4. [7; Example 9 (b)] Consider two stationary stochastic processes $y_{j,k}$, $k \in Z$, $j = 1,2$, with m-continuous spectral measures such that $\overline{sp}\{y_1\} \perp \overline{sp}\{y_2\}$ and

$$(y_{j,h} | y_{j,k}) = \int e^{i(h-k)\lambda} f_j(\lambda) d\lambda, \quad h,k \in Z;\ j = 1,2,$$

where

$$f_1 \equiv 1,\ f_2(\lambda) = \begin{cases} 0, & \lambda \in [0,\pi) \\ 1, & \lambda \in [\pi,2\pi). \end{cases}$$

Define a bivariate stationary stochastic process $\underline{\underline{x}}_k = (x_k^1, x_k^2)$, $k \in Z$, by

$$x_k^1 = y_{1,k} + y_{2,k}, \quad x_k^2 = y_{1,k} + 2y_{2,k}, \quad k \in Z.$$

The Wold-Cramér decomposition

$$S(\underline{\underline{x}}) = (y_2, 2y_2), \quad R(\underline{\underline{x}}) = (y_1, y_1)$$

cannot be formed by applying the algorithm (for the details see [7]).

References

[1] Cramér, H.: On some classes of non-stationary stochastic processes. - Proceedings of the Fourth Berkeley symposium on mathematical statistics and probability, Vol. II, pp. 57-76. University of California Press, Berkeley/Los Angeles, 1962.

[2] Jang Ze-Pei: The prediction theory of multivariate stationary processes, I. - Chinese Math. 4 (1963), 291-322; II. - Chinese Math. 5 (1964), 471-484.

[3] Masani, P.: Recent trends in multivariate prediction theory. - Multivariate Analysis I (ed. P.R. Krishnaiah), pp. 351-382. Academic Press, New York/London, 1966.

[4] Matveev, R.F.: On the regularity of one-dimensional stationary stochastic processes with discrete time. - Dokl.Akad.Nauk. SSSR 25 (1959), 277-280 (In Russian).

[5] Matveev, R.F.: On singular multidimensional stationary processes. - Theor. Probability Appl. 5 (1960), 33-39.

[6] Matveev, R.F.: On multidimensional regular stationary processes. - Theor. Probability Appl. 6 (1961), 149-165.

[7] Niemi, H.: On the construction of Wold decomposition for multivariate stationary processes. - J. Multivariate Anal. (to appear).

[8] Robertson, J.B.: Orthogonal decompositions of multivariate weakly stationary stochastic processes. - Canad. J. Math. 20 (1968), 368-383.

[9] Wiener, N., and P. Masani: The prediction theory of multivariate stationary processes I. - Acta Math. 98 (1957), 111-150.

[10] Wiener, N., and P. Masani: On bivariate stationary processes and the factorization of matrix-valued functions. - Theor. Probability Appl. 4 (1959), 300-308.

REPRESENTATION OF A BOUNDED OPERATOR AS A FINITE LINEAR COMBINATION OF PROJECTORS AND SOME INEQUALITIES FOR A FUNCTIONAL ON B(H).

A. Paszkiewicz

1. Notation and main results. Let H be a real or complex, separable Hilbert space of any dimension. Let $B(H)$ and $S(H)$ be the space of all bounded operators and the cone of all bounded self - adjoint operators acting in H, respectively. By $L(H)$ we denote the Logic of all projective operators belonging to $S(H)$. In non - commutative probability theory a linear functional f on $B(H)$ is treated as an extension of some orthogonally additive measure μ on $L(H)$, and the inequalities for the norm $\|f\|$, given in the following theorem, seem to be interesting. For the functional f, we denote

$$S(f) = \sup \{ |f(A)| ; A \in S(H)\} ,$$

$$P(f) = \sup \{ |f(P)| ; P \in L(H)\} .$$

Theorem 1. If the space H is complex, then the following is valid

$$(1.1) \quad \|f\| \leq \pi P(f),$$

$$(1.2) \quad S(f) \leq 2 P(f),$$

and, if $f(1_H) = 0$ (1_H = identity operator in H),

$$(1.3) \quad \|f\| \leq 2^{-1}\pi S(f).$$

The constans π, 2, $2^{-1}\pi$ in (1.1), (1.2), (1.3) cannot be diminished.

The question is whether inequalities (1.1) and (1.3) can become equalities for some functional f.

To prove Theorem 1, we shall show that the norm of the functional f is finite if only $P(f) < \infty$. This follows from the following.

<u>Theorem 2.</u> If H is a real or complex, separable Hilbert space, and $A \in S(H)$, then

$$(1.4) \quad A = \sum_{i=1}^{6} \alpha_i P_i ,$$

where $P_i \in L(H)$, $\alpha_i \in R$ (= the set of reals) and $|\alpha_i| \leq 5 \|A\|$ for $i = 1,\dots,6$.

This theorem states, in particular, that the set of projectors $L(H)$ is linearly complete in the linear space $S(H)$ over the field of real numbers.

In section 2. we shall give some remarks on the way our Theorem 2 was obtained in [1].

In sections 3 and 4 the proof of Theorem 1 will be shown. As an easy consequence of Lemma 3.3, given in section 3, one can obtain

Theorem 3. For any sequence of complex numbers $(z_j)_{j \in N}$,

$$(1.5) \sum_{j \in N} |z_n| \leq \pi \sup_{Z \subset N} \Big| \sum_{j \in Z} z_j \Big| - \sum_{j \in N} |z_j| .$$

and equality is impossible if one of the sides is finite.

N is, as umal, the set of positive integers, and Z ranges over all non - empty subsets of N.

2. Information on the proof of Theorem 2.

Let projectors $P_1, P_2 \in L(H)$ be mutually orthogonal, of the same dimension, and let I be a partial isometry carrying P_1 onto P_2, i.e.

$$I I^* = P_1, \quad I^* I = P_2,$$

and $I^2 = (I^*)^2 = 0$ since $P_1 \perp P_2$ (I^* is an operator adjoint to I). For any operator

$$A = \lambda P_1, \text{ where } 1 + \alpha \leq \lambda \leq 1 + \beta$$

$(-1 \leq \alpha < \beta \leq 1)$, one can define two operators

$$S = \tfrac{1}{2} [\lambda P_1 = \sqrt{\lambda(2-\lambda)} (I + I^*) + (2-\lambda) P_2] ,$$

$$T = \tfrac{1}{2} [\lambda P_1 - \sqrt{\lambda(2-\lambda)} (I + I^*) + (2-\lambda) P_2] .$$

It is obvious that $S,T \in L(H)$ since S,T are self - adjoint, and $S^2 = S$, $T^2 = T$. Moreover,

$$A - (S + T) = \lambda_2 P_2, \text{ where } -1 + \alpha \leq \lambda_2 \leq -1+\beta .$$

By help of the spectral theorem for a self - adjoint operator we can also prove

Lemma 2.1. If the operator $A \in S(H)$ satisfies

$$(2.1) \quad (a+\alpha) P_1 \leq A \leq (a+\beta) P_1$$

for some projector $P_1 \in L(H)$, $-a \leq \alpha < \beta \leq a$, and there exists a projector $P_2 \perp P_1$ such that $\dim P_2 = \dim P_1$, then there also exist projectors $S,T \leq P_1 + P_2$ such that

$$(2.2) \quad (-a+\alpha) P_2 \leq A-a(S+T) \leq (-a+\beta) P_2 .$$

The methods of proving Theorem 2 and the following remark are analogous.

Remark 2.1. If $\dim H = \infty$, and the operator $A \in S(H)$ satisfies

$$(2.3) \quad 0 \leq A \leq 2P_1, \quad \dim(I_H - P_1) = \infty ,$$

then

$$(2.4) \quad A = S' + T' - S'' - T''$$

for some projectors $S', T', S'', T'' \in L(H)$.

Proof. By (2.3), one can find a sequence of projectors $P_1, P_2, \ldots$ such that

$$(2.5) \quad P_i \perp P_j, \ \dim P_i = \dim P_j \quad (i \neq j, \ i,j = 1,2,\ldots)$$

Putting $a = \beta = 1$, $\alpha = -1$ in Lemma 2.1, we obtain (2.1) and thus, there exist projectors

$$S_1, T_1 \leq P_1 + P_2$$

such that

$$-2 P_2 \leq A - (S_1 + T_1) \leq 0.$$

Analogously we obtain projectors

$$S_2, T_2 \leq P_2 + P_3$$

such that

$$0 \leq A - (S_1 + T_1) + S_2 + T_2 \leq 2 P_3 .$$

By induction, there exist two sequences of projectors (S_i), (T_i) for which

$$(2.6) \quad S_i, T_i \leq P_i + P_{i+1}$$

$$0 \leq A + \sum_{i=1}^{2k} (-1)^i (S_i + T_i) \leq 2 P_{2k+1} ,$$

$$-2P_{2k} \leq A + \sum_{i=1}^{2k-1} (-1)^i (S_i+T_i) \leq 0 \quad (k = 1,2,\dots)$$

Conditions (2.5), (2.6) imply that the projectors

$$S' = S_2 + S_4 + \dots, \quad S'' = S_1 + S_3 + \dots$$

$$T' = T_2 + T_4 + \dots, \quad T'' = T_1 + T_2 + \dots$$

are well defined, and (2.4) holds. Indeed, for any $x \in H$,

$$\| A + \sum_{i=1}^{n} (-1)^i (S_i+T_i)x \| \leq 2 \| P_{n+1} x \| \to \text{ as } n \to \infty .$$

If $\dim H = \infty$, then, for any self - adjoint operator A with the norm $\|A\| < 1$, we can obtain in representation (1.4) standart (integer) coefficients independent of the operator A [1].

Remark 2.2. If $A \in S(H)$, $\|A\| < 1$, and $\dim H = \infty$, then

$$A = 5(Q_1+Q_2+Q_3+Q_4) - 5Q_5 - 8Q_6 - 12Q_7$$

for some projectors $Q_1,\dots,Q_7 \in L(H)$.

3. Auxiliary remarks on sequences of complex numbers.

Definition 1. We shall call the real number

$$\operatorname{ord}(w_1,\dots,w_n) = \max_Z \left| \sum_{l \in Z} w_l \right| ,$$

where Z ranges over all non - empty subsets of $\{1,\dots,n\}$, the ordering of the sequence $w_1,\dots,w_n$ of complex numbers.

Since the modulus is a subadditive function on C, we immediately obtain.

Lemma 3.1. For any numbers $w_{jl} \in C$ $(j = 1,\dots,m,\ l=1,\dots,n)$, the inequality

$$\operatorname{ord}(w_{11},\dots,w_{1n}; \dots; w_{m1},\dots,w_{mn}) \leq \sum_{j=1}^{m} \operatorname{ord}(w_{j1},\dots,w_{jn})$$

holds.

We shall make use of the following well-known

Lemma 3.2. If we put, for fixed real numbers α_m,

$$\varphi_j^m = \alpha_m + 2\pi \quad j\, m^{-1} \qquad (j = 1,\dots,m)$$

and

$$\Lambda_m = \{ j = 1,\dots,m; \quad \mathrm{Re}\exp i\,\varphi_j^m \geq 0 \},$$

then

$$(3.1) \qquad m^{-1} \sum_{j \in \Lambda_m} \mathrm{Re}\exp i \; \varphi_j^m \longrightarrow \pi^{-1} \quad \text{as } m \longrightarrow \infty .$$

A simple proof of the above can be obtained by help of Riemann's integral. Let us use the notation $\varphi_j^m = \alpha_m + 2\pi\, j\, m^{-1}$ for any integer j and observe that

$$\sum_{j \in \Lambda_m} \mathrm{Re}\exp i\,\varphi_j^m = \sum_{j=k}^{k+s} \cos \varphi_j^m$$

for some integers k,s, and the partition

$$(-\tfrac{\pi}{2}, \varphi_k^m, \varphi_{k+1}^m, \dots, \varphi_{k+s}^m, \tfrac{\pi}{2})$$

of the interval $(-\frac{\pi}{2}, \frac{\pi}{2})$ has the diameter $2\pi\, m^{-1}$,

Therefore

$$2\pi\, m^{-1} \sum_{l=k}^{k+s} \cos\varphi_j^m \longrightarrow \int_{-\frac{\pi}{2}}^{\frac{\pi}{2}} \cos\varphi \, d\varphi = 2 \text{ as } m \longrightarrow \infty ,$$

and (3.1) holds.

Lemma 3.3. For any sequence $(w_1,\dots,w_n)$ of complex numbers

$$\sum_{l=1}^{n} |w_l| \leq \pi \quad \mathrm{ord}\,(w_1,\dots,w_n).$$

Proof. Let us denote

$$(3.2) \qquad w_{jl} = w_l \exp i\, 2\,\pi\, jm^{-1}$$

for any positive integer m, and $j = 1,\dots,m$, $l=1,\dots,n$. By Definition 3.1

$$\mathrm{ord}\,(w_{11},\dots,w_{1n}; \dots; w_{m1},\dots,w_{mn}) = \Big|\sum_{(j,l) \in \Lambda_m} w_{jl}\Big|$$

for some set $\Lambda_m \subset \{1,\dots,m\} \times \{1,\dots,n\}$. Let, for any m, $\theta_m \in C$ be a number satisfying $|\theta_m| = 1$ and

$$\Big|\sum_{(j,l)\in\Lambda_m} w_{jl}\Big| = \theta_m \sum_{(j,l) \in \Lambda_m} w_{jl} .$$

It is easy to see that

$$\Lambda_m = \{(j,l)\ ;\ \operatorname{Re} \theta_m w_{jl} \geq 0\}\ .$$

For

$$\Lambda_m^l = \{j=1,\dots,m\ ;\ \operatorname{Re} \theta_m w_{jl} \geq 0\},\ l = 1,\dots,m\ ,$$

we obtain, by (3.2) and Lemma 3.2, that

$$n^{-1} \sum_{j\in\Lambda_m^l} \operatorname{Re}\ \theta_m w_{jl} \longrightarrow |w_l|\ \pi^{-1} \quad \text{as } n \longrightarrow \infty\ ,$$

and then,

$$n^{-1} \operatorname{ord}(w_{11},\dots,w_{1n};\dots;\ w_{m1},\dots,w_{mn}) =$$

$$= n^{-1} \sum_{l=1}^{n} \sum_{j\in\Lambda_m^l} \operatorname{Re}\ \theta_m\ w_{jl} \longrightarrow \pi^{-1} \sum_{l=1}^{n} |w_l| \text{ as } n \longrightarrow \infty$$

Moreover, we have

$$\operatorname{ord}(w_{j1},\dots,w_{jn}) = \operatorname{ord}(w_1,\dots,w_n),\quad j=1,\dots,m\ ,$$

which, by Lemma 3.1, ends the proof.

We shall use another characteristic of the sequence $(w_1,\dots,w_n)$ of complex numbers

$$S(w_1,\dots,w_n) = \max\{\sum_{l=1}^{n} \varepsilon_l\ w_l\ ;\ \ \varepsilon_l = \pm 1 (\ l=1,\dots,n)\ \}$$

(the coefficients ε_l take the values - 1 and 1 independently from one another).

<u>Lemma 3.4.</u> If, for complex numbers $w_1,\dots,w_n$,

$$(3.3) \qquad \sum_{l=o}^{n} w_l = 0.$$

then $S(w_1,\dots,w_n) = 2 \operatorname{ord}(w_1,\dots,w_n)$.

Proof. Really, by (3.3), the following conditions are equivalent

$$\operatorname{ord}(w_1,\dots,w_n) = |\sum_{l\in Z} w_l|\ ,$$

$$S(w_1,\dots,w_n) = |\sum_{l=1}^{n} \varepsilon_l\ w_l|,\ \varepsilon_l = 1 \quad \text{if } l \in Z,\ \varepsilon_l = -1 \text{ if } l \notin Z\ ,$$

for $Z \subset \{1,\dots,n\}$.

4. Functionals on B(H).

Lemma 4.1. If, for a linear functional f on $B(H)$, $P(f) < \infty$, then $\|f\| < \infty$.

Proof. For any operator $A \in S(H)$, according to (1.4),

$$|f(A)| \leq \sum_{i=1}^{6} |\alpha_i| \, |f(P_i)| \leq 5 \cdot 6 \cdot P(f) \cdot \|A\| ,$$

and thus, $|f(B)| \leq 60 \, P(f)$ if $B \in B(H)$, $\|B\| = 1$.

For a continuous linear functional f on $B(H)$, we shall use the notation

(4.1) $S_1(f) = \sup\{|f(P_1 - P_2)| \, ; \, P_1, P_2 \in L(H), \, P_1 \perp P_2 , \, P_1 + P_2 = 1_H\}$

(4.2) $U(f) = \sup\{|f(V)| \, ; \, V$ – a unitary operator in $H\}$.

The corollary in paper [1] states that (1.2) holds. In fact, the proof of this statement is given there.

Lemma 4.2. For any linear continuous functional f on $B(H)$.

$$S(f) = S_1(f).$$

Proof of Theorem 1. In virtue of our Lemma 4.1 and Corollary in [1], it is enough to prove inequalities (1.1), (1.3) for a continuous functional f.

For any operator $A \in B(H)$, $\|A\| \leq 1$, we have a polar decomposition $A = UW$ for some unitary operator U and self-adjoint operator W, $\|W\| \leq 1$. Using Lemma 4.2 for the functional $f_1(B) = f(UB)$ $(B \in B(H))$, we obtain, by (4.1), (4.2),

$$|f(A)| \leq S(f_1) = S_1(f_1) \leq U(f),$$

and thus,

(4.3) $\|f\| = U(f)$

By the spectral theorem for a unitary operator V, there exists an operator

$$V_\varepsilon = \sum_{i=1}^{n} \theta_i P_i$$

for some

$$\theta_i \in C, \; \|\theta_i\| = 1 \text{ and}$$

$$(4.4) \quad P_i \in L(H),\ P_i \perp P_j \ \text{ for } i \neq j,\ \sum_i P_i = 1 \qquad (i,j = 1,\dots,n),$$

such that

$$(4.5) \quad | f(V) - f(V_\varepsilon)| < \varepsilon \quad (\text{for any fixed } \varepsilon > 0).$$

Moreover, conditions (4.4) imply that

$$\operatorname{ord}(w_1,\dots,w_n) \leq P(f),$$
$$S(w_1,\dots,w_n) \leq S(f),$$

for $w_i = f(P_i)$ $(i = 1,\dots,n)$ and, by Lemma 3.3 ,

$$|f(V_\varepsilon)| \leq \sum_{i=1}^{n} |w_i| \leq \pi \ P(f).$$

If we assume, in addition, that

$$f(1_H) = \sum_{i=1}^{n} w_i = 0 ,$$

then, by Lemma 3.4 ,

$$|f(V_\varepsilon)| \leq \sum_{i=1}^{n} |w_i| \leq 2^{-1}\pi \ S(w_1,\dots,w_n) \leq$$
$$\leq 2^{-1}\pi \ S(f) .$$

In virtue of (4.3), (4.5), inequalities (1.1) and (1.3) (where $f(1_H) = 0$) take place.

It is obvious, by Lemma 4.2., that (1.2) becomes equality if only $f(1_H) = 0$. We shall now complete the proof of Theorem 1 by showing that the constants π, $2^{-1}\pi$ in (1.1), (1.3) cannot be diminished. Let $e_1,\dots,e_{2n}$ be an orthonormal basis in $H = C^{2n}$. For any operator $A \in B(H)$, we define

$$(4.5) \quad f_n(A) = (2n)^{-1} \sum_{l=1}^{2n} < Ae_l, e_l > \exp i \ \pi \ ln^{-1}.$$

Then, for a (unitary) operator U_n of the form

$$U_n e_l = e_l \exp(-i\pi \ ln^{-1}), \quad l = 1,\dots,2n,$$

we obtain $f_n(U_n) = 1$, and $\|f_n\| \geq 1$. On the other hand, by (4.5),

$$P(f_n) \leq (2n)^{-1} \sup \{ |\sum_{l=1}^{2n} \alpha_l \exp i\pi ln^{-1} | ;\ 0 \leq \alpha_l \leq 1 \quad (l = 1,\dots,2n) \}$$

and (cf. the proof of Corollary in [1])

$$P(f_n) \leq (2n)^{-1} \left| \sum_{l=1}^{n} \exp i \pi l n^{-1} \right| \longrightarrow$$

$$\longrightarrow (2\pi)^{-1} \left| \int_0^{\pi} \exp i x \, dx \right| = \pi^{-1} \quad \text{as} \quad n \longrightarrow \infty \; .$$

Basides, $f_n(1_H) = 0$ and, by Lemma 4.2 ,

$$S(f_n) = S_1(f_n) = 2P(f_n),$$

which ends the proof.

5. Proof of Theorem 3. We may assume that both sides of inequality (1.5) are finite. For

$$z_0 = - \sum_{j=1}^{\infty} z_j$$

and

$$a = \sup \{ | \sum_{j \in Z} z_j | ; \; \emptyset \neq Z \subset \{1,2,\dots\} \} ,$$

$$a_0 = \sup \{ | \sum_{j \in Z} z_j | ; \; \emptyset \neq Z \subset \{0,1,\dots\} \},$$

we have $a = a_0$. Indeed, $a \leq a_0$ and, for any $\varepsilon > 0$, one can find sets $Z_\varepsilon \subset \{0,1,\dots\}$, $Z'_\varepsilon = \{0,1,\dots\} \setminus Z_\varepsilon$, satysfying

$$(5.1) \qquad \left| \sum_{j \in Z_\varepsilon} z_j \right| = \left| \sum_{j \in Z'_\varepsilon} z_j \right| > a_0 - \varepsilon$$

since

$$\sum_{j=0}^{\infty} z_j = 0 .$$

Then $0 \notin Z_\varepsilon$ or $0 \notin Z'_\varepsilon$, and the left - hand side of inequality (5.1) is less or equal to a, which implies $a \geq a_0$.
Finally, note that

$$\sum_{j=0}^{n} |z_j| \leq \pi \; \mathrm{ord} \; (z_0,\dots,z_n) \leq \pi \, a_0 = \pi a \, ,$$

and inequality

$$\sum_{j=0}^{n} |z_j| \leq \pi \, a$$

is equivalent to (1.5).
The example of the sequence

$$z_j = \exp 2\pi i\, j\, n_o^{-1} \qquad \text{if} \quad j = 1,\dots,n_o$$

$$z_j = 0 \quad \text{if} \quad j > n_o \qquad (n_o = 1,2,\dots)$$

shows that no constant less than π can be used in (1.5).
On the other hand, equality in (1.5) is impossible if one of the sides is finite. Without loss of generality we may assume that

$$\sum_{j=1}^{\infty} |z_j| \leq 1 \quad \text{and} \quad z_1 \neq 0$$

and observe that for $v \in C$, $|v| = 1$

$$|v + 2^{-1} z_1| + \varepsilon_1 \leq \max (|v|, |v + z_1|)$$

if $\quad \varepsilon_1 = \sqrt{1+2^{-2} |z_1|^2} - 1$. Thus, for

$$w_o = 2^{-1} z_1 + i\, \varepsilon_1 |z_1|^{-1} z_1$$

$$(5.2) \qquad w_1 = 2^{-1} z_1 - i\, \varepsilon_1 |z_1|^{-1} z_1$$

$$w_j = z_j, \qquad j = 2,3,\dots,$$

the inequality

$$(5.3) \qquad \max (|v|, |v+w_o|, |v+w_1|) \leq \max (|v|, |v+w_o+w_1|)$$

holds. For $Z \subset \{0,1,2,\dots\}$, we denote $Z' = Z \cap \{2,3,\dots\}$,

$$v = \sum_{j \in Z'} w_j = \sum_{j \in Z'} z_j ,$$

then (5.3) implies

$$\sum_{j \in Z} w_j \leq \sum_{j \in Y} z_j$$

when $Y = Z'$ or $Y = Z' \cup \{1\}$. In consequence, by (5.2),

$$\sum_{j=1}^{\infty} |z_j| < \sum_{j=0}^{\infty} |w_j| \leq \pi \sup \{|\sum_{j \in Z} w_j |;\ \emptyset \neq Z \subset \{0,1,\dots,\}\} -$$

$$- |\sum_{j=0}^{\infty} w_j| \leq \pi \sup \{|\sum_{j \in Y} z_j|;\ \emptyset \neq Y \subset N\} - |\sum_{j=1}^{\infty} z_j|$$

holds, and Theorem 3 is proved.

References

[1] A.Paszkiewicz, Any self - adjoint operator is a finite linear combination of projectors, to appear in Bull. Acad. Polon. Sci.

Institute of Mathematics
Łódź University
90-238 Ł ó d ź
Banacha 22
P o l a n d

THE RATES OF CONVERGENCE IN THE CENTRAL LIMIT THEOREM IN BANACH SPACES

V. Paulauskas *)

In the paper we give short survey of recent results on estimates of the exactness of approximation in the central limit theorem in Banach spaces. Although at present there are results in the case of non-identically distibuted summands and in more general statement of a question with non-normal approximating law (see, for example [14], [1], [27]), for the sake of simplicity and clearness of presentation we confine ourselvs to the case of i.i.d. summands and Gaussian limit law. We give no proofs, since all results which we shall speak about are submitted or published.

In what follows B will stand for real separable Banach space with norm $\|\cdot\|$ and dual space B^*, H will denote real separable Hilbert space. Let ξ_i, $i \geq 1$ be i.i.d. B-valued random variables (B r.v.) with $E\xi_i = 0$, $E\|\xi_i\|^2 < \infty$, distribution F and covariance operator T. Let $F_n(A) = P\{\frac{1}{\sqrt{n}} \sum_{i=1}^{n} \xi_i \in A\}$ A-Borel set in B. If on B there exists mean zero Gaussian measure $\mu = \mu(0,T)$ with the same covariance operator T and if $F_n \Longrightarrow \mu$ ($\Longrightarrow$ denote weak convergence), we say that ξ_1 safisfies the central limit theorem (CLT). $F_n \Longrightarrow \mu$ means that (see [5]) $|F_n(A) - \mu(A)| \longrightarrow 0$ as $n \to \infty$ for $\forall$ sets A which are μ-continuity sets; or

$$\int_B f(x)(F_n(dx) - \mu(dx)) \longrightarrow 0 \quad \text{for every } f \in C^0(B)$$

where $C^0(B)$ denotes the class of bounded and continuous real-valued functions on B.

Starting from these two equivalent statements we are faced with two main directions in estimating the rate of convergence in CLT. Namely, one can consider the quantity

$$\Delta_n(\varepsilon) = \sup_{A \in \varepsilon} |F_n(A) - \mu(A)| \tag{1}$$

where ε is some subclass of μ-continuity class. The main examples of such classes, for which we should like to get estimate of (1) are the following classes:

---

*) To the 400 years anniversary of Vilnius University

$\varepsilon_1 = \{\{x \in B: \|x\| < r\}\ r > 0\}$, $\varepsilon_2(a) = \{\{x \in B: \|x-a\| < r\}, r>0\}$

$\varepsilon_3 = \{\varepsilon_2(a),\ a \in B\}$, $\varepsilon_4(f) = \{\{x \in B: f(x) < r\}, r>0\}$

where $f: B \longrightarrow R_1$.

In order to get estimate of $\Delta_n(\varepsilon)$, one must know that class ε is μ-uniformity class (for definition see [5]), and here we run into rather unpleasant fact that even in H for any mean zero Gaussian measure μ ε_3 is not μ-uniformity class [14], and this fact is due to infinite - dimensionality of H.

In the second direction one deals with the quantity:

$$\delta_n(\mathcal{F}) = \sup_{f \in \mathcal{F}} \left| \int_B f(x)\, (F_n-\mu)(dx) \right| \tag{2}$$

where $\mathcal{F}$ is some class of functions on B. For example one can put $\mathcal{F} = C^k(B)$(= class of k-times continuously differentiable functions) but here we are facing with another unpleasant fact, that in some important Banach spaces (for example $C[0,1]$, l_∞, C_0) the class of differentiable functions is rather pure (in the sense that there does not exist nontrivial differentiable function with bounded support) and the behaviour of differentiable functions is rather complicated (for details see [14]).

In the paper we shall deal only with the estimation of the quantity $\Delta_n(\varepsilon)$, for the second direction refering to the series of V.M.Zolotarev's papers [27], [28], [29] (some results of this kind are also in [15]; for finite-dimensional case see survey paper of P.L. Butzer and L.Hahn [6]).

In all paper we shall assume $E\|\xi_1\|^3 < \infty$ and discuss the following two problems, which to our opinion are the main at present in the first direction.

I. Let $B = H$. Does $E\|\xi_1\|^3 < \infty$ imply $\Delta_n(\varepsilon_1) = O(n^{-1/2})$?

(In other words, is the Berry-Esseen estimate valid in Hilbert space?) And if the answer is negative, then (Ia) how large β can we put in the implication $E\|\xi_1\|^3 < \infty \longrightarrow \Delta_n(\varepsilon) = O(n^{-\beta})$? Also one can ask (Ib) which stronger conditions on ξ_1 imply the rate of convergence $\Delta_n(\varepsilon_1) = O(n^{-1/2})$.

II. For which Banach spaces and which classes of sets ε we can obtain an estimate of $\Delta_n(\varepsilon)$?

Inspite the fact, that problem I is more than 20 years old, at present it is not solved and even the list of papers dealing with the rate of convergence in CLT in H rather short (see [10],

[26],[24],[25*], [11]. The paper of J.Kuelbs and T.Kurtz [11] gave strong impetus to the problem under consideration, and at present the best result in this direction is the implication
$E\|\xi_1\|^3 < \infty \Longrightarrow \Delta_n(\varepsilon_1) = O(n^{-1/6})$ [16], and it is a consequence of more general result which will be formulated bellow in theorem 4.

Now we shall turn to (Ib). The first result in this direction is due to S.V.Nagajev and V.J.Čebotarev [12],[13], obtained under rather strong assumption. Let $\xi_i = (\xi_{i1},\xi_{i2},\dots,\xi_{in},\dots)$, $i \geq 1$ be i.i.d. l_2r.v.with $E\xi_1 = 0$, $E\xi_{1i}^2 = \sigma_i^2$, $E|\xi_{1i}|^3 = \beta_i$.

<u>Theorem 1 [13].</u> Let ξ_{1j} be independent for all $j \geq 1$. Then there exists absolute constant C such that

$$\Delta_n(\varepsilon_1) = Cn^{-1/2}[(\prod_{j=1}^{4}\sigma_j)^{-3/4}\sum_{j=5}^{\infty}\beta_j + \sum_{j=1}^{4}\sigma_j^{-3}\beta_j] \qquad (3)$$

Later V.J.Čebotarev gave a generalization of (3) to spaces l_p, $2 \leq p < \infty$ [8].

Recently A.Račkauskas and V.V.Borovskich obtained the rate $O(n^{-1/2})$ under weaker conditions [4]. Let us denote
$\xi_1^N = (\xi_{1,N+1},\xi_{1,N+2},\dots)$ $\tilde{F}^N$ - distribution of ξ_1^N, $\beta^N = \int\|x\|^3\tilde{F}^N(dx)$.

<u>Theorem 2 [4].</u> If ξ_{1j}, $j = 1,2,\dots,N$, $N \geq 7$ are independent and independent from ξ_1^N, then there exists absolute constant C such that

$$\Delta_n(\varepsilon_1) \leq Cn^{-1/2}[\sum_{j=1}^{N}\frac{\beta_j}{\sigma_j^3} + (\prod_{j=1}^{N}\sigma_j^{-6})^{1/4}\max(1,(\sum_{j=N+1}^{\infty}\sigma_j^2)^{3/2})\,\beta^N] \qquad (4)$$

It is necessary to note, that estimate (4) is derived from more general result in Banach spaces of class $\mathcal{D}_3$ (for definition see below) and summands of the form $\xi_1 = \sum_{j=1}^{\infty}\zeta_j x_j$ where ζ_j are independent R_1 r.v. , $x_j \in B$ and series converges a.s. The estimate in [4] is expressed by means of pseudomoments.

Different kind of conditions, imposed on summands ξ_j were used in our recent paper [22]. The main goal of this paper was to adapt method of finite - dimensional approximation in CLT in l_p in general setting (in [24],[25]) this method was used in l_2 for specially constructed summands ξ_j, namely, for obtaining the rate

---

* Here and in what follows we do not mention the later papers dealing with the rate of convergence in ω^2-criterion,since they as a rule don't use the Hilbert space set-up and their results cannot be generalized to more general situation in H.

of convergence in ω^2-criterion). In this way we were able to obtain the rates of convergence, close to $O(n^{-1/2})$, and here we shall formulate some results, related to (Ib).

We shall use the notation, given before th 1, and let $T = \{t_{ij}\}$, $t_{ij} = E\,\xi_{1i}\,\xi_{1j}$. We assume that there exists a number $\gamma > 1$ such that

$$t_{ii} \sim i^{-\gamma} \tag{5}$$

(here $a_i \sim b_i$ means that for all i $C_1 b_i \le a_i \le C_2 b_i$, C_1, C_2 - some absolute constants). Let

$$\beta(\delta) = E\Big(\sum_{i=1}^{\infty} \xi_{1i}^2\, t_{ii}^{-\delta}\Big)^{3/2}, \quad \tilde\beta(\delta) = \sum_{i=1}^{\infty} E|\xi_{1i}|^3\, t_{ii}^{-\frac{3}{2}\delta}, \quad 0 \le \delta < 1$$

$$\beta_1(\varepsilon) = E\Big(\sum_{i=1}^{\infty} \xi_{1i}^2\, t_{ii}^{-(1-\frac{1}{\gamma}-\varepsilon)}\Big)^{3/2},$$

$$\tilde\beta_1(\varepsilon) = \sum_{i=1}^{\infty} E\,|\xi_{1i}|^3\, t_{ii}^{-3/2(1-\frac{1}{\gamma}-\varepsilon)} \qquad 0 < \varepsilon < 1 - \frac{1}{\gamma}$$

Theorem 3 [22] Let ξ_i, $i \ge 1$ be i.i.d. l_2 r.v. with $E\,\xi_1 = 0$ and (5).

If $\beta(\delta) < \infty$ then

$$\Delta_n(\varepsilon_1) = O(n^{-g(\delta,\gamma)}) \tag{6}$$

If $\tilde\beta(\delta) < \infty$, or $\beta_1(\varepsilon) < \infty$, or $\tilde\beta_1(\varepsilon) < \infty$, then respectively

$$\Delta_n(\varepsilon_1) = O(n^{-\tilde g(\delta,\gamma)}) \tag{7}$$

$$\Delta_n(\varepsilon_1) = O(n^{-g_1(\gamma,\varepsilon)}) \tag{8}$$

$$\Delta_n(\varepsilon_1) = O(n^{-\tilde g_1(\gamma,\varepsilon)}) \tag{9}$$

Where

$$g(\delta,\gamma) = \frac{1}{2}\,\frac{\gamma-1}{(\gamma-1)(4-3\delta)+5-3\delta},$$

$$\tilde g(\delta,\gamma) = \frac{1}{2}\,\frac{\gamma-1}{(\gamma-1)(4-3\delta)+3(2-\delta)},$$

$$g_1(\gamma,\varepsilon) = \frac{1}{2}\,\frac{\gamma-1}{(\gamma-1)(1+3/\gamma+3\varepsilon)+2+3(1/\gamma+\varepsilon)},$$

$$\tilde g_1(\gamma,\varepsilon) = \frac{1}{2}\,\frac{\gamma-1}{(\gamma-1)(1+3/\gamma+3\varepsilon)+3(1+1/\gamma+\varepsilon)}$$

From (6) - (9) it is easy to see that these estimates are good only for large values of patameter γ, more over, the diference between

(6) and (7) also (8) and (9) disappears when $\gamma \to \infty$. This is seen from the relations

$$g(\delta) = \lim_{\gamma\to\infty} g(\delta,\gamma) = \lim_{\gamma\to\infty} \tilde{g}(\delta,\gamma) = \frac{1}{2(4-3\delta)} ,$$

$$g(0) = \frac{1}{8}, \quad \lim_{\delta\to 1} g(\delta) = \frac{1}{2}$$

$$g_1(\varepsilon) = \lim_{\gamma\to\infty} g_1(\gamma,\varepsilon) = \lim_{\gamma\to\infty} \tilde{g}_1(\gamma,\varepsilon) = \frac{1}{2(1+3\varepsilon)}$$

$$\lim_{\varepsilon\to 0} g_1(\varepsilon) = \frac{1}{2}$$

Thus, if $\beta_1(\varepsilon)$ or $\tilde{\beta}_1(\varepsilon)$ are finite for sufficiently large γ then we get estimate

$$\Delta_n(\varepsilon_1) = O\,(n^{-1/2+\varepsilon_1})$$

where ε_1 depends on ε and γ. If instead of (5) we have exponential decrease ($t_{ii} \sim e^{-i\gamma}$, $\gamma > 0$) and $\beta_2(\varepsilon) =$
$= E(\sum_{i=1}^{\infty} \xi_{1i}^2 \; t_{ii}^{-1} \; |\ln t_{ii}|^{-\varepsilon})^{3/2}$, then

$$\Delta_n(\varepsilon_1) = O\,(n^{-1/2}(\ln n)^{\varkappa}) .$$

where $\varkappa$ depends on ε.

By means of the same method, in [22] there were obtained the estimates of quantity $\Delta_n(\varepsilon_4(p))$, $2 < p < \infty$, where $\varepsilon_4(p) =$

$= \{\{x \in l_2 : \|x\|_p < \tau\}\ \tau > 0\}$, $\|x\|_p = (\sum_{i=1}^{\infty} |x_i|^p)^{1/p}$. It is interesting to note that these sets are convex, but not bounded in l_2, so they are not included in the class of sets, which we shall describe below.

It can be mentioned that the existence of quantities $\tilde{\beta}_i(\varepsilon)$, $i = 1,2$ is not connected with the existence of moments of higher order than third. It is easy to give an example of l_2 r.v. ξ, for which $\tilde{\beta}_1(\varepsilon) < \infty$ but $E\,\|\xi\|^{3+\delta} = \infty$ for any $\delta > 0$.

Now we shall turn to the second problem. At present we know two classes of Banach spaces, for which it is possible to estimate the remainder term in CLT (at least $\Delta_n(\varepsilon_1)$).

The first class consists of spaces C(S), Banach spaces of continuous functions on metric compact S with supremum norm. The first result for these spaces belongs to E. Gine who for the estimating $\Delta_n(\varepsilon_1)$ in C(S) proposed the finite dimensional approximation with application of inequalities of metric entropy (see [9]). Later in the paper [18] (and these two papers till now remains only ones dealing with remainder term in CLT in C(s))the results of [9] were strenghtened and generalized - there were non-uniform estimates for non-identically distributed summands obtained, some new classes of processes, such as processes with sub-gaussian increments, were treated. We do not give here any results of this type, since they are rather complicated and for the statement we need many new notions.

The second class consists of Banach spaces with smooth norm, and in this case it is possible to obtain rather exact rates of convergence for more general sets than balls. We shall formulate some results, for more detail acquaintance refering to papers [15]-[17] [1]-[3], [23].

We say that Banach space B belongs to class $\mathcal{D}_k$, k - integer ≥ 1, if the map $\varphi : B\setminus\{0\} \to R_1$, $\varphi(x) = \|x\|$, is k times Freschet differentiable (for differentiation in normed spaces see, for example, [7]) and the following estimates for the derivatives hold

$$\|\mathcal{D}^{(i)}\varphi(x)\| \leq C(i,B)\ \|x\|^{1-i}, \quad i = 1,2,\ldots,k \quad k \neq 0$$

Here C(i,B) denotes constant depending on i and space B. It is known that l_p and $l_p \in \mathcal{D}_{[p]}$ and in [3] it is shown that if $B \in \mathcal{D}_k$ then $l_p(B) \in \mathcal{D}_k$ for $p \geq k$, where $l_p(B) = \{x = (x_1, x_2, \ldots,), x_i \in B, \sum_i^\infty \|x_i\|^p < \infty\}$ is separable Banach space with norm $\|x\|_{l_p(B)} = (\sum_i^\infty \|x_i\|_B^p)^{1/p}$. Also it is know that $\mathcal{D}_2$ is subclass of Banach spaces of type 2.

In Banach spaces of class $\mathcal{D}_3$ we are able to obtain the estimate of the remainder term for larger class of sets than balls. Now we shall give the definition of this class. Let A be Borel set in $B \in \mathcal{D}_3$. We introduce two following conditions:

(A1) the set A is conected, $0 \in A$ and every ray tx, $t > 0$, $\|x\| = 1$ intersects the boundary ∂A of A at one point.

(A2) the functional $d_A(x) = \sup\{t > 0 : \frac{tx}{\|x\|} \in A\}$, $x \neq 0$ is three times continuously differentiable for all $x \neq 0$ and

$$\|\mathcal{D}^{(i)} d_A(x)\| \leq M_i\ \|x\|^{-i}, \quad i = 1,2,3, \qquad x \neq 0$$

$$\inf_{\|x\|=1} d_A(x) = m_1 > 0, \qquad \sup_{\|x\|=1} d_A(x) = m_2 < \infty,$$

where M_i, $i = 1,2,3$ and m_1, m_2 are some positive constans. Let F be distribution on B with mean zero, strong second moment and covariance operator T; let $M = M(0,T)$ be Gaussian mean zero measure on B with the same covariana operator T. For formulation of our result we need more two conditions:

(A3) for all $x \neq 0$ $\int_B \mathcal{D}^{(2)} d_A(x)(y)^2\,(F-M)(dy) = 0$

(A4) for all $\varepsilon > 0$ there exist $\beta > 0$ such that

$$\sup_{\tau>0} M((\partial(A(\tau) + a))\varepsilon) \leq C_0(B,T)(1+\|a\|^{\beta})\varepsilon \,, \qquad (10)$$

and constant $C_0(B,T)$ has the property: for all $0 < \alpha \leq 1$

$$C_0(B,\alpha T) \leq \alpha^{-\gamma}\, C_0(B,T)$$

with some $0 < \gamma < \infty$. Here $A(\tau) = A \cdot \tau$, $(\partial A)_{\varepsilon} = A_{\varepsilon} \setminus A) \cup (A \setminus A_{-\varepsilon})$,

$$A_{\varepsilon} = (x: \|x - y\| < \varepsilon,\ y \in A)\ ,\ A_{-\varepsilon} = ((A^c)_{\varepsilon})^c,\quad A^c = B \setminus A$$

Let M_1, m_1, m_2, γ, C_0 fixed (C_0 fixed means that the dependence of C_0 on T is given) and ε_5 denote the class of sets A for which conditions (A1)-(A4) are satisfied. Denote

$$\varepsilon_6(a) = \{A(\tau) + a,\quad \tau > 0\ ,\quad A \in \varepsilon_5\ \}$$

<u>Theorem 4.</u> Let $B \in \mathcal{D}_3$ and let ξ_i, $i \geq 1$ be i.i.d. B r.v. with distribution F, $E\xi_1 = 0$, $E\|\xi_1\|^3 < \infty$ and covariance operator T. Then there exists constant C, depending on the class $\varepsilon_6(a)$ such that

$$\Delta_n(\varepsilon_6(a)) \leq C(1+\|a\|^{\beta})\, n^{-1/6}\ \vartheta_3^{\,1/3(1-\delta_n)}$$

where $\vartheta_3 = \int_B \|x\|^3 |F-M|(dx)$, $|F-M|$ (A) denotes the variation of $F - M$ on a set $A, \delta_n \sim \frac{1}{n^2}$.

In special case when ε_5 consists of unit ball in B this theorem was proved in [16] and result in this form (only with the restriction $0 \leq \beta \leq 3$) - in [20]. This restriction was removed by V.Bernotas [1], who considered non-identically distributed summands. Later non-uniform estimate for non-identically distributed summands was obtained in [3]. The class of sets ε_5 was introduced in [17] (in the case of H) and in [15] (in Banach spaces of class $\mathcal{D}_3$).

When considering sets, described as above, one rather natural way is to put $A \equiv A_f = \{x \in B:\ f(x) < 1\}$ where $f: B \to R_1$. Then the conditions (A1) and (A2) will be satisfied if one puts some restrictions on f. Using this approach results can be reformulated as the rates of convergence of distribution of functionals of sums of independent B r.v. to distribution of functional of Gaussian B r.v. Results of this type can be found in papers [17] and [15], the later one contains some examples of such functionals f, constructed by means of polynomial operators and operators of Hammerstein and Nemyckii.

Some remarks should be made on condition(A4). At present estimate (10) is proved only for balls in the following cases: in l_2 for any Gaussian measure and arbitraty $a \in l_2$ with $\beta = 1$ and this value of β is optimal (see [11], [14], [19]; in l_p, $1 \leq p < \infty$ for some Gaussian measures, a - arbitraty, $\beta = p - 1$ is optimal [16],[19]; in C(S) with a = 0 for mean zero Gaussian processes $\eta(t)$, $t \in S$ such that $E\eta^2(t) \geq \sigma > 0$ for all $t \in S$[18]; in l_2 for some stable laws with a = 0 [21] [23].

References

[1] V.Bernotas "On closeness of distributions of two sums of independent random variables with values in some Banach spaces". Liet.matem.rink. 18, 4 (1978), p. 5-12 (in Russian) English translation in "Transaction of Lithuanian Math.Soc."

[2] V.Bernotas "An estimate of closeness of two distributions of normed sums of random variables with values in some Banach spaces". Liet.matem.rink. 19, 3 (1979), p. 194-196 (In Russian)

[3] V.Bernotas, V.Paulauskaus "Non-uniform estimate in the Central Limit Theorem in some Banach spaces". Liet.matem.rink. 19, 2 (1979) p. 25-45 (in Russian)

[4] Yu.V.Borovskih, A.Račkauskas "Asymptotics of distributions in the Banach spaces". Liet.matem.rink. 19, 4 (1979), (in Russian)

[5] R.N.Bhattacharya, R.Rao "Normal approximation and asymptotic expansions, John Willey and Sons, N.Y. 1976

[6] P.L.Butzer, L.Hahn "General Theorems on Ratse of Convergence of Distribution of Random variables I, II. Jour. of Multivariate Anal. 8, 2 (1978) 181-220.

[7] A.Cartan, Differential calculus, Moscov 1971 (Russian translation)

[8] V.J.Čebotarev "On the estimates of the rate of convergence in the central limit theorem in l_p "Dokl.Akad.Nauk SSSR, 247, 2 (1979), 301-303 (in Russian)

[9] E.Gine "Bounds for the speed of convergence in the central limit theorem in C(S). Z.Wahrscheinlichkeits. Verw.Geb. 36 (1976) 317-331

[10] N.P.Kandelaki "On a limit theorem in Hilbert space, Trans.of Computer Center of AN.GSSR; 5, 1 (1965)46-55 (in Russian)

[11] J.Kuelbs, T.Kurtz "Berry - Eseen estimates in Hilbert space and on application to the law of the iterated logarithm", Ann.Probab. 2, 3(1974) 387-407

[12] S.V.Nagajev, V.J.Čebotarev, "Estimates of the rate of convergence in the central limit theorem in l_2 in the case of independent coordinates, Abstracts of communications, second Vilnius conf. on probab. and math.stat., 1977, 68-69 (in Russian)

[13] S.V.Nagajev, V.J.Čebotarev "On the rate of convergence in the central limit theorem for l_2-valued random variables", in book "Mathematical analysis and related problems of mathematics" Novosibirsk, "Nayka", 1978, p.153-182

[14] V.Paulauskas "On the closeness of distributions of sums of independent random variables with values in Hilbert space". Liet.matem.rink., 15, 3 (1975) 177-200, (in Russian)

[15] V.Paulauskas "On convergence of some functionals of sums of independence random variables in a Banach space", Liet. matem.rink. 16, 3 (1976) 103-121 (in Russian)

[16] V.Paulauskas "On the rate of convergence in the central limit theorem in some Banach spaces", Teor.verojat.i primen. 21, 4 (1976) 775-791 (in Russian)

[17] V.Paulauskas "Non-uniform estimate in the central limit theorem in a separable Hilbert space", Proc.of the Third Japan - USSR Symp. on Probab.,Lecture Notes in Math. 550 (1976), 475-499.

[18] V.Paulauskas, "The estimate of the rate of convergence in the central limit theorem in C(S)". Liet.matem.rink., 16, 4 (1976) 168-201 (in Russian)

[19] V.Paulauskas, letter to editors, Teor.verojat.i primen. 23, 2 (1978) p. 477 (in Russian)

[20] V.Paulauskas "Limit theorems for sums of independent random variables in Banach spaces", Doct.thesis, Vilnius, 1978 (in Russian)

[21] V.Paulauskas "The rates of convergence to stable laws and the law of the iterated logarithm in Hilbert space, Univ. of Goteborg, Depart. of Math. Nr.5 (1977)

[22] V.Paulauskas "The estimate of the rate of convergence in the central limit theorem in spaces l_p", submitted to Liet. matem.rink.

[23] V.Paulauskas, A.Rackauskas "Infinitely divirsible and stable laws in separable Banach spaces II", to appear in Liet. matem.rink.

[24] V.V.Sazonov, On ω^2 criterion, Sankhya, ser. A, 30, 2 (1968) 205-210.

[25] V.V.Sazonov "An improvement of one estimate of rate of convergence" Teor.verojat.i primen. 14, 4 (1969) 667-678 (in Russian).

[26] N.N. Vakhanija, N.P. Kandelaki, " On estimate of the rate of convergence in the central limit theorem in Hilbert space" Transactions of Computer Center of AN. GSSR, 9, 1 (1969) 150-160.

[27] V.M. Zolotarev " Ideal metrics in the problem of approximation of distributions of sums of independent random variables ", Teor.verojat.i primen. 22, 3 (1979) 449-465.

[28] V.M. Zolotarev, Metric distances in the spaces of random variables and their distributions, Matem.sbornik, 101 (143), 3, (1976) 104-141.

[29] V.M. Zolotarev, Approximation of the distribution of sums of independent random variables with values in infinite-dimensional spaces, Teor.verojat.i primen. 21, 4 (1976) 741-757.

Vilnius V. Kapsukas University
Department of Mathematics
USSR, Vilnius 232042

THE GENERALIZED ANSCOMBE CONDITION AND ITS APPLICATIONS IN RANDOM LIMIT THEOREMS

E.Rychlik and Z.Rychlik

1. Introduction. Let (S,d) be a complete separable metric linear space equipped with its Borel σ-field $\mathcal{B}$. Let $\{Y_n, n \geq 1\}$ be a sequence of S-valued random elements defined on a probability space $(\Omega, \mathcal{A}, P)$. Suppose that $\{N_t, t > 0\}$ is a positive integer-valued stochastic process defined on the same probability space $(\Omega, \mathcal{A}, P)$.

Following the classical work of Anscombe [3], many authors (see, e.g. [7],[11],[4],[12],[1]) have investigated the limit behaviour of the distribution of S_{N_t} as $t \to \infty$. The obtained results have the following form. Suppose $Y_n \Rightarrow \mu$, converges weakly on S, to a measure μ. If $N_t/a_t \xrightarrow{P} \lambda$, in probability, where here and in what follows λ is a positive random variable and a_t, $t > 0$, are constants going to infinity as $t \to \infty$, then under some additional assumptions $Y_{N_t} \Rightarrow \mu$. The condition $\lambda = 1$ was first discussed by Anscombe [3], who also introduced the following "uniform continuity" condition on $\{Y_n, n \geq 1\}$. For each $\varepsilon > 0$ there exists $\delta > 0$ such that

(1) $$\limsup_{n \to \infty} P\left[\max_{i \in D_n(\delta)} d(Y_i, Y_n) \geq \varepsilon\right] \leq \varepsilon,$$

where $D_n(\delta) = \{i : |i - n| \leq \delta n\}$.

Condition (1), in particular, known as "Anscombe's condition", has played a very important role in the proofs of the results in the so-called limit theorems with random indices.
Aldous [1] has pointed out that condition (1) is also axactly the right one when

(2) $$N_t/a_t \xrightarrow{P} 1 \quad \text{as } t \to \infty .$$

Let $\{X_k, k \geq 1\}$ be a sequence of independent S-valued random elements. Let us put $Y_n = S_n/k_n$, $n \geq 1$, where $S_n = X_1 + \ldots + X_n$ and $\{k_n, n \geq 1\}$ is a sequence of positive numbers. Then one can easily find conditions under which $S_n/k_n \Rightarrow \mu$. Thus in order to use Anscombe's [3] or Aldous'[1] results one must prove that $\{S_n/k_n, n \geq 1\}$ satisfies (1). But it is easy to see that in general $\{S_n/k_n, n \geq 1\}$ satisfies (1) in the case when $\{X_k, k \geq 1\}$ is stationary in the wide sense. Recently Csörgö and Rychlik [9] introduced much more useful version of "Anscombe's condition" as well as another assumption on $\{N_t, t > 0\}$. They have also given some applications of the introduced assumptions in the study of the limit behaviour of sequences of random elements with random indices.

In this paper we give applications of another version of Anscombe's condition, a more general one than that considered in [9]. The key concept employed is the so-called the generalized Anscombe condition introduced in [10].
The obtained theorems summarize and extend the results given in [1],[3],[4],[7],[9-11] and [12].

Throughout the paper we assume that the metric space (S,d) is such that $d(x+z,y+z) = d(x,y)$ for all $x,y,z \in S$.

2. Results. Let $\{k_n, n \geq 1\}$ and $\{w_n, n \geq 1\}$ be sequences of positive numbers. Suppose that $\{k_n, n \geq 1\}$ is nondecreasing.
Let $\{Y_n, n \geq 1\}$ be a sequence of S-valued random elements.

Definition 1. A sequence $\{Y_n, n \geq 1\}$ is said to satisfy the generalized Anscombe condition with norming sequences $\{k_n, n \geq 1\}$ and $\{w_n, n \geq 1\}$ if for every $\varepsilon > 0$ there exists $\delta > 0$ such that

$$(3) \quad \limsup_{n \to \infty} P\,[\max_{i \in D_n(\delta)} d(Y_i/w_n, Y_n/w_n) \geq \varepsilon\,] \leq \varepsilon ,$$

where here and in the sequel $D_n(\delta) = \{i: |k_i^2 - k_n^2| \leq \delta k_n^2\}$.

One can notice that in the special case $w_n = 1$ and $k_n^2 = n$, $n \geq 1$, the condition (3) reduces to (1).

Let $\{A_n, n \geq 1\}$ be a sequence of elements from S.

Theorem 1. The following conditions are equivalent:

(i) $\{Y_n, n \geq 1\}$ satisfies the generalized Anscombe condition (3) with norming sequences $\{k_n, n \geq 1\}$ and $\{w_n, n \geq 1\}$ and

$$(Y_n - A_n)/w_n \Longrightarrow \mu \quad \text{as } n \longrightarrow \infty ;$$

(ii) $(Y_{N_t} - A_{a_t})/w_{a_t} \Longrightarrow \mu$ as $t \longrightarrow \infty$ for every positive integer-valued stochastic process $\{N_t, t > 0\}$ such that

$$(4) \quad k_{N_t}/k_{a_t} \xrightarrow{P} 1 \quad \text{as } t \longrightarrow \infty ,$$

where $\{a_t, t > 0\}$ is a family of positive integers,

$$a_t \longrightarrow \infty \quad \text{as } t \longrightarrow \infty .$$

Proof. Write $D_{a_t}(\delta) = \{i : | k_i^2 - k_{a_t}^2 | \leq \delta\, k_{a_t}^2 \}$.
Let $\varepsilon > 0$ and a closed set $F \subset S$ be given. Then, for every $\delta > 0$, we get

$$(5) \quad P[(Y_{N_t} - A_{a_t})/w_{a_t} \varepsilon F] \leq P[\max_{i \varepsilon D_{a_t}(\delta)} d(Y_i/w_{a_t}, Y_{a_t}/w_{a_t}) \geq \varepsilon]$$
$$+ P[|k_{N_t}^2 - k_{a_t}^2| \geq \delta\, k_{a_t}^2] + P[(Y_{a_t} - A_{a_t})/w_{a_t} \varepsilon F^{\varepsilon}],$$

where $F^{\varepsilon} = \{ x \varepsilon S : \rho(x,F) \leq \varepsilon \}$ and $\rho(x,F) = \inf \{d(x,y): y \varepsilon F\}$. Thus, by (i), (4) and (5), we get

$$\limsup_{t \to \infty} P[(Y_{N_t} - A_{a_t})/w_{a_t} \varepsilon F] \leq \varepsilon + \mu(F) .$$

Since $\varepsilon > 0$ can be chosen arbitrarily small the last inequality and Theorem 1.1 [5, p.4] prove that (i) implies (ii).

For the converse, it is clear that by (ii) $(Y_n - A_n)/w_n \Longrightarrow \mu$ as $n \longrightarrow \infty$, so suppose that (3) fails. Thus there exist an $\varepsilon > 0$ and a family $\{a_t, t > 0\}$ of positive integers such that

$$(6) \quad P[\max_i d(Y_i/w_{a_t}, Y_{a_t}/w_{a_t}) \geq \varepsilon] > \varepsilon \quad \text{for every } t > 0,$$

where the maximum is taken over all i such that $k_{a_t}^2 \leq k_i^2 \leq (1 + 1/t)k_{a_t}^2$. Furthermore, by Lemma 2.2 [5, p.7], there exist B_k, $1 \leq k \leq M$, pairwise disjoint and open subsets of S such that the diameter of B_k is less than $\varepsilon/2$, $\mu(\bigcup_{i=1}^{M} B_i) > 1 - \varepsilon/2$ and every B_k is a μ-continuity set. So there must exist a set $B \varepsilon \{B_k, 1 \leq k \leq M\}$ and a subfamily $\{a_t', t > 0\}$ of $\{a_t, t > 0\}$ such that

$$(7) \quad P[(Y_{a_t'} - A_{a_t'})/w_{a_t'} \varepsilon B; \max_i d(Y_i/w_{a_t'}, Y_{a_t'}/w_{a_t'}) \geq \varepsilon] \geq \varepsilon/2M,$$

where maximum is taken over all i such that $k_{a_t'}^2 \leq k_i^2 \leq (1 + 1/t)\, k_{a_t'}^2$. Let $b_t^1 = \max \{i: k_{i_1}^2 \leq (1+1/t)k_{a_t'}^2\}$, $b_t^2 = \min \{ i \geq a_t' : (Y_i - A_{a_t'})/w_{a_t'} \notin B \}$, $N_t = \min(b_t^1, b_t^2)$. Then $k_{N_t}/k_{a_t} \xrightarrow{P} 1$ and

$$P[(Y_{N_t} - A_{a_t'})/w_{a_t'} \notin B] \geq P[(Y_{a_t'} - A_{a_t'})/w_{a_t'} \notin B] + \varepsilon/2M ,$$

ans so $(Y_{N_t} - A_{a_t'})/w_{a_t'} \not\Longrightarrow \mu$.

Let us observe that the random variables $\{N_t, t > 0\}$ constructed in the proof of Theorem 1 are stopping time random variables for $\{Y_n, n \geq 1\}$, and so Theorem 1 would be unchanged if we assumed $\{N_t, t > 0\}$ to be stopping time random variables for $\{Y_n, n \geq 1\}$.

Definition 2 [10]. A positive integer-valued stochastic processes $\{N_t, t > 0\}$ is said to satisfy the condition (Δ) with norming sequences $\{k_n, n \geq 1\}$ and $\{w_n, n \geq 1\}$ if for every $\varepsilon > 0$ and $\delta > 0$ there exist a finite and measurable partition $\{B_1, B_2, \ldots, B_m\}$ of Ω and a family $a_t(j)$, $1 \leq j \leq m$, $t > 0$, of positive integers such that $a_t(j) \longrightarrow \infty$ as $t \longrightarrow \infty$

$$(8) \quad \limsup_{t \to \infty} \sum_{j=1}^{m} P_{B_j} \left(\left| k^2_{N_t} - k^2_{a_t(j)} \right| > \delta\, k^2_{a_t(j)} \right) \leq \varepsilon$$

and

$$(9) \quad \limsup_{t \to \infty} \sum_{j=1}^{m} P_{B_j} \left(\left| w_{N_t} - w_{a_t(j)} \right| > \delta\, w_{a_t(j)} \right) \leq \varepsilon ,$$

where $P_A(B) = P(A \cap B)$.

Assume that $(S, \|\cdot\|)$ is a separable and normad space and let $d(x,y) = \|x - y\|$. Then we have the following extension of Theorem 3 [10].

Theorem 2. Let $\{Y_n, n \geq 1\}$ be a sequence of S-valued random elements such that $(Y_n - \theta)/w_n \Longrightarrow \mu$ (stably), [2], where θ is an element from S. If for every $\varepsilon > 0$ there exists $\delta > 0$ such that for each $B \in \mathcal{A}$ with $P(B) > 0$

$$(10) \quad \limsup_{t \to \infty} P_B \left[\max_{i \in D_n(\delta)} d(Y_i, Y_n) \geq \varepsilon\, w_n \right] \leq \varepsilon\, P(B),$$

then $(Y_{N_t} - \theta)/w_{N_t} \Longrightarrow \mu$ (stably) for every $\{N_t, t > 0\}$ satisfying the condition (Δ).

Proof. Given $\varepsilon > 0$ and a closed F choose $\delta > 0$ as in (10). Then there exist a measurable partition $\{B_1, B_2, \ldots, B_m\}$ of Ω and pósitive integers $a_t(j)$, $j \leq m$, $t > 0$, such that (8) and (9) hold true. So by (8), (9) and (10) we obtain

$$\limsup_{t \to \infty} P\left[(Y_{N_t} - \theta)/w_{N_t} \in F \right] \leq 3\varepsilon \; +$$

$$\limsup_{t \to \infty} \sum_{j=1}^{m} P_{B_j} \left[(Y_{N_t} - \theta)/w_{N_t} \in F;\ \left| k^2_{N_t} - k^2_{a_t(j)} \right| \leq \right.$$

$$\leq \delta\, k^2_{a_t(j)};\ |w_{N_t} - w_{a_t(j)}| \leq \delta\, w_{a_t(j)};\ \sup_{i \in D_{a_t(j)}(\delta)} d(Y_i, Y_{a_t(j)})$$

$$< w_{a_t(j)}\ ] \leq 3\varepsilon + \limsup_{t \to \infty} \sum_{j=1}^{m} P_{B_j}[(Y_{a_t(j)} - \theta)/w_{a_t(j)}$$

$$\in F^{\varepsilon_1} + \limsup_{t \to \infty} \sum_{j=1}^{m} P_{B_j}\ [\|Y_{a_t(j)} - \theta\| \geq K w_{a_t(j)}\ ],$$

where $\varepsilon_1 = (\varepsilon + \delta K)/(1 - \delta)$ and K is a positive number. Choosing an appropriately large K and then a sufficiently small $\delta > 0$ we get

$$\limsup_{t \to \infty} P[(Y_{N_t} - \theta)/w_{N_t} \in A] \leq 8\varepsilon + \mu(F) .$$

Thus Theorem 1.1 [5,p.4] and Remark 3 of Aldous [1] give the assertion.

We remark that if $(Y_n - \theta)/w_n \Longrightarrow \mu$ (mixing), then in the assertion of Theorem 2 we get $(Y_{N_t} - \theta)/w_{N_t} \Longrightarrow \mu$ (mixing) .

From Theorem 2 one can easily obtain following extension of results given in [4],[7],[11] and [12].

Theorem 3. Let $\{N_t, t > 0\}$ be a positive integer-valued stochastic process such that

$$(11) \qquad k_{N_t}/k_{a_t} \xrightarrow{P} \lambda \quad \text{as} \quad t \to \infty ,$$

where λ is a positive random variable. Assume that

$$(12) \qquad k_{n-1}/k_n \to 1, \quad k_n \to \infty \quad \text{as } n \to \infty ,$$

and for every $\varepsilon > 0$ there exists $\delta > 0$ such that for each $A \in \sigma(\lambda)$

$$(13) \qquad \limsup_{n \to \infty} P_A\ [\max_{i \in D_n(\delta)} d(Y_i, Y_n) \geq \varepsilon\, k_n] \leq \varepsilon\ P(A),$$

where $\sigma(\lambda)$ is the σ-field generated by λ . If for every $A \in \sigma(\lambda)$ with $P(A) > 0$ there exists a measure μ_A such that $(Y_n - \theta)/k_n \Longrightarrow \mu_A$ on the probability space $(\Omega, \mathcal{A}, P(\cdot\,|A))$, then $(Y_{N_t} - \theta)/k_{N_t} \Longrightarrow \mu_\Omega$.

Proof. One can easily check that (11) and (12) imply (8) and (9) with $B_j \in \sigma(\lambda)$, $1 \leq j \leq m$. So Theorem 3 is a consequence of Theorem 2.

3. Concluding remarks. Let $\{S_n, F_n, n \geq 0\}$ be a martingale on a probability space $(\Omega, \mathcal{A}, P)$ with $S_o = 0$. Define $X_n = S_n - S_{n-1}$, $n \geq 1$, $b_j^2 = E(X_j^2 | F_{j-1})$, $j \geq 1$, $V_n^2 = \sum_{j=1}^{n} b_j^2$, $B_n^{2'} = EV_n^2 = ES_n^2$, $n \geq 1$. Let us observe that, by Kolmogorov's inequality

$$P[\max_{i \in D_n(\delta)} d(S_i, S_n) \geq \varepsilon B_n] \leq 2\delta/\varepsilon^2,$$

where $D_n(\delta) = \{i : | B_i^2 - B_n^2| \leq \delta B_n^2\}$. Thus $\{S_n, n \geq 1\}$ satisfies (3) with $k_n = w_n = B_n$, $n \geq 1$. Let Y_n be the random element of $C[0,1]$ defined by interpolating between the points $(0,0)$, $(B_1^2/B_n^2, S_1/B_n), \ldots, (1, S_n/B_n)$. If, for example $V_n^2/B_n^2 \xrightarrow{P} 1$ and $\{X_k, k \geq 1\}$ satisfies Lindeberg's condition, then $Y_n \Longrightarrow W$(mixing), where W is the Wiener measure on $C[0,1]$. So by Theorem 2 $Y_{N_t} \Longrightarrow W$ (mixing] for every $\{N_t, t > 0\}$ satisfying (8) with $k_n = B_n$, $n \geq 1$, and this generalize Theorem 17.2 [6] and the main result of Babu and Ghosh [4].

One can easily check that functional random limit theorems given in [8] can also be obtained, and even extended, by our Theorems 1,2 and 3.

To give a better illustration of the meaning of Theorem 1 let us note that from a very special case of it we immediately obtain the following.

<u>Corollary</u>. Let $\{X_k, k \geq 1\}$ be a sequence of independent random variables with zero means and finite variances. Let $S_n = X_1 + \ldots + X_n$, $B_n^2 = D^2 S_n$, $n \geq 1$. Then the following conditions are equivalent

(a) $S_n/B_n \Longrightarrow \mu$;

(b) $S_{N_t}/B_{N_t} \Longrightarrow \mu$ (mixing) for every $\{N_t, t > 0\}$ satisfying (4) with $k_n = B_n$, $n \geq 1$.

Let us observe that the sequence $\{S_n, n \geq 1\}$, given in Corollary, satisfies (10) as well as (13) with $k_n^2 = w_n = B_n^2$, $n \geq 1$.

Theorem 1, especialy, might be also of some use in statistical applications where we want to use sequential estimation of an unknown parameter with given required accuracy. Some remarks concerning this problem can be found in [3] and [10].

References

[1] Aldous, D.J. Weak convergence of randomly indexed sequences of random variables. Math.Proc.Camb.Phil.Soc. 83(1978),117-126

[2] Aldous, D.J. and Eagleson, G.K. On mixing and stability of limit theorems. Ann.Probability 6(1978), 325-331.

[3] Anscombe, F.J. Large-sample theory of sequential estimation. Proc.Cambridge Philos.Soc. 48 (1952), 600-607.

[4] Babu, G.J. and Ghosh, M. A random functional central limit theorems for margingales. Acta Math.Acad.Sci.Hung. 27(1976), 301-306.

[5] Bhattacharya, R.N. and Ranga Rao R. Normal Approximation and Asymptotic Expansions. John Wiley 1976.

[6] Billingsley, P. Convergence of probability measures. New York: Wiley 1968.

[7] Blum, J.R., Hanson, D.I. and Rosenblatt, J.I. On the central limit theorem for the sum of a random number of independent random variables. Z. Wahrscheinlichkeitstheorie verw. Gebiete 1(1963), 389-393.

[8] Byczkowski, T. and Inglot, T. The invariance principle for vector-valued random variables with applications to functional random limit theorems. (to appear)

[9] Csörgő, M. and Rychlik, Z. Weak convergence of sequences of random elements with random indices. Math.Proc.Camb. Phil. Soc.(submitted)

[10] Csörgő, M. and Rychlik, Z. Asymptotic properties of randomly indexed sequences of random variables. Carleton Mathematical Lecture Note No. 23, July 1979.

[11] Guiasu, S. On the asymptotic distribution of sequences of random variables with random indices. Ann.Math.Statist.42 (1971), 2018-2028.

[12] Prakasa Rao, B.L.S. Limit theorems for random number of random elements on complete separable metric spaces. Acta Math.Acad.Sci. Hung. 24(1973), 1-4 .

Instytut Matematyki
Uniwersytet Warszawski
Pałac Kultury i Nauki
00-901 Warszawa Poland

Instytut Matematyki UMCS
20-031 Lublin
Nowotki 10
Poland

ON MOVING AVERAGE REPRESENTATIONS OF BANACH-SPACE VALUED STATIONARY PROCESSES OVER LCA-GROUPS

F. Schmidt

Let $\underline{F}$ be a (complex) Banach space, and let $\underline{F}^*$ be the space of all bounded linear functionals on $\underline{F}$. Further, let $\underline{H}$ and $\underline{K}$ be (complex) Hilbert spaces, and let $\underline{C}$ be the set of all complex numbers. For Banach spaces $\underline{F}_0$, $\underline{F}_1$, $\underline{F}_2$ we denote by $BL(\underline{F}_0,\underline{F}_0)$ the space of all bounded linear operators in $\underline{F}_0$ and by $BL(\underline{F}_1,\underline{F}_2)$ the space of all bounded linear operators from $\underline{F}_1$ into $\underline{F}_2$.

1. A representation theorem for weak densities

Let G be a LCA-group which satisfies the T_0- separation axiom, and let λ be the Haar measure on G.

1.1. The quadruple $(\underline{F},G,\lambda,w)$ is called a weak density if $\underline{F}$, G, λ are as above and

$$w:\ \underline{F}\times\underline{F}\ni(f,f')\longrightarrow w(\cdot,f,f')\in\underline{L}_1(G,\lambda)$$

is a mapping with the properties

$$\left.\begin{array}{ll}\text{(i)} & w(\cdot,f,f')\geq 0\\ \text{(ii)} & \int_G w(x,f,f)\,\lambda(dx)\leq C\|f\|_{\underline{F}}^2<\infty\end{array}\right\}(f\in\underline{F})$$

$$\text{(iii)}\quad w(\cdot,\alpha_1f_1+\alpha_2f_2,f')=\alpha_1\,w(\cdot,f_1,f')+\alpha_2\,w(\cdot,f_2,f')\quad(\alpha_i\in\underline{C},\ f_i,f'\in\underline{F}\ (i=1,2))$$

$$\text{(iv)}\quad \overline{w(\cdot,f',f)}=w(\cdot,f,f')\quad(f,f'\in\underline{F}).$$

The quadruple $(\underline{F},G,\lambda,W)$ is called a positive operator function if $\underline{F}$, G, λ are as above and

$$W:\ G\longrightarrow BL(\underline{F},\underline{F}^*)$$

is a function with the properties

$$\text{(i)}\ \langle f,W(x)f\rangle\ \geq 0\quad(x\in G,\ f\in\underline{F})$$

$$\text{(ii)}\ \langle f,W(\cdot)f'\rangle\in\underline{L}_1(G,\lambda)\quad(f,f'\in\underline{F}).$$

Each positive operator function $(\underline{F},G,\lambda,W)$ defines a weak density $(\underline{F},G,\lambda,w)$ by

(1) $w(\cdot,f,f') := \langle f, W(\cdot)f' \rangle \quad (f,f' \in \underline{F})$.

However, there exist weak densities which are not representable by positive operator functions ([8], 1.5., Beispiele 1 und 2).

1.2. Obviously, for each $A \in BL(\underline{F},\underline{L}_2(\underline{K},G,\lambda))$ by

(2) $w(\cdot,f,f') := ((Af)(\cdot),(Af')(\cdot))_{\underline{K}} \quad (f,f' \in \underline{F})$

a weak density $(\underline{F},G,\lambda,w)$ is defined. Conversely, we can prove the following

Theorem 1. Let $(\underline{F},G,\lambda,w)$ be a weak density. Then there exist a (complex) Hilbert space $\underline{K}$ and an operator $A \in BL(\underline{F},\underline{L}_2(\underline{K},G,\lambda))$ such that (2) holds.

Proof. cf. [2], another proof was given in [8], Satz 3.1., a third proof for the case of the circle group in [7], Theorem 2.

Corollary 1. Let $\underline{F}$ be separable, and let $(\underline{F},G,\lambda,W)$ be a positive operator function. Then there exist a (complex) Hilbert space $\underline{K}$ and a (strongly) measurable function $Q: G \longrightarrow BL(\underline{F},\underline{K})$ such that

(3) $W(\cdot) = Q(\cdot)^* Q(\cdot)$

holds.

Proof. cf. [8], Folgerung 3.2.

2. Banach-space valued stationary processes over LCA-groups

Let Γ be an abelian group.

2.1. The quadruple $(\Gamma,\underline{F},\underline{H},X)$ is called a (Banach-space valued) stationary process if Γ, $\underline{F}$, $\underline{H}$ are as above and

(4) $X: \Gamma \longrightarrow BL(\underline{F},\underline{H})$

is a function with the property

(5) $X(\gamma)^*X(\gamma') = X(0)^*X(\gamma'-\gamma) \quad (\gamma, \gamma' \in \Gamma)$.

For a stationary process $(\Gamma,\underline{F},\underline{H},X)$ let

$$\underline{H}(X,\Delta) := \bigvee_{\gamma \in \Delta} (X(\gamma)\underline{F}) \quad (\Delta \subseteq \Gamma), \quad \underline{H}(X) := \underline{H}(X,\Gamma).$$

If Γ is a topological group, then the stationary process $(\Gamma,\underline{F},\underline{H},X)$ is said to be continuous if the mapping (4) is continuous w.r.t. the strong operator topology.

2.2. If Γ is the character group of the group G (cf. 1.) and $(\Gamma,\underline{F},\underline{H},X)$ is a continuous stationary process we have the spectral representation

(6) $$(X(\gamma)f, X(\gamma')f')_{\underline{H}} = \int_G (\gamma'-\gamma)(x)\ \mu_X[f,f'](dx)$$
$$(\gamma,\gamma' \in \Gamma,\ f,f' \in \underline{F})$$

where $\mu_X[f,f']$ $(f,f' \in \underline{F})$ are σ-additive regular complex-valued measures on G uniquely determined by X ([8] (2.1.)). If the measures $\mu_X[f,f'[$ $(f,f' \in \underline{F})$ are λ-absolutely continuous we denote by $w_X(\cdot,f,f')$ the Radon-Nikodym derivative of $\mu_X[f,f']$ w.r.t. λ. Obviously, $(\underline{F},G,\lambda,w_X)$ is a weak density. Conversely, for each weak density $(\underline{F},G,\lambda,w)$ there exists a continuous stationary process $(\Gamma,\underline{F},\underline{H},X)$ for which the measure $\mu_X[f,f']$ $(f,f' \in \underline{F})$ are λ-absolutely continuous and the corresponding Radon-Nikodym derivatives $w_X(\cdot,f,f')$ are equal to $w(\cdot,f,f')$ $(f,f' \in \underline{F})$ ([8],2.3.)

3. Moving average representations

Let Γ be the character group of the group G (cf.1.), and let $\varkappa$ be the Haar measure on Γ, normalized as in [1] §31.

3.1. We denote by $F(\underline{K}) \in BL(\underline{L}_2(\underline{K},G,\lambda),\underline{L}_2(\underline{K},\Gamma,\varkappa))$ the (isometric) operator of the Fourier transformation,

$$(F(\underline{K})\underline{v})(\gamma) := \int_G \overline{\gamma(x)}\ \underline{v}(x)\ \lambda\ (dx)$$
$$(\gamma \in \Gamma, \underline{v} \in \underline{L}_2(\underline{K},G,\lambda) \cap \underline{L}_1(\underline{K},G,\lambda))$$

and by $S(\underline{K},\gamma) \in BL(\underline{L}_2(\underline{K},\Gamma,\varkappa))$ and $U(\underline{K},\gamma) \in BL(\underline{L}_2(\underline{K},G,\lambda))$ $(\gamma \in \Gamma)$ the operators defined by

$$S(\underline{K},\gamma)\underline{w}(\cdot) := \underline{w}(\gamma+\cdot) \quad (\gamma \in \Gamma,\ \underline{w} \in \underline{L}_2(\underline{K},\Gamma,\varkappa))$$

and

$$U(\underline{K},\gamma)\underline{v}(\cdot) := \overline{\gamma(\cdot)}\underline{v}(\cdot) \ (\gamma \in \Gamma, \underline{v} \in L_2(\underline{K},G,\lambda))$$

respectively. Then we have the equality

(7) $$S(\underline{K},\gamma)\ \ F(\underline{K}) = F(\underline{K})\ \ U(\underline{K},\gamma) \quad (\gamma \in \Gamma).$$

3.2. The quintuple $(\Gamma,\underline{\underline{J}}_\varkappa,\underline{K},\underline{H},Y)$ is called a quasi-isometric measure ([3], Def. 8.2.) if $\Gamma,\varkappa,\underline{K},\underline{H}$ are as above, $\underline{\underline{J}}_\varkappa$ is the δ-ring of the $\varkappa$-integrable subsets of Γ, and

(8) $$Y:\ \underline{\underline{J}}_\varkappa \longrightarrow BL(\underline{K},\underline{H})$$

is a mapping with the property

(9) $$Y(\Delta)^*Y(\Delta') = \varkappa(\Delta \cap \Delta')\ \ I_{\underline{K}} \quad (\Delta,\Delta' \in \underline{\underline{J}}_\varkappa)$$

($I_{\underline{K}}$: identity operator in $\underline{K}$). For a quasi-isometric measure $(\Gamma,\underline{J}_{\varkappa},\underline{K},\underline{H},Y)$ let

$$\underline{H}(Y,\Delta) := \bigvee_{\substack{\Delta' \in \underline{J}_{\varkappa} \\ \Delta' \subsetneq \Delta}} (Y(\Delta')\underline{K}) \quad (\Delta \subseteq \Gamma),\ \underline{H}(Y) := \underline{H}(Y,\Gamma).$$

3.3. Let $(\Gamma,\underline{J}_{\varkappa},\underline{K},\underline{H},Y)$ be a quasi-isometric measure and let $B \in BL(\underline{F},\underline{L}_2(K,\Gamma,\varkappa))$. Then the "moving averages"

$$(10) \quad X(\gamma)f := \int_{\Gamma} Y(d\xi)(Bf)(\gamma-\xi) \quad (\gamma \in \Gamma,\ f \in \underline{F})$$

define a continuous stationary process $(\Gamma,\underline{F},\underline{H},X)$ with $\underline{H}(X) \subseteq \underline{H}(Y)$ ([4], Satz 2.3.1. and Hilfssatz 2.3.).

Theorem 2. The continuous stationary process $(\Gamma,\underline{F},\underline{H},X)$ has a representation in form of moving averages iff all measures $\mu_X[f,f']$ $(f,f' \in \underline{F})$ are λ-absolutely continuous.

Proof. "only if" (For details cf.[4], Folgerung 2.3.1.):
Let X be as in (10), and let $Af := F(\underline{K})^{-1}Bf$ $(f \in \underline{F})$. Then we have by (7)

$$\begin{aligned}(X(\gamma)f,X(\gamma')f')_{\underline{H}} &= \Big(\int_{\Gamma} Y(d\xi)(Bf)(\gamma-\xi), \int_{\Gamma} Y(d\xi')(Bf')(\gamma'-\xi')\Big)_{\underline{H}} = \\ &= \int_{\Gamma}((Bf)(\gamma-\xi),(Bf')(\gamma'-\xi))_{\underline{K}}\,\varkappa(d\xi) = \\ &= \int_{G}(\gamma'-\gamma)(x)((Af)(x),(Af')(x))_{\underline{K}}\,\lambda(dx) \\ &\qquad (\gamma,\gamma' \in \Gamma,\quad f,f' \in \underline{F}).\end{aligned}$$

It follows (cf.(6)) that all measures $\mu_X[f,f']$ $(f,f' \in \underline{F})$ are λ-absolutely continuous and that

$$(11) \qquad w_X(\cdot,f,f') = ((Af)(\cdot),(Af')(\cdot))_{\underline{K}} \quad (f,f' \in \underline{F})$$

holds.

"if" (For details cf.[8], Satz 4.2.):
Let all measures $\mu_X[f,f']$ $(f,f' \in \underline{F})$ be λ-absolutely continuous. Then by Theorem 1 there exist a (complex) Hilbert space $\underline{K}$ and an operator $A \in BL(\underline{F},\underline{L}_2(K,G,\lambda))$ such that (11) holds. Let $(\Gamma,\underline{J}_{\varkappa},\underline{K},\underline{H}',Y)$ be a quasi-isometric measure. We set $Bf := F(\underline{K})\,Af$ $(f \in \underline{F})$ and define the stationary process $(\Gamma,\underline{F},\underline{H}'X')$ by

$$X'(\gamma)f := \int_{\Gamma} Y'(d\xi)(Bf)(\gamma-\xi) \quad (\gamma \in \Gamma,\ f \in \underline{F})$$

Then we have

$$(X'(\gamma)f, X'(\gamma')f')_{\underline{H}'} = (X(\gamma)f, X(\gamma')f')_{\underline{H}} \qquad (\gamma,\gamma' \in \Gamma,\ f,f' \in \underline{F}).$$

Hence there exists an isometric operator $V' \in BL(\underline{H}X'), \underline{H}(X))$ such that $V'X'(\gamma) = X(\gamma)$ $(\gamma \in \Gamma)$. Let $V \in BL(\underline{H}(Y'), \underline{H})$ be an isometric extension of V'($\underline{H}$ denotes some Hilbert space containing $\underline{H}$), and let $Y(\Delta) := VY'(\Delta)$ $(\Delta \in \underline{J}_{\varkappa})$. Then we have

$$X(\gamma)f = VX'(\gamma)f = V\int_{\Gamma} Y'(d\xi)(Bf)(\gamma-\xi) = \int_{\Gamma} Y(d\xi)(Bf)(\gamma-\xi) \quad (\gamma \in \Gamma,\ f \in \underline{F}).$$

From the proof of Theorem 2 we get the

Corollary 2. The continuous stationary process $(\Gamma, \underline{F}, \underline{H}, X)$ has a representation in form of moving averages with a given Hilbert space $\underline{K}$ and a given operator $B \in BL(\underline{F}, \underline{L}_2(K,\Gamma,\varkappa))$ iff all measures $\mu_X[f,f']$ $(f,f' \in \underline{F})$ are λ-absolutely continuous and the operator $A := F(\underline{K})^{-1}B$ satisfies the equality (11).

4. Regularity and Singularity

Let Γ be a partially ordered abelian group (order relation compatible with the addition). Then $\Pi := \{\gamma \in \Gamma \mid \gamma \geq 0\}$ is a subsemigroup of Γ with

(12) $$\Pi \cap (-\Pi) = \{0\}$$

(Conversely, each subsemigroup Π which satisfies (12) induces by $(\gamma \geq \gamma') :\Longleftrightarrow \gamma - \gamma' \in \Pi$ a partial order on Π, compatible with the addition.)

4.1. For the stationary process $(\Gamma, \underline{F}, \underline{H}, X)$ we define the subspaces

$$\underline{H}_-(X) := \bigcap_{\gamma \in \Gamma} \underline{H}(X, -\Pi+\gamma)$$

$$\underline{H}_+(X) := \bigvee_{\gamma \in \Gamma} (\underline{H}(X) \ominus \underline{H}(X, -\Pi+\gamma)).$$

Then we have

$$\underline{H}(X) = \underline{H}_+(X) + \underline{H}_-(X).$$

The stationary process $(\Gamma, \underline{F}, \underline{H}, X)$ is said be regular, if $\underline{H}(X) = \underline{H}_+(X)$, and singular, if $\underline{H}(X) = \underline{H}_-(X)$. Each stationary process $(\Gamma, \underline{F}, \underline{H}, X)$ admits a Wold decomposition into its regular part $(\Gamma, \underline{F}, \underline{H}, X)$ and its singular part $(\Gamma, \underline{F}, \underline{H}, X_-)$ defined by $X_{\pm}(\gamma) = P_{\pm}X(\gamma)$ $(\gamma \in \Gamma)$ where P_+ and P_- are the orthogonal projection operators onto $\underline{H}_+(X)$

and $\underline{H}_-(X)$, respectively. (For details cf. [9], Theorem 5.5.).

4.2. Let Γ be the character group of the group G (cf. 1.) and moreover, let Γ be partially ordered. By 1_Δ we denote the indicator function of the set $\Delta \subseteq \Gamma$. Let

$$\underline{L}_2^+(\underline{K},\Gamma,\varkappa) := \{\underline{w} \in \underline{L}_2(\underline{K},\Gamma,\varkappa) \mid \underline{w}(\cdot) = \underline{w}(\cdot)\, 1_\Pi(\cdot)\}$$

$$\underline{L}_2^+(\underline{K},G,\lambda) := F(\underline{K}^{-1}\, \underline{L}_2^+(\underline{K},\Gamma,\varkappa) .$$

The weak density $(\underline{F},G,\lambda,w)$ is called "factorable" if the Hilbert space $\underline{K}$ and the operator A in (2) can be chosen such that $A\underline{F} \subseteq L_2^+(\underline{K},G,\lambda)$. If $\underline{F}$ is separable and $(\underline{F},G,\lambda,w)$ has the form (1) with a positive operator function $(\underline{F},G,\lambda,W)$, then $(\underline{F},G,\lambda,w)$ is factorable iff there exist a Hilbert space $\underline{K}$ and a (strongly) measurable function $Q: G \longrightarrow BL(\underline{F},\underline{K})$ such that (3) and $Q(\cdot)f \in L_2^+(\underline{K},G,\lambda)$ $(f \in \underline{F})$ hold.

The moving average representation (10) of the stationary process $(\Gamma,\underline{F},\underline{H},X)$ is called "one-sided" if $B\underline{F} \subseteq L_2^+(\underline{K},\Gamma,\varkappa)$.

Theorem 3. The continuous stationary process $(\Gamma,\underline{F},\underline{H},X)$ has a representation in form of one-sided moving averages iff all measures $\mu_X[f,f']$ $(f,f' \in \underline{F})$ are λ-absolutely continuous and $(\underline{F},G,\lambda,w_X)$ is factorable.

Proof. Follows immediately from Corollary 2.

Theorem 4. If the continuous stationary process $(\Gamma,\underline{F},\underline{H},X)$ $(\Gamma \neq \{0\})$ has a representation in form of one-sided moving averages then it is regular.

Proof. Let X be as in (10), with $B\underline{F} \subseteq L_2^+(\underline{K},\Gamma,\varkappa)$. There exists an isometric operator $V(X,B) \in BL(\underline{H}(X),L_2(K,\Gamma,\varkappa))$ such that

$$V(X,B)\; X(\gamma)f = S(\underline{K},\gamma)Bf \quad (\gamma \in \Gamma\;,\;\; f \in \underline{F}).$$

Obviously,

$$V(X,B)\;\; \underline{H}(X,-\Pi+\gamma) \subseteq S(\underline{K},\gamma)L_2^+(\underline{K},\Gamma,\varkappa) \quad (\gamma \in \Gamma).$$

We have

$$V(X,B)\underline{H}_-(X) = \bigcap_{\gamma \in \Gamma} V(X,B)\; \underline{H}(K,\, -\Pi+\gamma) \subseteq \bigcap_{\gamma \in \Gamma} S(\underline{K},\gamma)L_2^+(\underline{K},\Gamma,\varkappa) =$$

$$= \bigcap_{\gamma \in \Gamma} \{\underline{w} \in \underline{L}_2(\underline{K},\Gamma,\varkappa) \mid \underline{w}(\cdot)=w(\cdot)\; 1_{\Pi-\gamma}(\cdot)\} = \{0\}$$

since $\bigcap_{\gamma \in \Gamma} (\Pi-\gamma) = \emptyset$. It follows that $\underline{H}(X) = \{0\}$, $\underline{H}_+(X) = \underline{H}(X)$ i.e., $(\Gamma,\underline{F},\underline{H},X)$ is regular.

We remark that Theorem 4 was proved in [4], Satz 3.1.1. under the additional condition " Γ is archimedean ordered".

4.3. In the cases $\Gamma = Z$ (group of the integers) and $\Gamma = R$ (group of the reells) the condition in Theorem 4 is also necessary for the regularity of $(\Gamma,\underline{F},\underline{H},X)$ (cf. [5], Folgerung 3.3.1. and [6], Satz 4.3.1.). However, in general the condition in Theorem 4 is not necessary for the regularity. We give an example of a regular continuous stationary process $(\Gamma,\underline{F},\underline{H},X)$ which cannot have a representation in form of moving averages.

Example. Let Γ be the group $Z \times Z$. Then Γ is the character group of the group $G = T \times T$ (T: circle group). In the usual manner, we identify T with the group $[-\pi,+\pi)$ (addition modulo 2π). Then the Haar measure λ on G is the Lebesque measure on $[-\pi,+\pi)\times[-\pi,+\pi)$ (normalized by $\lambda([-\pi,+\pi)\times[-\pi,+\pi)) = 1$ if the Haar measure $\varkappa$ on Γ is normalized by $\varkappa((0,0)) = 1$). Let $(Z,\underline{F},\underline{H},X')$ be a stationary process. Then we define tha stationary process $(Z \times Z,\underline{F},\underline{H},X)$ by

$$(13) \qquad X(m,n) := X'(m) \qquad (m,n \in Z).$$

We hawe $\mu_X[f,f'] = \mu_{X'}[f,f'] \times \varepsilon_0$ $(f,f' \in \underline{F})$ where ε_0 is the measure defined by $\varepsilon_0(D) = 1 (0 \in D)$ and $= 0$ $(0 \notin D)(D = [-\pi,+\pi))$. This shows that the measures $\mu_X[f,f']$ $(f,f'\ \underline{F})$ are supported by the set $[-\pi,+\pi)\times\{o\}$ which has λ-measure zero. Hence they are λ-absolutely continuous only if they are zero, i.e. only if $\underline{H}(X) = \{o\}$. By Theorem 2 it follows that the stationary process $(Z \times Z,\underline{F},\underline{H},X)$ defined by (13) has a representation in form of moving averages only if $\underline{H}(X') = \underline{H}(X) = \{o\}$. Let now Γ be endowed with the lexicographic order, and let $(Z,\underline{F},\underline{H},X')$ be a nontrivial $(\underline{H}(X') \neq \{o\})$ regular stationary process. Then $(Z \times Z,\underline{F},\underline{H},X)$ is nontrivial and regular, too. Howeover, $(Z \times Z,\underline{F},\underline{H},X)$ is not representable by moving averages.

References

[1] E.Hewitt, K.A. Ross , Abstract harmonic analysis I/II, Berlin - Göttingen-Heidelberg 1963/ Berlin-Heidelberg -New York 1970 .

[2] A.Makagon, F. Schmidt, A decomposition theorem for densities of positive operator-valued measures. Bull.Acad.Polon.Sci., Ser. Sci.Math. Astronom.Phys (to appear).

[3] P.Masani, Quasi-isometric measures and their applications, Bull.Amer.Math.Soc. 76, 3 (1970) 427-528.

[4] F.Schmidt, Über die Darstellung einer Klasse von stationaren stochastischen Prozessen mit Hilfe von verallgemeinerten zufälligen Masen, Math. Nachr. 56 (1973) 21-41.

[5] -------- Verallgemeinerte stationäre stochastische Prozesse auf Gruppen der Form $Z \times G^{-}$, Math. Nachr. 57 (1973) 337-357.

[6] -------- Verallgemeinerte stationare stochastische Prozesse auf Gruppen der Form $R \times G^{-}$, Math.Nachr. 68 (1975) 29-48.

[7] -------- Banach-space valued stationary processes with absolutely continuous spectral function, Probability Theory on Vector Spaces, Lecture Notes in Math. 656 (1978) 237-244.

[8] -------- Positive operatorwertige Mase und banachraumwertige stationare Prozesse auf LCA-Gruppen, Studia Mathematica (to appear).

[9] A.Weron, Prediction Theory in Banach Spaces, Probability - Winter School, Lecture Notes in Math. 472 (1975) 207-228.

Sektion Mathematik
Technische Universität
DDR - 8027 D r e s d e n
Mommsenstrasse 13

DILATIONS OF REPRODUCING KERNELS

R. Shonkwiler

§1 Introduction.

Recently there has been some activity toward extending the reproducing kernel theory so elegantly reported by Aronszajn [2] for Hilbert spaces to the more general setting of Banach spaces and beyond; see especially Weron and Gorniak [6]. However it is the intent of this paper to relate an approach to reproducing kernel theory put forth by the writer, Shonkwiler [5], which is more general than Aronszajn's but still within Hilbert space theory with the setting established in the papers of Masani [3] and [4]. As the latter works are extensive, we only begin this task here. However, it is observed that when restricted to Hilbert spaces, our kernels are directly comparable to those of Masani's. In addition, we extend here our Kernel Dilation Theorem of [5] to a new case - that of logarithmic kernels. Also we show that any positive definite commuting kernel can be dilated. It is observed that the utility of the dilation concept lies in the possibility of dilating a Hilbertian variety so that its spread (see §3) is cartesian (see §2).

Let H be a Hilbert space over the complex numbers $\mathbb{C}$, $H_0 \subset H$ be a dense linear manifold of H, and $L(H_0)$ denote the linear (possibly unbounded) operators in H whose domains contain H_0. Let R be any set having an idempotent unary operation *, $\rho^{**} = \rho$ for $\rho \in R$, and at least one *-fixed element ε, $\varepsilon^* = \varepsilon$.

Remark. Any set R may be endowed with such an operation *, namely set $\rho^* = \rho$ for each $\rho \in R$.

Definitions. By a Moore space with reproducing kernel K we mean a Hilbert space $\mathcal{H}$ having a dense linear manifold $\mathcal{H}_0 \subset \mathcal{H}$ of functions $\phi : R \to H$ along with a function $K : R \times R \to L(H_0)$ such that for $\rho \in R$ and $x \in H_0$, $K(\cdot,\rho)X \in \mathcal{H}_0$ and for $\phi \in \mathcal{H}_0$ the reproducing property holds,

$$(\phi(\cdot), K(\cdot,\rho)x)_{\mathcal{H}} = \langle\phi(\rho), x\rangle_H .$$

If $\mathcal{H}_0 = \mathcal{H}$, then we say $\mathcal{H}$ is a reproducing kernel Hilbert space or an Aronszajn space.

Remark. Often we take $H = \mathbb{C}$ in which case $K(\rho,\sigma)$ is just a complex number.

<u>Example</u> 1. Let μ be a non-negative bounded measure on $[a,b] \subset \mathbb{R}$, $H = \mathbb{C}$, $R = \{0,1,2,\cdots\}$, $\rho^* = \rho$, $\varepsilon = 0$ and

$$K(\rho,\sigma) = \int_a^b t^{\rho+\sigma} d\mu(t) = \int_a^b t^{\rho}\overline{t^{\sigma}} d\mu(t) .$$

If $H_0 = \text{span}\{K(\cdot,\rho)\alpha : \rho\in R, \alpha\in\mathbb{C}\}$ and $H = \overline{H_0}$ with inner-product given by

$$(\sum_i K(\cdot,\rho_i)\alpha_i, \sum_j K(\cdot,\sigma_j)\beta_j) = \sum_{i,j} \alpha_i \overline{\beta_j} K(\sigma_j,\rho_i) ,$$

then H is a Moore space (in fact an Aronszajn space).

<u>Theorem</u>. An Aronszajn space H is minimal in the sense that

$$H = \text{closure}(\text{span}\{K(\cdot,\rho)X : \rho\in R, x\in H_0\}) .$$

<u>Proof</u>. By the reproducing property, if ϕ is orthogonal to the second member of the above equation, then for all $x \in H_0$,

$$0 = (\phi(\cdot), K(\cdot,\rho)X) = \langle\phi(\rho), X\rangle ,$$

so that $\phi(\rho) = 0$ for all ρ.

<u>Theorem</u>. Evaluation is weakly continuous in a Moore space, i.e. for ϕ, $\psi \in H_0$, $\phi(\tau) \to \psi(\tau)$ in H as $\phi(\cdot) \to \psi(\cdot)$ in H.

<u>Proof</u>. By the reproducing property, for all $x \in H_0$,

$$\langle\phi(\tau) - \psi(\tau), x\rangle = (\phi(\cdot) - \psi(\cdot), K(\cdot,\tau)x) \to 0$$

as $\phi(\cdot) \to \psi(\cdot)$ in H.

<u>Corollary</u>. If $K(\rho,\rho) \in B(H)$ for $\rho \in R$, then evaluation is strongly continuous.

<u>Proof</u>.

$$||\phi(\tau) - \psi(\tau)|| \le \sup_{||x||=1} |\langle\rho(\tau) - \psi(\tau), x\rangle|$$

$$\le \sup_{||x||=1} ||\rho(\cdot) - \psi(\cdot)||\cdot||K(\cdot,\tau)x|| .$$

But

$$||K(\cdot,\tau)x||^2 = (K(\cdot,\tau)x, K(\cdot,\tau)x) = \langle K(\tau,\tau)x, x\rangle$$

$$\le ||K(\tau,\tau)||\cdot||x||^2$$

by hypothesis.

Definition. A positive definite kernel $K : R \times R \to L(H_0)$ is one satisfying

$$\sum_{i=1}^{n} \sum_{j=1}^{n} \langle K(\rho_i,\rho_j)x_j,x_i\rangle \geq 0$$

for all $n = 1,2,\cdots$, $\{\rho_1,\cdots,\rho_n\} \subset R$ and $\{x_1,\cdots,x_n\} \subset H_0$.

Theorem. If K is a reproducing kernel, then K is positive definite.

Proof. Given n, $\rho_1,\cdots,\rho_n$, and $x_1,\cdots,x_n$ put $\phi(\cdot) = \sum_{i=1}^{n} K(\cdot,\rho_i)x_i \in \mathcal{H}_0$. Then

$$0 \leq (\phi(\cdot),\phi(\cdot)) = \sum_{i,j} \langle K(\rho_i,\rho_j)x_j,x_i\rangle .$$

Theorem. If the kernel K is positive definite, then K conjugates, i.e. dom $K^*(\rho,\sigma) \supset H_0$ and on H_0 $K^*(\rho,\sigma) = K(\sigma,\rho)$.

Proof. For all x and ρ, $\langle K(\rho,\rho)x,x\rangle$ is real. Since for x,y and i $y \in H_0$,

$$0 \leq \langle K(\rho,\rho)x,x\rangle + \langle K(\rho,\sigma)y,x\rangle + \langle K(\sigma,\rho)x,y\rangle + \langle K(\sigma,\sigma)y,y\rangle$$

and

$$0 \leq \langle K(\rho,\rho)x,x\rangle + i\langle K(\rho,\sigma)y,x\rangle - i\langle K(\sigma,\rho)x,y\rangle + \langle K(\sigma,\sigma)y,y\rangle ,$$

it follows that $\langle K(\rho,\sigma)y,x\rangle + \langle K(\sigma,\rho)x,y\rangle$ and $i\langle K(\rho,\sigma)y,x) - i\langle K(\rho,\sigma)x,y\rangle$ are both real. This implies $\langle K(\sigma,\rho)x,y\rangle = \overline{\langle K(\rho,\sigma)y,x\rangle} = \langle x,K(\rho,\sigma)y\rangle$.

Remark. The argument in the following theorem is essentailly due to Moore.

Theorem. If the kernel K is positive definite, then there exists a Moore space $\mathcal{H}$ whose kernel is K. In addition, if for $\rho \in R$, $K(\rho,\rho)$ is bounded, then $\mathcal{H}$ is a reproducing kernel Hilbert space.

Proof. As is well-known the collection of functions $\phi : R \to H$ defined by $\phi(\cdot) = \sum_{i=1}^{n} K(\cdot,\rho_i)x_i$ for $n = 1,2,\cdots$, $\rho_1,\cdots,\rho_n \in R$ and $x_1,\cdots,x_n \in H_0$ can be made into an inner-product space $\mathcal{H}_0$ with inner-product

$$(\sum_{i=1}^{n} K(\cdot,\rho_i)x_i, \sum_{j=1}^{m} K(\cdot,\sigma_j)y_j) = \sum_{i}\sum_{j}\langle K(\sigma_j,\rho_i)x_i,y_j\rangle .$$

This proves the first conclusion. If in addition $K(\rho,\rho)$ is bounded, then letting $\{\psi_n\}^{\infty}$ be a fundamental sequence in $\mathcal{H}_0$ converging to the ideal element $[\{\psi_n\}]$, we have

$$G_0(y) = \lim_n \langle y,\psi_n(\rho)\rangle = \lim_n(K(\cdot,\rho)y,\psi_n(\cdot)) = (K(\cdot,\rho)y,[\{\psi_n\}])$$

for $y \in H_0$ and $\rho \in R$. But as above the fourth number of this chain of equalities is bounded by

$$||K(\cdot,\rho)y||\cdot||[\{\psi_n\}]|| \le ||K(\rho,\rho)||^{\frac{1}{2}}||y|| \quad ||[\{\psi_n\}]|| \quad .$$

Thus G_0 is a bounded linear functional on H_0 and has a unique extension G to all of H by continuity. By the Riesz theorem there is a vector $\psi(\rho)$ in H so that

$$<y,\psi(\rho)> = G(y) = \lim_n <y,\psi_n(\rho)>_n$$

for all $y \in H_0$ and $\rho \in R$. It follows that $[\{\psi_n\}]$ may be identified with $\psi(\cdot)$ and

$$(\psi(\cdot),K(\cdot,\rho)y) = <\psi(\rho),y> \ .$$

§2 Special Kernels.

Definition. Let (Ω,A,μ) be a bounded non-negative B(H) valued measure space and let $X(\rho,\cdot) : \Omega \to \mathbb{C}$ be square summable for each $\rho \in R$. The kernel given by

$$K(\rho,\sigma) = \int_\Omega X(\rho,\omega)\overline{X(\sigma,\omega)}d\mu(\omega)$$

is called a covariance kernel.

Example. (Stochastic process). Let (Ω,A,P) be a probability space, $H = \mathbb{C}$, and $X : R \to L_2(\Omega,A,P)$ be a stochastic process. Then the complex valued kernel

$$K(\rho,\sigma) = \int_\Omega \overline{X(\rho)}X(\sigma)dP$$

is a covariance kernel.

Definition. Let E_t, $-\infty < t < \infty$, be a resolution of the identity. (Also called a spectral family or an orthogonally scattered measure, E_t is an orthogonal projection for each t and one has $\lim_{t\to-\infty} E_t = 0$, $\lim_{t\to\infty} E_t = I$, and $E_tE_s = E_{\min(s,t)}$.) Let $f : R \times \mathbb{R} \to \mathbb{C}$ be square summable as a function of $\mathbb{R}$ and put

$$K(\rho,\sigma) = \int_{-\infty}^{\infty} f(\rho,t)\overline{f(\sigma,t)}dE_t \ .$$

This covariance kernel K is called a spectral kernel.

Theorem. If the covariance kernel K is either complex valued or spectral, then K is positive definite.

Proof. Assume K is complex valued; the proof for a spectral kernel is done in a theorem below. Given n, $\alpha_1,\cdots,\alpha_n \in \mathbb{C}$ and $\rho_1,\cdots,\rho_n \in R$,

$$\sum_{i=1}^{n}\sum_{j=1}^{n} K(\rho_i,\rho_j)\alpha_j\overline{\alpha_i} = \sum_{i,j}\alpha_j\overline{\alpha_i}\int_\Omega X(\rho_i,\omega)\overline{X(\rho_j,\omega)}d\mu(\omega)$$
$$= \int_\Omega \left|\sum_{j=1}^{n}\alpha_j X(\rho_j,\omega)\right|^2 d\mu(\omega) \geq 0 .$$

Definition. An arbitrary function $T : R \to L(H_0)$ is called a Hilbertian variety in $L(H_0)$. If there exists a Hilbertian variety T in $L(H_0)$ such that for $x,y \in H_0$,

$$\langle K(\rho,\sigma)x,y\rangle = \langle T(\sigma)x,T(\rho)y\rangle \quad ,$$

then K is a cartesian kernel with generator T.

Theorem. If the cartesian kernel K is generated by T, then $T(\sigma)x \in \mathrm{dom}T^*(\rho)$ for $\sigma \in R$ and $x \in H_0$ and

$$K(\rho,\sigma) = T^*(\rho)T(\sigma)$$

on H_0.

Proof. For $x \in H_0$ put $y^* = K(\rho,\sigma)x$. Then $T(\sigma)x \in \mathrm{dom}T^*(\rho)$ since for all $y \in H_0$,

$$\langle T(\rho)y,T(\sigma)x\rangle = \langle y,y^*\rangle \ .$$

Hence $T^*(\rho)T(\sigma)x = y^* = K(\rho,\sigma)x$.

Remarks. From this result, it is seen that our cartesian kernel is known as a covariance kernel in Masani [3]. Trivially a cartesian kernel is positive definite. Also the converse is true by Masani's Kernel Theorem, loc. cit. pp. 241.

Theorem. Every cartesian kernel is positive definite and conversely every B(H) valued positive definite kernel is a cartesian kernel.

Theorem. Every spectral kernel is a cartesian kernel.

Proof. Put $T(\rho) = \int\overline{f(\rho,t)}dE_t$. Then

$$\langle T(\sigma)x,T(\rho)y\rangle = \int_{-\infty}^{\infty} f(\rho,t)d_t \int_{-\infty}^{\infty}\overline{f(\sigma,s)}d_s\langle E_s x,E_t y\rangle$$
$$= \int_{-\infty}^{\infty} f(\rho,t)d_t \int_{-\infty}^{t}\overline{f(\sigma,s)}d_s\langle E_s x,y\rangle$$

$$= \int_{-\infty}^{\infty} f(\rho,t)\overline{f(\sigma,t)}<d_t E_t x,y> = <K(\rho,\sigma)x,y> .$$

Theorem. Every complex valued positive definite kernel is a covariance kernel.

Proof. Let H be the Moore space of K and let $\{e_\alpha(\cdot)\}$ be an orthonormal basis of H. By the Fourier Series Theorem

$$K(\cdot,\sigma) = \sum_\alpha \overline{(e_\alpha(\cdot),K(\cdot,\sigma))} e_\alpha(\cdot) = \sum_\alpha \overline{e_\alpha(\sigma)} e_\alpha(\cdot)$$

convergent in H. Since evaluation is continuous here, for $\rho \in R$, we also have

$$K(\rho,\sigma) = \sum_\alpha e_\alpha(\rho)\overline{e_\alpha(\sigma)} .$$

But the latter may be realized as an integral

$$\int_\Omega e(\alpha,\rho)\overline{e(\alpha,\sigma)}d\mu(\alpha)$$

over some measure space (Ω,A,μ).

Definition. If for some *-fixed element ε a kernel satisfies

$$x \in H_0 \text{ implies } K(\varepsilon,\sigma)x \in \operatorname{dom} K(\rho,\varepsilon)$$

and

$$K(\rho,\varepsilon)K(\varepsilon,\sigma)x = K(\rho,\sigma)x ,$$

for all $\rho,\sigma \in R$, then K splits.

Remark. In Example 1 one sees that not every positive definite kernel splits. But, as seen next, splitting, conjugating kernels are positive definite.

Theorem. If the kernel K splits and conjugates for some *-fixed element ε, then it conjugates for all $\rho,\sigma \in R$. Moreover, in this case K is a cartesian kernel with generator

$$T(\sigma) = K(\varepsilon,\sigma) .$$

Proof. For $x,y \in H_0$, and T as above

$$\langle K(\rho,\sigma)x,y\rangle = \langle K(\rho,\varepsilon)K(\varepsilon,\sigma)x,y\rangle = \langle K(\varepsilon,\sigma)x,K(\varepsilon,\rho)y\rangle$$
$$= \langle T(\sigma)x,T(\rho)y\rangle$$
$$= \langle x,K(\sigma,\rho)y\rangle$$

by symmetry.

Theorem. If K splits and conjugates and $K(\rho,\rho)$ is bounded for each $\rho \in R$, then $K(\rho,\sigma)$ is bounded for all $\rho,\sigma \in R$.

Proof. Since $K(\rho,\sigma) = K(\rho,\varepsilon)K(\varepsilon,\sigma) = K^*(\varepsilon,\rho)K(\varepsilon,\sigma)$ on H_0, it suffices to show that $K(\varepsilon,\rho)$ is bounded for each $\rho \in R$. But for $x \in H_0$,

$$||K(\varepsilon,\rho)x||^2 = \langle K(\varepsilon,\rho)x,K(\varepsilon,\rho)x\rangle = \langle K(\rho,\rho)x,x\rangle \leq M_\rho^2||x||^2$$

for some M_ρ depending only on ρ.

§3 Dilations.

Definitions. A kernel $\tilde{K} : R \times R \to L(\tilde{H}_0)$ is a dilation of the kernel $K : R \times R \to L(H_0)$, and we write $K = \mathrm{pr}\tilde{K}$, if $\tilde{H}_0 \supset H_0$ and for $x \in H_0$, $\rho \in R$,

$$P\tilde{K}(\varepsilon,\rho)x = K(\varepsilon,\rho)x$$

where P is the orthogonal projection P: closure $(\tilde{H}_0) \to$ closure(H_0).

Theorem. If $K = \mathrm{pr}\tilde{K}$, then $K(\varepsilon,\rho)$ and $\tilde{K}(\varepsilon,\rho)$ have the same weak values on H_0. Further, if $\tilde{K}(\varepsilon,\rho)$ is bounded, so is $K(\varepsilon,\rho)$.

Proof. For $x \in H_0$ and $y \in H$

$$\langle K(\varepsilon,\rho)x,y\rangle = \langle P\tilde{K}(\varepsilon,\rho)x,y\rangle = \langle\tilde{K}(\varepsilon,\rho)x,y\rangle .$$

And

$$||K(\varepsilon,\rho)x|| = ||P\tilde{K}(\varepsilon,\rho)x|| \leq ||P||\ ||\tilde{K}(\varepsilon,\rho)||\ ||x|| .$$

Theorem. Let $K : R \times R \to L(H_0)$ be positive definite, commute with itself on H_0, and satisfy $K(\varepsilon,\varepsilon) = I$. Then K can be dilated to a splitting, conjugating kernel $\tilde{K}$.

Proof. Let H be the Moore space of K and on H_0 define K formally by

$$\tilde{K}(\rho,\sigma)\sum_{i=1}^{n} K(\cdot,\tau_i)x_i = \sum_{i=1}^{n} K(\cdot,\tau_i)K(\rho,\varepsilon)K(\varepsilon,\sigma)x_i .$$

By direct computation one shows that formally $\tilde{K}(\rho,\sigma) \supset \tilde{K}(\sigma,\rho)$ and $\tilde{K}(\rho,\sigma) = \tilde{K}(\rho,\varepsilon)\tilde{K}(\varepsilon,\sigma)$. Then to prove $\tilde{K}$ is well-defined, let $\sum_i K(\cdot,\tau_i)x_i$ be a representation of zero. For $\psi(\cdot) \in \mathcal{H}_0$,

$$(\tilde{K}(\rho,\sigma)\sum_i K(\cdot,\tau_i)x_i, \psi(\cdot)) = (\sum_i K(\cdot,\tau_i)x_i, \tilde{K}(\sigma,\rho)\psi(\cdot)) = 0 \; .$$

Therefore $\tilde{K}(\rho,\sigma)\sum_i K(\cdot,\tau_i)x_i = 0$.

Embed H_0 in $\mathcal{H}_0$ by identifying x with $K(\cdot,\varepsilon)x$. Then the orthogonal projection P of $\bar{\mathcal{H}}_0$ onto $\bar{H}_0$ satisfies

$$P\psi(\cdot) = \psi(\varepsilon) \quad , \quad \psi(\cdot) \in \mathcal{H}_0 \; .$$

Finally, by direct computation, one sees that

$$P\tilde{K}(\varepsilon,\rho) = K(\varepsilon,\rho) \quad \text{on} \quad H_0 \; .$$

<u>Remarks</u>. While interesting, this result has not proved useful and shows therefore that the importance of dilation theory stems from the additional structure which kernels appearing in the literature may possess. We give below the 5 different additional structures known to us (including 3 due to us). Moreover we attempt to present a unifying framework for this additional structure to help search for new cases. Interestingly, this unifying format provides the key insight into the correct dilation for these kernels. In the definition that follows if $n > 0$, then $T^{(n)}(\rho)$ is understood to be the nth derivative of $T(\rho)$ with respect to ρ and in this case R will possess appropriate topological structure.

<u>Definition</u>. Let $T : R \to L(H_0)$ be a Hilbertian variety in $L(H_0)$. We say that the kernel $K : R \times R \to L(H_0)$ is a <u>spread</u> of T if there exists an integer $n = 0,1,2,\cdots$, a bounded complex valued measure space $(W,\mathcal{W},\lambda)$ and functions $C : R \times R \times W \to \mathbb{C}$, $g : R \times R \times W \to R$ such that the following is defined

$$F(T^{(n)}(\cdot);\rho,\sigma) = \int_W C(\rho,\sigma,\omega)T^{(n)}(g(\rho,\sigma,\omega))d\lambda(\omega)$$

and

$$K(\rho,\sigma) = F(T^{(n)}(\cdot);\rho^*,\sigma) \; .$$

Cases. 1) (Semi-group) Let R be a $*$-semi-group with associative binary operation $\cdot$, $(\rho\cdot\sigma)^* = \sigma^*\cdot\rho^*$. Put $K(\rho,\sigma) = T(\rho^*\cdot\sigma)$.

2) (Cosine) Let $R = \mathbb{R}$, $\rho^* = -\rho$, $\varepsilon = 0$ and $K(\rho,\sigma) = \frac{1}{2}T(\sigma+\rho) + \frac{1}{2}T(\sigma-\rho)$.

3) (Resolvent) $R = (\mathbb{C}-\mathbb{R}) \cup \{0\}$, $\rho^* = \bar{\rho}$, $\varepsilon = 0$, $K(\rho,\sigma) = \frac{\rho^*}{\rho^*-\sigma}T(\rho^*) + \frac{\sigma}{\sigma-\rho^*}T(\sigma)$. Here assume $\rho T(\rho)$ differentiable.

4) (Hankel) See Shonkwiler[5].

5) (∞-Log) Let $\bar{\mathbb{C}} = \mathbb{C} \cup \{\infty\}$ be the complex sphere with the chordal metric and put $R = \bar{\mathbb{C}} - \mathbb{R}$, $\rho^* = \bar{\rho}$, $\varepsilon = \infty$, and $K(\rho,\sigma) = \frac{\sigma}{\sigma^*-\rho^*}T'(\rho^*) - \frac{\rho^*}{\sigma^*-\rho^*}T'(\sigma)$.

Remark. The cosine example shows that $K^*(\varepsilon,\rho) \neq K(\varepsilon,\rho^*)$ in general unless assumed to hold.

Theorem. (Kernel Dilation) If K is the spread of some variety $T(\cdot)$ according to one of the cases above and satisfies:

(0) $K(\varepsilon,\rho) = T(\rho)$

(1) $K(\varepsilon,\varepsilon) = I$

(2) K is positive definite

and (3) $K(\rho^*,\varepsilon) = K(\varepsilon,\rho)$,

then K has a splitting conjugating dilation also satisfying (1) - (3).

Proof. Cases (1) - (4) are proved in Shonkwiler [5]. As in those proofs, for Case (5) let $\tilde{H}$ be the Moore space of K and on $\tilde{H}_0 = \text{span}\{K(\cdot,\rho)x : \rho \in R,\ x \in H_0\}$ define $\tilde{K}$ by extending the following:

$$\tilde{K}(\infty,\rho)K(\tau,\sigma)x = F(K(\tau,\cdot);\rho,\sigma)x$$
$$= (\sigma K(\tau,\rho) - \rho K(\tau,\sigma))x/(\sigma-\rho),$$

and

$$\tilde{K}(\rho,\sigma)K(\tau,\omega)x = F(\tilde{K}(\infty,\cdot);\rho^*,\sigma)K(\tau,\omega)x$$
$$= (\sigma\tilde{K}(\infty,\bar{\rho}) - \bar{\rho}\tilde{K}(\infty,\sigma))K(\tau,\omega)x/(\sigma-\bar{\rho}) .$$

Singularities which occur in these assignments are removable since T has derivatives of all orders here. By tedious, but straightforward calculation, it can be shown that: (a) the second assignment above is consistent with the first, (b) (1) and (3) above hold, (c) $\tilde{K}$ splits and conjugates, and (d) $\tilde{K}(\varepsilon,\rho)$ projects to $K(\varepsilon,\rho)$ on H_0. Then (2) follows by a previous theorem and the proof is complete.

Remark. By techniques similar to those in the last section of Shonkwiler [5], it can be shown that if the hypothesis of the Kernel Dilation Theorem are satisfied for case (5), then

$$T'(\rho) = \int_{-\infty}^{\infty} \frac{\rho}{\rho-t}\, dE_t \ , \ \rho \text{ complex}$$

for some spectral family. Thus for any contour C remaining entirely in the upper or lower half-planes joining 0 to ρ, we have

$$T(\rho) = \int_C T'(\zeta)d\zeta = \int_{-\infty}^{\infty} (\rho+t\ \ell n\ (\rho-t))dE_t \ .$$

REFERENCES

[1] N. Aronszajn, La Théorie générale des royaux reproduisants et ses applications, Première Partie, Proc. Cambridge Philos. Soc. Vol. 39, (1944), 133-153.

[2] ____________, Theory of reproducing kernels, Trans. Amer. Math. Soc., 68(1950), 337-404.

[3] P. Masani, Dilations as Propagators of Hilbertian Varieties, S.I.A.M. Math. Anal., 9, No. 3(1978), 414-456.

[4] _________, Propagators and dilations, Probability theory on vector spaces (Proc. Conf. Trzebieszowice, 1977), p. 95-117, Lecture Notes in Math., 656, Springer, Berlin, 1978.

[5] R. Shonkwiler and G. Faulkner, Kernel Dilation in Reproducing Kernel Hilbert Space and its Application to Moment Problems, Pac. J. Math., Vol. 77(1978), No. 1, 103-115.

[6] A. Weron and J. Gorniak, An analogue of Sz.-Nagy's dilation theorem, Bull. Acad. Polonaise Sci. 24(1976), 867-872.

[7] A. Weron, Remarks on positive definite operator valued functions in Banach spaces, Bull. Acad. Polonaise Sci. 24(1976), 873-876.

REMARKS ON PETTIS INTEGRABILITY OF CYLINDRICAL PROCESSES

Z. Suchanecki

Let E be a locally convex space (l.c.s.) and $(\Omega, \mathcal{A}, P)$ a probability space. The linear mapping T from E' (the topological dual to E) into $L^0(\Omega, \mathcal{A}, P)$ is called the cylindrical process. We say that the cylindrical process T is Pettis integrable if (1) $Tx' \in L^1$ (2) for each $A \in \mathcal{A}$ there is a $x_A \in E$ such that $\langle x_A, x' \rangle = \int_A Tx' dP$ for each $x' \in E'$.

In this note we study the correspondence between the Pettis integrability and continuity of cylindrical processes. It is easy to see that, if the cylindrical process $T: E' \to L^1$ is $(\tau(E', E), \|\cdot\|_{L^1})$ continuous (here $\tau(E', E)$ denotes the Mackey topology on E'), then T is Pettis integrable.

S.D. Chatterji show (see [1] Prop.1) that, if $f: \Omega \to E$ is weakly measurable and Pettis integrable function and E is a complete l.c.s., then the corresponding to f cylindrical process T_f, $T_f x' = \langle f, x' \rangle$, is $(\tau(E', E), \|\cdot\|_{L^1})$ continuous. In Theorem 1 we show that this is also true for cylindrical processes and under a weaker assumption. We need only that E is a sequentially complete l.c.s. . This assumption on E is also nessesary if E is a normed space(see Proposition 1), but it is not true if E in non-normed space (see Example 1).

In the sequel, by a locally convex space we shall mean a real Hausdorff l.c.s. . If A is a subset of E then A^0 denotes the polar with respect to the duality $\langle E, E' \rangle$. For duality $\langle E, F \rangle$ (E, F are l.c.s.), by $\sigma(E,F)$ and $\tau(E,F)$ we denote the weak and Mackey topologies respectively.

<u>Theorem 1</u>. Let E be a l.c.s. and $T: E' \to L^1(\Omega, \mathcal{A}, P)$ a Pettis integrable cylindrical process. Then T is $(\tau(E', E''), \|\cdot\|_{L^1})$ continuous. Moreover, if E is sequentially complete, then T is $(\tau(E', E), \|\cdot\|_{L^1})$ continuous.

In the proof of Theorem 1 we shall have need of two Lemmas.

<u>Lemma 1.</u> If E and F are l.c.s. , $S: E \to F$ is a continuous linear operator and $T: E' \to L^1$ is a Pettis integrable

cylindrical process, then $T.S^* : F' \longrightarrow L^1$ is Pettis integrable and $S(\int_A T\,dP) = \int_A T.S^*\,dP$ for each $A \in \mathcal{A}$.

<u>Proof.</u> Let $A \in \mathcal{A}$ and $x_A = \int_A T\,dP$. Setting $y_A = Sx_A$, we obtain

$$< y_A, y' > = < Sx_A, y' > = < x_A, S^*y' > = \int_A T(S^*y')dP$$

for each $y' \in F'$.

<u>Lemma 2.</u> Let E be a l.c.s. . If $T: E' \longrightarrow L^1$ is a Pettis integrable cylindrical process, then for each absolutely convex neighborhood U of zero in E

$$\sup_{x' \in U^\circ} \int_\Omega |Tx'|\,dP < \infty .$$

In particular if E is a normed linear space, then T is continuous with the operator norm

$$\|T\| = \sup_{\|x'\| \leq 1} \int_\Omega |Tx'|\,dP$$

<u>Proof.</u> First we show the lemma in the case E is a Banach space. Let $\|x'_n - x'_o\|_{E'} \longrightarrow 0$ and $\|Tx'_n - \varphi\|_{L^1} \longrightarrow 0$.

Then for each $A \in \mathcal{A}$

$$\int_A \varphi dP = \lim_n \int_A Tx'_n\,dP = \lim_n < x'_n, \int_A T\,dP > = < x'_o, \int_A T\,dP > =$$

$$= \int_A Tx'_o\,dP.$$

Therefore $\varphi = Tx'_o$ and by the Banach closed graph theorem T is bounded.

Now let E be a l.c.s. , U an absolutely convex neighborhood of zero in E and $p_U(.)$ the Minkowski functional of U. Denote by $\hat{E}_U$ the completion of the normed space $(E/p_U^{-1}(0), p_U(.))$ and by $\Phi_U : E \longrightarrow \hat{E}_U$ the quotient map.

By Lemma 1 $T.\Phi_U^* : (\hat{E}_U)' \longrightarrow L^1$ is Pettis integrable.

Since U° is the unit ball in $(\hat{E}_U)'$, we have

$$\sup_{x' \in U^\circ} \int_\Omega |Tx'|\,dP = \sup_{\substack{y' \in (\hat{E}_U)' \\ \|y'\| \leq 1}} \int_\Omega |T \circ \Phi_U^* y'|\,dP < \infty .$$

Proof of Theorem 1. Consider the operator T^* defined on L^∞_s (the space of all simple functions in L^∞) as follows

$$T^*\left(\sum_{i=1}^{n} a_i \mathbb{1}_{A_i}\right) = \sum_{i=1}^{n} a_i \int_{A_i} T\,dP.$$

Since T is Pettis integrable, $T^* : L^\infty_s \longrightarrow E$. Let $\{\varphi_n\} \subset L^\infty_s$ be a Cauchy sequence in L^∞. Then for each absolutely convex neighborhood U of zero in E

$$\sup_{x' \in U^\circ} |\langle T^*\varphi_n - T^*\varphi_m, x'\rangle| \leq \|\varphi_n - \varphi_m\|_{L^\infty} \sup_{x' \in U^\circ} \int_\Omega |Tx'|\,dP.$$

Because $\sup_{x' \in U^\circ} \int_\Omega |Tx'|\,dP < \infty$ (Lemma 2), then $\{T^*\varphi_n\}$ is the Cauchy sequence in E.

If E is sequentially complete, then T^* extends to the linear operator from L^∞ into E. By ([2], 7.4., p.158), this proves the second part of Theorem 1.

If E is not sequentially complete then we consider the algebraical embeddings $E \subset E'' \subset E'^*$, where E'' denotes the topological dual to E'_β (E' with the strong topology) and E'^* the algebraical dual to E'. Since $\{T^*\varphi_n\}$ defined above is the Cauchy sequence in E, then it is Cauchy in $\sigma(E, E')$ and therefore in $\sigma(E'^*, E')$. Because E'^* is weakly complete, then $\{T^*\varphi_n\}$ converges in $\sigma(E'^*, E')$ to some $x_o \in E'^*$. Using the fact, that E'' is the union of the $\sigma(E'^*, E')$ - closures in E'^* of all bounded subsets of E (cf.[2], 5.4. p.143) we obtain that $x_o \in E''$. Hence T^* extends to the linear operator from L^∞ into E''. By ([2], 7.4, p.158) this proves the first part of Theorem 1.

Corollary 1. Let E be a sequentially complete l.c.s. . A cylindrical process $T: E' \longrightarrow L^1$ is Pettis integrable if and only if T is $(\tau(E', E), \|\cdot\|_{L^1})$ continuous.

Proof. If T is $(\tau(E', E), \|\cdot\|_{L^1})$ continuous then T^* maps L^∞ into E. Hence for $A \in \mathcal{A}$ we put $\int_A T\,dP = T^*(\mathbb{1}_A)$.

In the sequel we shall use the following lemma, which is a slight modification of a part of Proposition 1 in [1].

Lemma 3. Let E be a l.c.s. and $T : E' \longrightarrow L^1$ a cylindrical process. The following are equivalent

(a) T is $(\tau(E', E), \|\cdot\|_{L^1})$ continuous

(b) for each $\Psi \in L^\infty$, there is $x_\Psi \in E$ such that
$$< x_\Psi, x' > = \int_\Omega \Psi . Tx' \, dP \quad \text{for each } x' \in E'.$$

One may prove Lemma 3 in the same way as the equivalence of (ii) and (iii) in([1], Proposition 1).

<u>Proposition 1</u>. Let E be a normed space. Suppose that for each probability space $(\Omega, \mathcal{A}, P)$ and each Pettis integrable function $f: \Omega \longrightarrow E$, the cylindrical process T_f, related to f, is $(\tau(E',E), \|\cdot\|_{L^1})$ continuous. Then E is complete.

<u>Proof</u>. Suppose that E is not complete. Then there is a sequence $\{x_n\} \subset E$ such that $\sum_{n=1}^{\infty} \| x_n \| < \infty$ and $\sum_{n=1}^{\infty} x_n$ does not converges in E.
Let $\{\alpha_n\}$ be a non decreasing sequence of positive numbers such that $\alpha_n \nearrow \infty$ and $\sum_{n=1}^{\infty} \alpha_n \|x_n\| < \infty$.
We consider a subspace E_1 of $\hat{E}$ (the completion of E), defined as follows

$$E_1 = \text{lin}\,(\{ \sum_{k_1 < k_2 < \dots} \alpha_{k_n} x_{k_n}\} \cup E),$$

where the summation is taken over all subsets $\{k_1, k_2, \dots, \}$, $k_1 < k_2 < \dots$, of the natural numbers.
Obviously $E \subset E_1 \subset \hat{E}$, $E' = (E_1)'$ and $\tau(E', E) = \tau(E', E_1)$.

Let $\Omega = \mathbb{N}$, $\mathcal{A} = 2^{\mathbb{N}}$ and $P(\{n\}) = \| x_n \| \cdot (\sum_{n=1}^{\infty} \| x_n \|)^{-1}$.

Define a function $f: \Omega \longrightarrow E_1$, $f(n) = \alpha_n \| x_n \|^{-1} . x_n$ for each $n \in N$. It is obvious that f is Pettis integrable.
Let $\Psi(n) = \alpha_n^{-1}$, $n = 1,2,\dots$. Then $\Psi \in L^\infty$, but $\int_\Omega \Psi f \, dP \notin E_1$, because the series $\sum x_n$ does not converges in E_1. Using the Lemma 3 we obtain, that the cylindrical process $T_f x' = < f, x' >$ is not $\tau(E', E_1)$ continuous, what gives the contradiction.

Now we shall show, that Proposition 1 does not hold for non-normed spaces.

<u>Example 1.</u> Let E be the space of all bounded real sequences with topology of pointwisse convergence. So E is metrizable and separable l.c.s. . If $(\Omega, \mathcal{A}, P)$ is a probability space, then each measurable function $f: \Omega \longrightarrow E$ is of the form $f = (\varphi_1, \varphi_2, \dots)$

where φ_n, $n = 1,2,\dots$ are real measurable functions. It is easy to see, that f is Pettis integrable if and only if

$$\sup_n \int_\Omega |\Psi\varphi_n| dP < \infty \text{ . Therefore } \int_\Omega \Psi f \, dP \in E \quad \text{for each } \Psi \in L^\infty \text{ .}$$

This shows that $T_f x' = \langle f, x' \rangle$ is $(\tau(E',E), \|\cdot\|_{L^1})$ continuous. On the other hand the space E is not complete, because for example the sequence $x_n = (1,\dots,n,0,\dots)$ is Cauchy in E, but not converges.

References

[1] S.D. Chatterji - Sur l'intégrabilite de Pettis, Math. Z. 136, 53-58 (1974)

[2] H.H. Schaefer - Topological vector spaces, Springer-Verlag: New York - Heidelberg - Berlin (1971)

Institute of Mathematics
Wrocław Technical University
50-370 Wrocław
Poland

A PROBABILITISTIC CHARACTERIZATION OF UNCONDITIONALLY SUMMING OPERATORS

Rafał Sztencel

In this note it is proved that an operator is unconditionally summing if and only if it maps a.s. bounded symmetric random series into a.s. convergent random series. This is a generalization of the theorem conjectured by Hoffmann-Jørgensen [1] and proved by Kwapień [2] which asserts that a Banach space E does not contain subspaces isomorphic to c_0 if and only if the a.s. boundedness of sums of independent, symmetric E-valued random variables implies the a.s. convergence of the sums.

We begin with definitions. E, F are Banach spaces.

<u>Definition 1.</u> A sequence of vectors (x_i) from a Banach space E is weakly summable iff for every $x^* \in E^*, \Sigma|x^*(x_i)| < \infty$.

<u>Definition 2.</u> (Pełczyński [3], Pietsch [4]). An operator $T:E \longrightarrow F$ is unconditionally summing iff every weakly summable sequence (x_i) is mapped into a sequence (Tx_i) summable in the norm topology.

In the sequel (ε_i) will denote a Bernoulli sequence on a probability space $(\Omega , \mathcal{F} , P)$, $L_p(E)$ will denote $L_p(\Omega , \mathcal{F} ,P;E)$. If T is an operator from E into F, then $\tilde{T}: L_0(E) \longrightarrow L_0(F), (\tilde{T}f)(\omega)= Tf(\omega)$.

Now we can formulate our result.

<u>Theorem.</u> The following conditions are equivalent:

(i) $T: E \longrightarrow F$ is unconditionally summing

(ii) The a.s. boundedness of the series $\Sigma x_i\varepsilon_i$ implies the a.s convergence of the series $\Sigma Tx_i\varepsilon_i$

(iii) If (ξ_i) is a symmetric sequence of E-valued r.v. then the a.s. boundedness of $\Sigma \xi_i$ implies the a.s. convergence of $\Sigma T\xi_i$

(iv) $\forall$ $p\in[1,\infty)$ $\tilde{T}: L_p(E) \longrightarrow L_p(F)$ is unconditionally summing

(v) $\exists$ $p\in[1,\infty)$ $\tilde{T}: L_p(E) \longrightarrow L_p(F)$ is unconditionally summing

Proof. (i) $\Longrightarrow$ (ii). Suppose that $\Sigma x_i \varepsilon_i$ is a.s. bounded and $\Sigma Tx_i \varepsilon_i$ does not convergence a.s. Then it does not converge in probability and there is an increasing sequence of indices (n_i) such that for some $\varepsilon > 0$ $P(\|\sum_{i=n_k+1}^{n_{k+1}} Tx_i \varepsilon_i\| > \varepsilon) > \varepsilon$. Put $U_k = \sum_{i=n_k+1}^{n_{k+1}} x_i \varepsilon_i$, $Y_k = TU_k$, $A_k = \{\|Y_k\| > \varepsilon\}$. Let (ε_i') be a Bernoulli sequence on a probability space $(\Omega', \mathcal{F}', P')$. The series $\Sigma \varepsilon_k' U_k$ is $P' \times P$-a.s. bounded. Hence $P(B) = 1$, where $B = \{\omega \in \Omega: \Sigma \varepsilon_k' U_k(\omega)$ is P'-a.s. bounded $\}$. Let $C = \bigcap_n \bigcup_{k \geq n} A_k$. By the Borel-Cantelli lemma $P(C)=1$ Now choose $\omega \in B \cap C$, put $u_k = U_k(\omega)$. The series $\Sigma u_n \varepsilon_n$ is a.s. bounded, On the other hand, $\limsup \|Tu_n\| \geq \varepsilon$, hence there are $z_i = u_{n_i}$ such that $\|Tz_i\| > \varepsilon/2$, while $\Sigma z_i \varepsilon_i$ is a.s. bounded. Thus by the result of [2] we can choose (z_{i_k}) which is weakly summable. Since ΣTz_{i_k} does not converge, T is not unconditionally summing - a contradiction. This completes the proof of (i) $\Longrightarrow$ (ii)

(ii) $\Longrightarrow$ (iii). Suppose that there is a symmetric sequence (ξ_i) such that $\Sigma \xi_i$ is a.s. bounded and $\Sigma \tilde{T}\xi_i$ is not convergent a.s. The same is true for $\Sigma \varepsilon_i' \xi_i$ and $\Sigma \varepsilon_i' T\xi_i$. To get a contradiction take $\omega \in \Omega$ such that $\Sigma \varepsilon_i' \xi_i(\omega)$ is P' a.s. bounded and $\Sigma \varepsilon_i' T \xi_i(\omega)$ does not converge P'- a.s. This completes the proof of (ii) $\Longrightarrow$ (iii).

(iii) $\Longrightarrow$ (iv). We shall need the following known

Lemma. If (f_n) is a symmetric sequence, Σf_n converges a.s., $\Sigma f_n \in L_p(E)$, then Σf_n converges in $L_p(E)$.

It is easy to see that T is bounded, hence $\tilde{T}$ maps $L_p(E)$ into $L_p(F)$. Suppose that for some $1 \leq p < \infty$ there is a weakly summable sequence $(\eta_i) \subset L_p(E)$ such that $\Sigma \tilde{T}\eta_i$ is not convergent in norm. We can assume that $\|\tilde{T}\eta_i\|_{L_p} > \delta > 0$. Put $\xi_i(\omega,\omega') = \varepsilon_i'(\omega') \eta_i(\omega)$. Then (ξ_i) is a symmetric sequence which is weakly summable, hence its sums are bounded in $L_p(E)$ and $P' \times P$-a.s. bounded. Then $\Sigma \tilde{T} \xi_i$ converges a.s. and the sum is in $L_p(F)$ by the boundedness of $\tilde{T}$ and the Fatou lemma. Now we conclude by the lemma that $\Sigma \tilde{T}\xi_i$ converges in $L_p(F)$; but $\|\tilde{T}\xi_i\|_{L_p} = \|\tilde{T}\eta_i\|_{L_p} > \delta$ - a contradiction. This completes the proof of (iii) $\Longrightarrow$ (iv) $\Longrightarrow$ (v) $\Longrightarrow$ (i) obvious. This completes the proof of the theorem.

It is known (cf.[4]) that the identity map $I_E: E \longrightarrow E$ is unconditionally summing iff E does not contain subspaces isomorphic to c_0. Thus as a corollary we obtain the result of [2].

References

[1] J.Hoffmann-Jørgensen, "Sums of independent Banach space valued random variables", Studia Math.52 (1974),159-186.

[2] S.Kwapień, "On Banach spaces containing c_0 ", ibidem, 187-188.

[3] A.Pełczyński, "Banach spaces on which every unconditionally converging operator is weakly compact", Bull. Ac.Polon.Sci. 10 (1962), 641-648.

[4] A.Pietsch, "Operator ideals", Berlin 1978.

Department of Mathematics
University of Warsaw
00-901 Warszawa, PKiN

ON OPERATOR CHARACTERIZATION OF AM- AND AL-SPACES

J. Szulga (Wrocław)

Let Y be a infinite dimensional Banach space. We characterize a Banach lattice E as an AM- or AL-space by dealing with the equality $A(E,Y) = B(E,Y)$ where A and B are classes of operators from E into Y.

Notations and basic facts.

In the paper E denotes a Banach lattice with the norm dual E' and the positive cone E_+. E is said to be an AM-space (we write $E \in AM$) if there exists an equivalent lattice norm such that $\|\sup(|x|,|y|)\| = \max(\|x\|,\|y\|)$. Note that $E \in AM$ iff for some constant $K > 0$ $\|\sup|x_i|\| \leq K \max\|x_i\|$ for all $x_1,\dots,x_n \in E$. (cf. [4], Th. IV. 2.8.) E is said to be an AL-space (we write $E \in AL$) if $\|x+y\| = \|x\| + \|y\|$ for all $x,y \in E_+$. By the Kakutani representation theorem (cf. [4], Th. II.8.5.) $E \in AL$ iff E isomorphic to some $L_1(\mu)$. It is known that $E \in AL$ iff $E' \in AM$ (cf. [4], Prop. II.9.1.).

Let $1 \leq p \leq \infty$ and $1/p + 1/q = 1$. Following J. Krivine we consider the function

$$E^n \ni (x_1,\dots,x_n) \to (\Sigma|x_i|^p)^{1/p} \in E$$

where $(\Sigma|x_i|^p)^{1/p} = \sup\{\Sigma \alpha_i x_i : \Sigma|\alpha_i|^q \leq 1,\ \alpha_i \in R\}$ (we refer to [2] for details)

Fact 1. ([2]). If $\{f_i\}$ is a finite sequence of standard independent gaussian random variables then for some constant a_1

$$E|\Sigma x_i f_i| = a_1(\Sigma |x_i|^2)^{\frac{1}{2}}$$

for all $x_1,\dots,x_n \in E$.

Remark. For real x_i we have

$$(E|\Sigma x_i f_i|^p)^{1/p} = a_p(\Sigma |x_i|^2)^{\frac{1}{2}}$$

for some constant a_p. Here $1 \le p < \infty$.

For $x_1,\dots,x_n \in E$ we put $\omega_p(\{x_i\}) = \sup\{(\Sigma|\langle x',x_i\rangle|^p)^{1/p} : \|x'\| \le 1\}$.
An operator $T: E \to F$, where F is a Banach, is said to be p-concave if for some $c > 0$

(1) $$(\Sigma \|Tx_i\|^p)^{1/p} \le c\|(\Sigma |x_i|^p)^{1/p}\| \quad \text{for all } x_1,\dots,x_n \in E.$$

We put $K_p(T) = \inf\{c: c \text{ satisfies } (1)\}$. 1-concave operator is called also cone-absolutely summing (see [4], p.244).
An operator $T: E \to F$ is said to be p-absolutely summing if for some $c > 0$

(2) $$(\Sigma \|Tx_i\|^p)^{1/p} \le c\omega_p(\{x_i\}) \quad \text{for all } x_1,\dots,x_n \in E.$$

We put $\pi_p(T) = \inf\{c: c \text{ satisfies } (2)\}$. Note that $\omega_p(\{x_i\}) \le$ $\le \|(\Sigma|x_i|^p)^{1/p}\|$ and if $E \in AM$ then also $\|(\Sigma|x_i|^p)^{1/p}\| \le K\omega_p(\{x_i\})$ for some $K > 0$. Hence operators p-concave are p-absolitely summing and for $E \in AM$ the both notion coincide. One can ask whether this is the neccessary condition. We give the positive answer.

Fact 2. (cf. [3], Prop.1 and [4] Prop.II.9.1) The following conditions are equivalent:

(i) Each operator from E into L_1 is 2-absolutely summing;
(ii) Each operator from E' into l_2 is 1-absolutely summing;
(iii) $E \in AM$.

Characterizations of AM-space

Lemma. $E \in AM$ iff $\|(\Sigma|x_i|^2)^{\frac{1}{2}}\| \le c\omega_2\{x_i\}$ for all $x_1,\dots,x_n \in E$.

Proof: The statement follows from Theorem 3 in [4] and Fact 2.

Theorem 1. Let $1 \le p < \infty$. The following conditions characterize AM-spaces:

(i) $\|(\Sigma|x_i|^p)^{1/p}\| \le C\omega_p\{x_i\}$ for all $x_1,\ldots,x_n \in E$;

(ii) For all $T: E \to L_1$ p-concavity is equivalent to p-absolute summability;

(iii) $\omega_p(\{|x_i|\}) \le C\omega_p\{x_i\}$ for all $x_1,\ldots,x_n \in E$;

(iv) $\|\,|X|\,\|_p^* \le C\|X\|_p^*$ for all Bochner integrable random vectors X.

Here $\|\cdot\|_p^*$ denotes p'th Pettis norm: $\|X\|_p^* = \sup\{(E|\langle x',X\rangle|^p)^{1/p}: \|x'\| \le 1\}$.

Proof: Immediately (i) is satisfied for an AM-space and (i) => (ii). (ii) => (iii). Let $\Pi_p(T) \le CK_p(T)$ for all $T: E \to L_1$. Let $x' \in E_+'$. Then one can define the L-seminorm on E: $N(x) = \langle x', |x|\rangle$. The completion of $E/N^{-1}(\{0\})$, which we denote by (E,x') is an AL-space (cf. [4], II, 8.Ex.1). Since the operator $T: E \to (E,x')$ given by $Tx = [x]$ is p-concave, hence by hypothesis T is p-absolutely summing. That is

$$(\Sigma|\langle x',|x_i|\rangle|^p)^{1/p} \le C\omega_p\{x_i\} \quad \text{for all } x_1,\ldots,x_n \in E.$$

Since C doesn't depend on x', thus (iii) follows. (iii) => (iv). The statement follows by classical arguments of the theory of the integral. Now we will show that $E \in AM$ provided (iv). We have

$$\|E|X|\,\| = \|\,|X|\,\|_1^* \le C\|X\|_p^*.$$

In particular, taking $X = \Sigma x_i f_i$, where f_i are independent standart guassian random variables, we obtain

$$\|(\Sigma|x_i|^2)^{\frac{1}{2}}\| \le C'\,\omega_2(\{x_i\}) \quad \text{for all } x_1,\ldots,x_n \in E.$$

by Fact 1. In view of Lemma the proof is complete.

AL-space and regular operators.

An operator $T: E \to F$, where E,F are Banach lattices, is said to be regular if $T = T_1 - T_2$, where $T_1, T_2 \ge 0$. It is easy to see that T

is regular iff $|T|$ exists in the canonical order of the space of operators. If $E \in AL$ and F is the range of a positive contractive projection from F'' then all $T: E \to F$ are regular (cf. [4], Prop.IV.1.5). The following result shows that for some F this situation is possible only for AL-spaces.

Theorem 2. Let $1 \le p < \infty$. All $T: E \to L^p$ are regular iff $E \in AL$.

Proof: We have to prove just the part "only if". By hypothesis for all $T: E \to L^p$ $|T|$ exists and $\| |T| \| \le C\|T\|$ for some $C > 0$ (cf. [4], Prop.IV.1.4) $|T|$ can be found by the formula

$$|T|x = \sup\{|Ty| : |y| \le x\} ,$$

where $x \in E_+$.
Let $X' \in L^p(E')$. Consider T of the form

$$Tx = \langle x, X' \rangle .$$

Then $|T| = \langle x, |X'| \rangle$ since

$$\langle x, |X'| \rangle = \sup\{\langle y, X' \rangle : |y| \le x\} \qquad \text{a.e.}$$

Hence by Theorem 1 $E' \in AM$, so $E \in AL$. The proof is complete.

Corollary. Let $1 \le p < \infty$ and E be a Banach space with unconditionall basis (which makes E the Banach lattice under the coordinatewise order). Then for some $C > 0$

$$(3) \qquad \| \Sigma |Tx_i| \| \le C \| \Sigma |x_i| \| \qquad \text{for all } x_1, \dots, x_n \in E$$

iff E is isomorphic to l_1.

Proof: The part "if" follows immediately. Let us assume (3) is satisfied. Let (e_i) denotes the unconditionall basis of E. If $T: E \to L^p$ then for $x = \Sigma \xi_i e_i \in E_+$, $\Sigma \xi_i |Te_i|$ converges in L^p and $\| \Sigma \xi_i |Te_i| \| \le C\|x\|$ by hypothesis. Hence T is regular, since for $|y| \le x \in E_+$

$$|Ty| \le \Sigma \xi_i |Te_i| .$$

Thus $E \in AL$ by Theorem 2 and so E is isomorphic to l_1.

Problem. Whether (3) characterizes AL-spaces for arbitrary Banach lattices?

Corollary 2. Let E be a Banach space with u.b.
Then E is isomorphic to l_1 iff for all $T: E \to L_p$, T is 1-concave. The last statement can be generalized.

Theorem 3. Let E be a Banach lattice. The following condition are equivalent:

(1) There exists $1 \le p < \infty$ such that all $T: E \to L_p$ are 1-concave;

(2) For some infinite dimensional Banach space Y all $T: E \to Y$ are 1-concave;

(3) All $T: E \to L_2$ are 1-concave;

(4) $E \in AL$.

Proof: We will show (2) => (3) and (3) => (4). The other implications follows immediately. (2) => (3) It sufficies to show that for all operators $T: E \to l_2$ of the form $T = \Sigma\, x_i' \times e_i$, $K_1(T) \le C\|T\|$, where (e_i) denotes the standard basis of l_2, so let $T = \Sigma\, x_i' \times e_i$. By Dvoretzky--Roqers lemma there exists a subspace $Y_n \subset Y$ and isomorphism $I_n : l_2^n \to Y_n$ such that $\|I_n\| \le K$ (K doesn't depend on Y_n). Then

$$\Pi_1(T) \le K\Pi_1(I_n \circ T) \le CK\,\|I_n \circ T\| \le C^1\|T\|$$

(2) => (4) If all $T: E \to L_2$ are 1-concave, so its possese factorization through L_1 (cf. [4], Th.IV.3.3). Hence all T are 1-absolutely summing and $E \in AL$ by Fact 2. The proof is complete.

Remark. The equivalence (2) and (4) is also proved in [1].

References

[1] L.P. Janovskiĭ, *Summing, order summing operators and characterization of AL-spaces*, Sibirskiĭ Matematiceskii Žurnal, 1979, T.XX, No 2, 401-408 (in Russian).

[2] J. Krivine, *Théorèmes de factorization dans les espaces réticules*, Séminaire Maurey-Schwartz, 1973/74, Exp.XXII, XXIII.

[3] G. Pisier, *Une novelle classe de Banach vérifiant le théorème de Grothendieck*, to appear in Annales de l'Institut Fourier.

[4] H.H. Schaeffer, *Banach Lattices and Positive Operators*, Springer Verlag Berlin-Heidelberg-New York 1974.

Jerzy Szulga
Wrocław University
Institute of Mathematics
Pl. Grunwaldzki 2/4
50-384 Wrocław, Poland

ON NUCLEAR COVARIANCE OPERATORS

V.I. Tarieladze

It is well-known that the covariance operator of strong second order probability measure in separable Hilbert space is nuclear. The analoguous fact for Banach spaces was proved in [1]. Here we give a proof of the more strong assertion and note two related problems.

Let Y and Z be Banach spaces. A linear operator $R: Y \to Z$ is called nuclear if it admits the representation

$$R\, y = \sum_{k=1}^{\infty} < y, y_k^* > Z_k \quad , \qquad y \in Y$$

where $(y_k^*) \subset Y^*$ (Y^* is the topological dual of Y), $(Z_k) \subset Z$ and $\sum_{k=1}^{\infty} \|y_k^*\| \, \|Z_k\| < \infty$. If H is a real separable Hilbert space, then a symmetric and positive linear operator $R: H \to H$ is nuclear if and only if for some (or for any) orthonormal basis (e_k) of H we have $\Sigma(R\, e_k, e_k) < \infty$. The immediate consequence of this fact is that if we have a strong second order measure μ on H, then its covariance operator $R_\mu : H \to H$ is nuclear.

Let X be a real Banach space, μ be a strong second order Radon probability measure on X (i.e. $\int_X \|x\|^2 \, d\mu(x) < \infty$). We can define the operator $R_\mu : X^* \to X$ (the covariance operator of μ) by the equality

$$R_\mu\, x^* = \int_X < x, x^* > x \, d\mu(x), \qquad x^* \in X^*$$

(Here the integral is obvious Bochner or Pettis integral).

In [1], using Grothendieck's theorem on representation of nuclear operators on L_∞, was proved that symmetric and positive linear operator R_μ is nuclear. By the definition, nuclearity of R_μ means that $R_\mu\, x^* = \Sigma < x^*, x_k^{**} > x_k$, where $(x_k^{**}) \subset X^{**}$, $(x_k) \subset X$ and $\Sigma \|x_k^{**}\| \, \|x_k\| < \infty$. The following result gives more precise structure of R_μ.

<u>Theorem.</u> Let X be a real Banach space and μ be a strong second order Radon probability measure on X. Then the covariance operator R_μ of μ admits the representation

$$R_\mu x^* = \sum_{k=1}^{\infty} < x_k, x^* > x_k \ , \quad x^* \in X^* ,$$

where $(x_k) \subset X$ and $\Sigma \|x_k\|^2 < \infty$. In particular $R_\mu : X^* \longrightarrow X$ is the nuclear operator.

Proof. Let $(\Omega, \mathcal{A}, P)$ be a probability space and $\xi \in L_2(\Omega, \mathcal{A}, P; X)$ be a function with $P\xi^{-1} = \mu$. Then we have

$$R_\mu x^* = \int_\Omega <\xi, x^*> \xi \; dP \; , \qquad x^* \in X^* .$$

We can suppose, that $\mathcal{A} = \sigma(\bigcup_{n=1}^\infty \mathcal{A}_n)$, where $\mathcal{A}_1 \subset \mathcal{A}_2 \subset \ldots$ are finite sub-algebras of $\mathcal{A}$. Let $\mathbb{E}^{\mathcal{A}_n}$ be the conditional expectation operator defined by $\mathcal{A}_n$. Denote $\mathbb{E}^{\mathcal{A}_n} \xi = \xi_n$.
Then ξ_n is a simple function. By martingale convergence theorem (see [2], p. 126, corollary 2) we have

$$\int_\Omega \|\xi - \xi_n\|^2 \, dP \longrightarrow 0 \quad (n \longrightarrow \infty).$$

We can choose subsequence (ξ_{k_n}) such that $\int_\Omega \|\xi - \xi_{k_n}\|^2 \, dP < \frac{1}{2^n}$.
For the simplicity we assume that $k_n = n$. $\xi_n - \xi_{n-1}$ also is simple function ($\xi_0 = 0$) : $\xi_n - \xi_{n-1} = \sum_{k=1}^{s_n} x_{nk} I_{A_{nk}}$, where $x_{n1}, \ldots, x_{n,s_n} \in X$, $A_{n,1}, \ldots, A_{n,k_n} \in \mathcal{A}$ are disjoint elements of $\mathcal{A}_n$, $s_n \in N$.

We have

$$\sum_{n=1}^\infty \sum_{k=1}^{s_n} \|x_{nk}\|^2 P(A_{nk}) = \sum_{n=1}^\infty \int_\Omega \|\xi_n - \xi_{n-1}\|^2 \, dP \leq$$

$$\leq 6 \sum_{n=1}^\infty \frac{1}{2^n} \leq 6 < +\infty .$$

An another hand $(\xi_n, \mathcal{A}_n)$ is martingale and this implies that for all $x^* \in X^*$ the sequence $<\xi_n - \xi_{n-1}, x^*>$ is orthogonal sequence in $L_2(\Omega, \mathcal{A}, P)$. We have also

$$<\xi, x^*> = \sum_{n=1}^\infty <\xi_n - \xi_{n-1} \; x^*> \qquad x^* \in X^*$$

where the series converges in $L_2(\Omega, \mathcal{A}, P)$. By the orthogonality we obtain

$$\int_\Omega <\xi, x^*>^2 \, dP = \sum_{n=1}^\infty \int_\Omega <\xi_n - \xi_{n-1}, x^*>^2 \, dP =$$

$$= \sum_{n=1}^\infty \sum_{k=1}^{s_n} <x_{nk}, x^*>^2 P(A_{nk}), \quad x^* \in X^* .$$

Thus, we have

$$< R_\mu x^*, x^* > \sum_{n=1}^{\infty} \sum_{k=1}^{s_n} < x_{nk}, x^* >^2 P(A_{nk}) \qquad x^* \varepsilon X^*$$

This equality implies that

$$R_\mu x^* = \sum_{n=1}^{\infty} \sum_{k=1}^{s_n} < x_{nk}, x^* > x_{nk} P(A_{nk}), \qquad x^* \varepsilon X^*$$

and this representation can be rewrited as required one, since $\sum_n \sum_{k=1}^{s_n} \| x_{nk} \|^2 P(A_{nk}) < \infty$.

In connection of this theorem two problems arize.

Problem 1. Is it possible in the theorem to choose the topologically independent sequence $(x_n) \subset X$ (i.e. the sequence (x_n) with $x_n \notin$ closed linear span of $\{x_1, \ldots, x_{n-1}, x_{n+1}, \ldots\}$) ?

The proof of theorem 1 in [3] shows that the answer is positive if μ is symmetric Gaussian measure.

Problem 2. Let X be a real Banach space. Has every symmetric and positive nuclear linear operator $R: X^* \longrightarrow X$ the representation $Rx^* = \sum < x_k, x^* > x_k$, where $(x_k) \subset X$ and $\sum \|x_k\|^2 < \infty$?

References

[1] S.Chevet, S.A. Chobanjan, W.Linde, V.I. Tarieladze, caracterization de certaines classes d'espaces de Banach par les measures gaussiennes, C.R. Acad.Sc.t-285 (1977), 793-796.

[2] J.Diestel and J.J. Uhl, Iz.Vector measures. American Math.Soc. Providence, 1977.

[3] S. Kwapień, B.Szymański, Some remarks on Gaussian measures in Banach spaces. Preprint.

Academy of Sciences of the Georgian SSR
Computing Center, Tbilisi, 380093 USSR

ON SYMMETRIC STABLE MEASURES WITH DISCRETE SPECTRAL MEASURE ON BANACH SPACES

Dang-Hung Thang and Nguyen Zui Tien

In this paper we consider stable measures on Banach spaces. There are given some results concerning operator representation, norm convergent expansions, the 0-1 law for subgroups, the set of possible translations and the structure of linear measurable functionals. The similar problems were intensively studied, especially for Gaussian case, in many papers, see detailed references.

§1. Operator representation

Let μ be a symmetric stable measure on a separable Banach space. It is known (see [18],[21]), that the characteristic functional (ch.f.) of μ is of the following form

$$\chi_\mu(x^*) = \exp\{-\int_S |< x,x^* >|^\alpha \ \Gamma(dx)\} \qquad (1)$$

where $S = \{ x \in X : \|x\| = 1 \}$, Γ is a finite measure on S, α is the index of μ : $0 < \alpha \leq 2$. Γ is called the spectral measure of μ and formula (1) the Levy representation of μ.

Now, let Γ be a discrete measure concentrated on points $(a_n \in S)_{n=1}^\infty$:

$$\Gamma(\{ a_n \}) = \lambda_n > 0 \qquad n = 1,2,\ldots$$

$$\sum_{n=1}^\infty \lambda_n = \Gamma(S) < +\infty$$

Let us put

$$x_n = \lambda_n^{1/\alpha} a_n$$

Then it is easy to see that

$$\chi_\mu(x^*) = \exp\{-\sum_{n=1}^\infty |< x_n,x^* >|^\alpha \},$$

where $(x_n)_{n=1}^\infty$ is a sequence of elements of X such that

$$\sum_{n=1}^\infty \|x_n\|^\alpha < +\infty .$$

Consider the mapping

$$T : X^* \longrightarrow l_\alpha$$

$$x^* \longrightarrow (< x_n, x^* >)_{n=1}^{\infty}$$

Evidently $T \in L(X^*, l\alpha)$,

$$\chi_\mu(x^*) = \exp\{- \|Tx^*\|^{\alpha}\} \qquad (2)$$

and

$$T^* : l_\beta = l_\alpha^* \longrightarrow X$$

$$b = (b_n)_{n=1}^{\infty} \longrightarrow \sum_{n=1}^{\infty} b_n x_n ,$$

The formula is called the operator representation of χ_μ.

Conversely, if χ_μ has the operator representation of the following form

$$\chi_\mu(x^*) = \exp\{-\|Ax^*\|^{\alpha}\}, \quad A \in L(x^*, l_\alpha), \quad 0 < \alpha < 2,$$

then χ_μ can be presented in the form (1) with a discrete spectral measure. Indeed, it is known (see [8]) that in this case $A^* \in L(l_\beta, X)$ and the series

$$\sum_{n=1}^{\infty} \gamma_n^{(\alpha)} A^* e_n^* \qquad (3)$$

is convergent with probability 1, where $(\gamma_n^{(\alpha)})_{n=1}^{\infty}$ is a standard α-stable sequence (i.e. a sequence of independent identically distributed random variables with the ch.f. $\exp\{-|t|^{\alpha}\}$, $(e_n)_{n=1}^{\infty}$ - the natural basis in $l\alpha$, $(e_n^*)_{n=1}^{\infty}$ - the dual basis. Let us put

$$x_n = A^* e_n^*.$$

Since each Banach space is of p-stable cotype for $0 < p < 2$ (see[20]) it follows from the convergence of the series (3) that

$$\sum_{n=1}^{\infty} \|x_n\|^{\alpha} < +\infty .$$

Taking $a_n = \|x_n\|^{-1} x_n$, $\lambda_n = \|x_n\|^{\alpha}$, $\Gamma = \sum_{n=1}^{\infty} \lambda_n \delta(a_n)$

we get

$$\Gamma(S) = \sum_{n=1}^{\infty} \|x_n\|^{\alpha} < +\infty ,$$

$$\chi_\mu(x^*) = \exp\{-\int_S | < x^*, x_n >|^{\alpha} \Gamma(dx)\}$$

Thus we have proved the following fact

<u>Proposition 1.1</u>

(a) A ch.f. of an arbitrary symmetric stable measure μ with a discrete spectral measure has the following operator representation

$$\chi_\mu(x^*) = \exp\{-\|Tx^*\|^\alpha\}, \quad 0 < \alpha \leq 2,$$

where T is the mapping defined as follows

$$\begin{aligned} T : X^* &\longrightarrow l\alpha \\ x^* &\longrightarrow (\langle x_n, x^* \rangle)_{n=1}^\infty \\ (x_n)_{n=1}^\infty \subset X, &\quad (\|x_n\|)_{n=1}^\infty \in l\alpha \end{aligned} \tag{4}$$

(b) If a ch.f. of measure μ has the operator representation of the form

$$\chi_\mu(x^*) = \exp\{-\|Ax^*\|^\alpha\}, \ A \in L(X^*, l\alpha), \quad 0 < \alpha < 2,$$

then μ is a symmetric stable measure with a discrete spectral measure. Moreover in this case we can choose a mapping $T \in L(X^*, l\alpha)$ which satisfies the conditions (4) and $\|Ax^*\| = \|Tx^*\|$.

<u>Theorem 1</u>

Let $\mathfrak{X}$ be a Banach space and $0 < \alpha < 2$. Then the following conditions are equivalent

(a) $\mathfrak{X}$ is of α-stable type,

(b) for every $A \in L(X^*, l_\alpha)$ the formula $\exp\{-\|Ax^*\|^\alpha\}$ is a.ch.f. of a Radon measure on $\mathfrak{X}$ if and only if there exists $T \in L(X^*, l\alpha)$ which satisfies the conditions (4) and $\|Ax^*\| = \|Tx^*\|$

<u>Proof.</u> (a) $\Rightarrow$ (b) By virtue of Proposition 1.1 it suffices to prove that for every $T \in L(X^*, l\alpha)$ satisfying the conditions (4), $\exp\{-\|Tx^*\|^\alpha\}$ is a ch.f. of Radon measure in X. Since X is of α-stable type, from the last formula in (4) it follows that

$$\xi = \sum_{n=1}^\infty \gamma_n^{(\alpha)} x_n \in X$$

with probability 1.
It is easy to verify that

$$\chi_\mu(x^*) = \exp\{-\|Tx^*\|^\alpha\},$$

where μ is the distribution of ξ.

(b) $\Longrightarrow$ (a) Let $(x_n)_{n=1}^\infty \subset X$ and $\sum_{n=1}^\infty \|x_n\|^\alpha < +\infty$

Put

$$\begin{aligned} T: X^* &\longrightarrow l\alpha \\ x^* &\longrightarrow (\langle x_n, x^* \rangle)_{n=1}^\infty \end{aligned}$$

Then (b) implies that there exists a Radon measure such that

$$\chi_\mu(x^*) = \exp\{-\|Tx^*\|^\alpha\}$$

On the other hand for every $x^* \varepsilon X^*, \sum_{n=1}^{\infty} |<x_n,x^*>|^\alpha < +\infty$, and the series

$$\sum_{n=1}^{\infty} < \gamma_n^{(\alpha)} x_n, x^* >$$

is convergent with probability 1.

Applying the Ito-Nisio theorem (see [12]) we get that the series

$$\sum_{n=1}^{\infty} \gamma_n^{(\alpha)} x_n$$

is convergent with probability 1. This means that X is of α-stable type.

§ 2. Norm convergent expansions

Throughout this paper we shall assume that μ is a symmetric stable measure with a discrete spectral measure defined on a separable Banach space X. It has been shown in § 1 , that for such measure

$$\chi_\mu(x^*) = \exp\{-\sum_{n=1}^{\infty} |< x_n,x^*>|^\alpha\} \qquad 0 < \alpha \leq 2$$

where $(x_n)_{n=1}^{\infty} \subset X$ and $(\|x_n\|)_{n=1}^{\infty} \varepsilon\ l\alpha$

Without loss of the generality we can assume that $x_n \notin \overline{\text{span}}$ $(x_m: m = 1,2,\dots, m \neq n)$, where $\overline{\text{span}}$ (A) is the smallest linear closed subspace X which contains A. By the Hahn-Banach theorem there exists a sequence $(x_n)_{n=1}^{\infty} \subset X^*$ such that:

$$< x_n,x_m^* > = \begin{cases} 1 & \text{when } m = n \\ 0 & \text{when } m \neq n . \end{cases}$$

Since $(x_n)_{n=1}^{\infty}$ is a sequence of real random variables defined on probability space $(X,\mathcal{B}(X),\mu)$ we get the following

<u>Theorem 2</u>.

(i) $(x_n^*)_{n=1}^{\infty}$ - standard α-stable sequence

(ii) $x = \sum_{n=1}^{\infty} < x,x^*_n> x_n \mod \mu$.

Proof.

We have

$$\int_X \exp\{i\, t< x,x_n^* > \}\ \mu(dx) = \chi_\mu(t\ x_n^*)$$

$$= \exp\{-|t|^\alpha \sum_{m=1}^{\infty} |< x_m,x_n^* >|^\alpha\} = \exp\{-|t|^\alpha\}$$

where x_n^* - standard α-stable variable ;

$$\int_X \exp\{i \sum_{n=1}^{N} tn < x,x_n^* >\}\mu(dx) = \chi_\mu(\sum_{n=1}^{N} tn\ x_n^*)$$

$$= \exp\{-\sum_{m=1}^{\infty} |< x_m, \sum_{n=1}^{N} tn\ x_n^* >|^\alpha\} = \exp\{-\sum_{n=1}^{N} |tn|^\alpha\}$$

where $(x_n^*)_{n=1}^{\infty}$ is a sequence of independent random variables

(ii) For all $x^* \varepsilon X^*$ we have :

$$\int_X \exp\{it < x,x^* >\}\mu(dx) = \exp\{-|t|^\alpha \sum_n^{\infty} |< x_n,x >|^\alpha\}$$

By virtue of (i) and Lévy theorem we receive that for all $x^* \varepsilon X^*$

$$< x,x^* > = \sum_n^{\infty} < x,x_n^* > < x_n,x^* > \qquad \text{mod } \mu .$$

So (ii) is a simple consequence of Ito-Nisio theorem ([12]).

In a further paper, for the sake of brevity and convenience, we will use the terminology introduced in the begiming of § 1 and §2 : $\Gamma, T, T^*, (x_n), (x_n^*)$.

It has been shown (see [1], [5],[23]) that support S_μ of symmetric stable measure μ is a closed linear subspace. As a consequence of [8] we have

Proposition 2.1.

$S_\mu = \overline{T^* 1\beta} = \overline{\text{span}}\ (x_n:\ n = 1,2,\ldots) = \overline{\text{span}}\ (S_\Gamma)$ where S_Γ is a support of spectral measure Γ.

By 2.1 we receive immediately

Proposition 2.2. (see [26])

The following conditions are equivalent:

(i) μ- mondegenerate $(< x,x^*> \not\sim 0,\quad x^* \varepsilon X^*)$

(ii) $S_\mu = X$,

(iii) $\mu(O) > 0$ fòr all open sets $O \subset X$,

(iv) Span (x_n, $n = 1,2,\ldots$) dense in X.

(v) $S_T = S$

Finally, considering then mapping

$$\varphi : X \longrightarrow R^{\infty}$$

$$x \longrightarrow (< x, x_n^* >)_{n=1}^{\infty}$$

and set

$$E = \{ x \in X : x = \sum_{n=1}^{\infty} < x, x_n^* > x_n \text{ converges } \}$$

on the basis of theorem 2 we obtain

Proposition 2.3.

(i) φ -continuous linear map

(ii) E - linear subspace such that $\mu(E) = 1$

(iii) φ/E is 1 - 1 unique.

§ 3 0-1 law for subgroups

Let $\mathcal{B}_{\mu}(X)$ be a completetion in measure μ Borel σ-algebra $\mathcal{B}(X)$.

Theorem 3

Let G be a μ-measurable subgroup (for generality) then $\mu(G) = 0$ or 1.

Proof. (see also [27])

Let $\varphi(\mu) = \mu \circ \varphi^{-1}$ be a image μ under mapping φ
On the basis theorem 2 we receive that

$$\varphi(\mu) = \prod_{n=1}^{\infty} \vartheta_n$$

where ϑ_n is a distribution function (on real axis) of random real variable X_n^* and consistently

$$X_{\vartheta_n}(t) = \exp\{-|t|^{\alpha}\} \qquad \forall\, n = 1,2, \ldots \quad .$$

Let assume now that $\mu(G) > 0$. In this case $E \cap G$ - subgroup and $\mu(E \cap G) > 0$ (Proposition 2.3). Since μ-Radon measure, there exists σ-compact subgroup H such that

$$\mu(H) = \mu(E \cap G) > 0$$

Hence, $\varphi(H)$ - subgroup, $\varphi(H) \in \mathcal{B}(R^\infty)$ and

$$\mu(\varphi)(\varphi(H)) = \mu(H) > 0$$

and $\varphi(\mu)(\varphi(H)) = 1$ (the 0-1 law in R^∞).

From part b) of Proposition 2.3 we obtain

$$H = E \cap \varphi^{-1}(\varphi(H)).$$

The theorem is prooved.

§ 4. Set of admissible translations

For each $x \in X$

$$\mu_x(B) = \mu(B-x), \qquad \forall B \in \mathcal{B}(X)$$

is called a admissible translation related to μ if $\mu_x \ll \mu$. We denote by A_μ the set of all admissible translations related to μ.

Proposition 4.1. (see [28])

(i) A_μ - linear set

(ii) If G - subgroup and $\mu(G) > 0 (=1)$, then $G \supset A_\mu$ and $\mu_x \perp \mu$ for all $x \notin G$

(iii) $A_\mu \in \mathcal{B}(X)$.

Corollary 4.2.

If $\mu_x \ll \mu$ then $\mu_x \sim \mu$.

Proof.

Let $\mu_x(B) = 0$, $\mu(B-x) = 0$. By condition $x \in A\mu$ we receive $-x \in A\mu$ and $\mu(B) = \mu(B-x+x) = 0$.

Theorem 4. $\mu_x \sim \mu$ if and only if $x \in T\, l_2^*$, i.e. $A_\mu = T^* l_2$.

Proof. Let E, φ be as in Proposition 2.3. Then there exists σ-compact subgroup $E_0 \subset E$ such that $\mu(E_0) = 1$. Let $E_0 = \bigcup_{n=1}^{\infty} K_n$ where K_n are compacts in X. We have

$$\varphi(B) = \bigcup_{n=1}^{\infty} \varphi(B\, K_n)$$

for every $B \in \mathcal{B}(E_o)$. Since φ/E_o is one - to - one, K_n is compact and $\varphi(B) \in \mathcal{B}(R^\infty)$ for $\forall\ B \in \mathcal{B}(E_o)$ we obtain by the Kuratowski-Suslin Theorem (see [17]) $\varphi(B\ K_n) \in \mathcal{B}(R^\infty)$. This shows that φ^{-1} is a measurable mapping from R^∞ into E_o. Now it is easy to check that

$$A_\mu = E_o \cap A_\mu = \varphi^{-1}(A_{\varphi(\mu)}) .$$

On the other hand it is known (see [19], [28]) that

$$A_{\varphi(\mu)} = \{(a_n) \in R^\infty : \sum_{n=1}^{\infty} |a_n|^2 < \infty \} = l_2$$

and consequently

$$A_\mu = \{x \in X : \sum_{n=1}^{\infty} |< x, x_n^* >|^2 < \infty \}$$

in 1 we have shown that

$$T^* : l_\beta \longrightarrow X$$
$$b = (b_k) \longrightarrow \sum_{k=1}^{\infty} b_k x_k$$

Note that $\beta \geq 2$, so $l_2 \subset l_\beta$. Take $b \in l_2$. Now it is clear that

$$\sum_{n=1}^{\infty} |< \sum_{k=1}^{\infty} b_k x_k, x_n^* >|^2 = \sum_{n=1}^{\infty} |b_n|^2 < \infty .$$

Conversly let $x \in A_\mu$. Taking $b = (< x, x_k^* >)_{k=1}^{\infty} \in l_2$ we obtain

$$T^* b = \sum_{k=1}^{\infty} < x, x_k^* > x_k = x$$

(as $A_\mu \in E$) i.e. $A_\mu \subset T^* l_2$.

§ 5. Measurable linear functionals

Let f be a mapping from X into R^1. f is called linear μ-measurable functional if f is a μ-measurable mapping and there exists a set $D_f \in \mathcal{B}_\mu(X)$ such that D_f is a linear set, $\mu(D_f) = 1$ and f is linear on D_f.

Lemma 5.1. If $f(x_n) = 0$ for all $n = 1,2,\ldots$, then $f(x) = 0$ μ - a.e.

Proof. First note that

$$(x_n)_{n=1}^{\infty} \subset A_\mu \subset G = E \cap D_f, \quad \mu(G) = 1$$

Moreover for every $x \in G$ we have

$$f(x) = f(\sum_{n=1}^{N} < x,x_n^* > x_n) + f(\sum_{N+1}^{\infty} < x,x_n^* > x_n) =$$

$$= \sum_{n=1}^{\infty} < x,x_n^* > f(x_n) + f(\sum_{N+1}^{\infty} < x,x_n^* > x_n) =$$

$$= 0 \qquad + f(\sum_{N+1}^{\infty} < x,x_n^* > x_n).$$

Hence by Theorem 2 and Kołmogorov 0-1 law we claim that f takes constant value μ-a.e. As f is linear on G $f = 0$ μ- a.e.

<u>Lemma 5.2.</u> $(f(x_n))_{n=1}^{\infty} \in l_\alpha$

<u>Proof</u>. By Theorem 2 it is easy to see that

$$\int_G \exp\{it\, f(x)\}\mu(dx) =$$

$$\exp\{-|t|^\alpha \sum_{n=1} |f(x_n)|^\alpha \cdot \int_G \exp\{it\, f(\sum_{N+1}^{\infty} < x,x_n^* > x_n)\}\mu(dx)$$

and consequently for all $N \geqslant 1$

$$\int_X \exp\{it\, f(x)\}\mu(dx) \leqslant \exp\{-|t|^\alpha \sum_{n=1}^{N} |f(x_n)|^\alpha\}$$

Therefore if $(f(x_n))_{n=1}^{\infty} \notin l_\alpha$ then

$$\int_X \exp\{it\, f(x)\}\mu(dx) = 0 \quad \forall\, t \neq 0 ,$$

the contradiction. This proves the lemma.

<u>Lemma 5.3.</u> If $(a_n)_{n=1}^{\infty} \in l_\alpha$ then

$$\mu\{ x : \sum_{n=1}^{\infty} a_n < x_n, x_n^* > \text{ is convergent } \} = 1$$

and consequently there exists a μ-measurable linear functional f such that

$$f(x) = \sum_{n=1}^{\infty} a_n < x,x_n^* > \quad \mu\text{-a.e.}$$

<u>Proof.</u> Put

$$U = \{ x : \sum_{n=1}^{\infty} a_n < x, x_n^* > \text{ converges } \}$$

Obviously U is a linear set. Hence by Theorem 2 and Theorem Lévy we have $\mu(U) = 1$ (because $(a_n)_{n=1}^{\infty} \in l_\alpha$) . Now consider an algebraic decomposition of X in the form

$$X = U \oplus V$$

and put

$$f(x) = \sum_{n=1}^{\infty} e_n < x_U, x_n^* > \quad , \quad \forall\, x \in X$$

where $x = x_U + x_V$, $x_U \in U$, $x_V \in V$. It is easy to check that f is a required functional.

From Lemmas 5.1 - 5.3 it follows

Proposition 5.4. If f is a μ-measurable linear functional then f has unique representation in the form

$$f(x) = \sum_{n=1}^{\infty} a_n < x, x_n^* > \qquad \text{mod. } \mu \qquad (5)$$

where $(a_n)_{n=1}^{\infty} = (f(x_n))_{n=1}^{\infty} \in l_\alpha$.

Denote by $\mathcal{L}_\mu(X)$ the set of all μ-measurable linear functionals. $\mathcal{L}_\mu(X)$ is a linear set (with natural addition and scalar multiplication). By Theorem 2 and Proposition 5.4 every element of $\mathcal{L}_\mu(X)$ is a stable random variable of the same index α. Therefore if

$$0 < p < \bar{\alpha} = \begin{cases} \alpha & \text{for } 0 < \alpha < 2 \\ +\infty & \text{for } \alpha = 2, \end{cases}$$

then $\mathcal{L}_\mu(X) \subset L_p(X,\mu,R^1)$.

Theorem 5. (a) For every $0 < p < \bar{\alpha}$ $\mathcal{L}_\mu(X)$ is a closed linear subspace of $L_p(X,\mu,R^1)$.

(b) The set of all linear continuous functional is dence in $\mathcal{L}_\mu(X)$.

(c) $\mathcal{L}_\mu(X)$ isomorphic with l_α .

Proof.(a) Let $(f_n)_{n=1}^{\infty} \subset \mathcal{L}_\mu(X)$ and $f_n \longrightarrow f$ in the norm of $L_p(X,\mu,R^1)$. Then there exists a subsequence $(f_{n_k})_{n=1}^{\infty}$ such that $f_{n_k} \longrightarrow f$ mod μ . This follows $f \in \mathcal{L}_\mu(X)$.

(b) This is a corollary of Proposition 5.4 and the fact that the series in (5) converges in the norm of $L_p(X,\mu,R^1)$.

(c) Consider the mapping

$$A : \mathcal{L}_\mu(X) \longrightarrow l_\alpha$$

$$f \longrightarrow (f(x_n))_{n=1}^{\infty}$$

It is clear that A is linear. From Lemmas 5.1 - 5.3 it follows that A is one - to - one and $A(\mathcal{L}_\mu(X)) = l_\alpha$. Moreover, we recall that f is stable random variable with characteristic function

$$\chi_f(t) = \int_X \exp\{it\, f(x)\mu(dx) = \exp\{-|t|^\alpha \sum_{n=1}^{\infty} |f(x_n)|^\alpha \} =$$

$$= \exp\{-|t|^\alpha \|Af\|^\alpha\} .$$

Hence for $0 < p < \bar{\alpha}$ we have

$$\int_X |f(x)|^p \mu(dx) = A_\alpha(p) \| Af \|^p ,$$

where $A_\alpha(p)$ is the constant depending only on α and p.

So the theorem is proved.

§ 6. Measurable linear operators

Let X,Y be a Banach spaces and $A: X \rightarrow Y$. Suppose μ is a given measure on X. A is called μ-measurable linear operator if A is a measurable mapping with respect to $(\mathcal{B}_\mu(X), \mathcal{B}(Y))$ and there exists a linear set $D_A \in \mathcal{B}_\mu(X)$ such that $\mu(D_A) = 1$ and A is linear on D_A.

Denote by $\mathcal{L}_\mu(X,Y)$ the set of all μ-measurable linear operators from X into Y. As befoer, μ denotes here a symmetric stable measure with discrete spectral measure.

Proposition 6.1. $A \in \mathcal{L}_\mu(X,Y)$ may be uniquely represented in the form :

$$Ax = \sum_{n=1}^{\infty} < x, x_n^* > A\, x_n \qquad (\mathrm{mod.}\ \mu) \tag{6}$$

Proof. For every $y_0^* \in Y^*$, $y_0^* A : X \rightarrow R^1$ a μ-measurable linear functional. Therefore by Proposition 5.4 we obtain

$$(y_0^* A)(x) = \sum_{n=1}^{\infty} < x, x_n^* > y_0^*(Ax_n) \qquad (\mathrm{mod.}\ \mu)$$

Regording A as a random element with values in Y we have that $(< \cdot, x_n^* > Ax_n)_{n=1}^{\infty}$ is a sequence of independent symmetric random elements and by Theorem Ito-Nisio (see [12]) we obtain (6) The uniqueness of (6) follows also by Proposition 5.4.

Let $A \in \mathcal{L}_\mu(X,Y)$. Then A is a stable random element with index α . Hence by the result of de Acosta [1] we obtain

$$\int_X \| Ax\|^p \mu(dx) < \infty \qquad \text{for} \quad 0 < p < \bar{\alpha} ,$$

i.e. $\mathcal{L}_\mu(X,Y) \subset L_p(X,\mu,Y)$

By Proposition 6.1 and the result of Hoffman-Jørgensen [11] it follows.

<u>Proposition 6.2.</u> (a) $\mathcal{L}_\mu(X,Y)$ is a closed linear subspace of $L_p(X,\mu,Y)$

(b) The set of all linear continuous operators from X into Y is dense in $\mathcal{L}_\mu(X,Y)$

Now let $0 < \alpha < 2$. Then by Theorem 2 and Proposition 6.1 we have

$$(Ax_n)_{n=1}^\infty \in l_\alpha(Y) = \{(y_n)_{n=1} \subset Y : \sum_{n=1}^\infty \|y_n\|^\alpha < \infty\}$$

(as every Banach space has stable cotype α for $0 < \alpha < 2$).
Consider the mapping

$$Q : \mathcal{L}_\mu(X,Y) \longrightarrow l_\alpha(Y)$$

$$A \longrightarrow (Ax_n)_{n=1}^\infty$$

<u>Theorem 6.</u> Let $0 < \alpha < 2$. The following conditions are equivalent

(a) Y has stable type α

(b) Q is an isomorphism of $\mathcal{L}_\mu(X,Y)$ onto $l_\alpha(Y)$.

<u>Proof.</u> (a) $\Longrightarrow$ (b) . Since $0 < \alpha < 2$ and Y has stable type α there exist constants C_1, C_2 such that

$$C_1(\sum_{n=1}^\infty \|Ax_n\|^\alpha)^{p/\alpha} \leq \int_X \|\sum_n < x,x_n^* > Ax_n\|^p \mu(dx) \leq$$

$$\leq C_2(\sum_{n=1}^\infty \|Ax_n\|^\alpha)^{p/\alpha}$$

(see [20]). Now let $(y_n)_{n=1}^\infty \in l_\alpha(Y)$. Then

$$\sum_{n=1}^\infty < x,x_n^* > y_n \in Y \qquad \text{mod } \mu \tag{7}$$

(as Y has stable type α). Put

$$Ax = \begin{cases} \sum_{n=1}^\infty < x,x_n^* > y_n & \text{if the series (7) converges} \\ 0 & \text{otherwise} \end{cases}$$

It is clear that $A\, x_n = y_n$.

(b) ==> (a) . Let $(y_n) \in l_\alpha(Y)$ and Q is an isomorphism between $\mathcal{L}_\mu(X,Y)$ and $l_\alpha(Y)$. Then there exists $A \in \mathcal{L}_\mu(X,Y)$ such that $A x_n = y_n$. By Proposition 6.1 it follows

$$\sum_{n=1}^{\infty} < x, x_n^* > y_n \in Y \qquad \text{mod } \mu ,$$

i.e. Y has stable type α .

For the case $\alpha = 2$ **we can show** the following proposition.

<u>Proposition 6.3</u>. The following conditions are equivalent:

(a) Y is isomorphic to a Hilbert space,

(b) Q is an isomorphism $\mathcal{L}_\mu(X,Y)$ onto $l_2(Y)$.

Finally consider the case $Y = l_p$, $1 \leq p < \bar{\alpha}$. Note that for $p < \bar{\alpha}$ l_p does not have stable typ α.
Therefore Theorem 6 does not apply in this case.

<u>Proposition 6.4.</u> For $1 \leq p < \bar{\alpha}$ $\mathcal{L}_\mu(X,l_p)$ is isomorphic to Banach space of matrices (a_{i_j}) satisfying the condition

$$\sum_{j=1}^{\infty} \Big(\sum_{i=1}^{\infty} |a_{i_j}|^\alpha\Big)^{p/\alpha} < \infty , \tag{8}$$

<u>Proof</u>. First, is is easy to check that the set $\mathcal{A}$ of matrices satisfying (8) is a Banach space with the norm

$$\Big[\sum_j \Big(\sum_i |a_{i_j}|^\alpha\Big)^{p/\alpha}\Big]^{\frac{1}{p}} .$$

Now let $A \in \mathcal{L}_\mu(X,l_p)$. Then by Proposition 6.1 and the result form the paper [7] we have

$$(< A x_i, b_j^* >)_{i j=1}^{\infty} \in \mathcal{A} ,$$

where $(b_j^*)_{j=1}^{\infty}$ is a natural basis in l_p^* . Consider the mapping

$$\Phi : \mathcal{L}_\mu(X,l_p) \longrightarrow \mathcal{A}$$

$$A \longrightarrow (< A x_i, b_j^* >)_{i,j=1}^{\infty}$$

We have

$$\int_X \|Ax\|^p \mu(dx) = \sum_{j=1}^{\infty} \int_X | < Ax, b_j^* > |^p \mu(dx) =$$

$$= A_\alpha^p(p) \sum_{j=1}^{\infty} \Big(\sum_{i=1}^{\infty} | < A x_i, b_j^* > |^\alpha\Big)^{p/\alpha}$$

Now let $(a_{i_j}) \in \mathcal{A}$. Then $y_i = (a_{i_j})_{j=1}^{\infty} \in l_p$ and by the result from [7] we obtain

$$\sum_{n=1}^{\infty} < x, x_n^* > y_n \in l_p \quad \text{mod } \mu .$$

Put

$$Ax = \sum_{n=1}^{\infty} <x, x_n^* > y_n .$$

It is easy to see that $A \in \mathcal{L}_{\mu}(X,Y)$ and $\Phi A = (a_{i_j})_{i,j=1}^{\infty}$, this proves the proposition.

Acknowledgment: The final form of this paper was prepared dusing the second author's stay at the Technical University of Wrocław. The author wants to express his gratitude to Professors S.Gładysz, C.Ryll-Nardzewski and A.Weron for the hospitality offered to him during his stay in Wrocław.

References

[1] A.De Acosta, Stable measures and seminorms, Ann.Prob.3 (1975), 865-875.

[2] C.Borell, Gaussian Radon measures on locally convex spaces, Math.Scad. 38 (1976), 265-283,

[3] P.L.Brockett, Support of infinitely devissible measures on Hilbert space, Ann.Prob.5 (1977), 1012-1017.

[4] T.Byczkowski, Some results concerning Gaussian measures on matric linear spaces, Lect.Notes in Math. 656 (1978), 1-16.

[5] Dang Hung Thang, Nguyen Zui Tien, Support of probability measure in linear topological spaces (in Vietnamese), Proc. of the Second Math. conference, Hanoi 1977.

[6] Dang Hung Thang, Nguyen Zui Tien, Linear measurable functional with respect to stable measure in Banach Spaces (in Vietnamese), Proc.of the Second Math.Conference, Hanoi 1977.

[7] Dang Hung Thang, Nguyen Zui Tien, On symmetric stable measures on space l_p, $1 \leq p < +\infty$, Teor.Verojat i Primen.

[8] Dang Hung Thang, Nguyen Zui Tien, On stable measures on Banach spaces (Preprint)

[9] R.M. Dudley, Singularity of measures on linear spaces, Z.Wahr. 6(1966), 129-132.

[10] R.M. Dudley and M.Kanter, Zero-one laws for stable measures, Proc.AMS 45(1974), 245-252.

[11] T.Hoffmann - Jørgensen, Sums of independent Banach space valued random variables, Studia Math.52(1974), 159-186.

[12] K.Ito, M.Nisio, On the convergence of sums of independent Banach space valued random variables, Osaka T.Math.5 (1968) 35-48.

[13] N.C. Jain, A zero-one law for Gaussian processes, Proc.AMS 29 (1971), 585-587.

[14] N.C. Jain, and G.Kallianpur, Norm convergent expansions for Gaussian processes in Banach spaces, Proc. AMS 25(1970), 890-895.

[15] G.Kallianpur, Zero-one laws for Gaussian processes Trans.AMS 149(1970), 199-211.

[16] Nguyen Zui Tien, The structure of linear measurable functionals with respect to Gaussian measure in Banach spaces, Teor. Verojat i Primen 23(1978)

[17] K.R.Parthasarathy, Probability Measures on Metric spaces, New York-London, 1967.

[18] V.Paulauskas, On stable distributions on separable Banach spaces, Second Vilnius conference on Probability theory and Mathematical statistics. Abstracts of communications, Vilnius, 1977, 166-167.

[19] L.A.Shepp, Distinguishing a sequence of random variables from a translate of itself, Ann.Math.Stat.36 (1965),1107-1112.

[20] L.Schwartz, Les espaces de type et cotype 2, d'apres Bernard Maurey el leurs applications, Ann.Inst.Fourier, Grenoble 24(1974), 179-188.

[21] A.V.Skorokhod, On admissible translations of measures in Hilbert space. Teor.Verojat i Primen, 15(1970), 577-598 in Russian.

[22] V.N. Sudakov, Linear sets with quasi-invariant measures, Dokl.Akad.Nauk. SSR 127(1959),524-525 (In Russian)

[23] B.S.Rajput, On the support of symmetric infinitely divisible and stable probability on LCTVS, Proc.AMS 66(1977),331-334.

[24] A.Tortrat, Sur les lois e(λ) dans les espaces vectorielles Applications aux lois stables, Z.Wahr.27(1976), 175-182.

[25] K.Urbanik, Random linear functionals and random integrals, Collog.Math., 33(1975), 255-263.

[26] N.N.Vakhania, Nguyen Zui Tien, On the probability measures in Banach spaces (in Vietnamese), Tap San Toan Hoe II, 3-4 (1974), 1-19.

[27] T. Zinn, Zero-one laws for no-Gaussian measures, Proc.AMS 44(1974), 179-185.

[28] T.Zinn, Admissible translations of stable measures, Studia Math. 54(1976), 245-257.

University of Hanoi
Department of Mathematics
Vietnam

A CHARACTERIZATION OF SOME PROBABILITY DISTRIBUTIONS

by

Nguyen Van Thu (Hanoi)

Abstract. The aim of the present paper is to give a charcterization of a probability distribution μ_{n+1} $(n=1,2,\ldots)$ on a real separable Banach space $(X, \|\cdot\|)$ such that for some probability distributions $\mu_1,\ldots,\mu_n$ and $c_1,\ldots,c_n \in (0,1)$ the following convolution equations hold:

$$\mu_k = T_{c_k}\mu_k * \mu_{k+1} \qquad (k=1,2,\ldots,n) . \tag{1}$$

The study of decomposability semigroups associated with probability distributions on linear spaces (see [3]) leads to the problem of characterization of distributions μ_{n+1} satisfying the equations (1).

Let (Ω,F,P) be a fixed probability system. For any X-valued random variable Z with distribution μ and for every real number c let $T_c\mu$ denote the distribution of cZ. Throughout the paper we shall fix a sequence Z, $Z_{k_1,\ldots,k_n}$ $(k_1,\ldots,k_n=0,1,2,\ldots)$ of i.i.d. X-valued random variables with distribution μ_{n+1}. Without loss of generality one may assume that $\mu_{n+1} \neq \delta_o$.

Theorem. The equations (1) hold for some probability distributions $\mu_1,\ldots,\mu_n$ on X and $c_1,\ldots,c_n \in (0,1)$ if and only if the random series

$$\sum_{k_1,\ldots,k_n=0}^{\infty} c_1^{k_1} \cdots c_n^{k_n} Z_{k_1,\ldots,k_n} \tag{2}$$

is convergent (in the norm topology) with probability 1 or equivalently, if and only if

(3) $$E \log^n \max(1, \|Z\|) < \infty$$

Proof. We first suppose that (2) is convergent with probability 1. Put $b = \min(c_1, \dots, c_n)$ then we have $0 < b < 1$ and the random series

(4) $$\sum_{k=0}^{\infty} c^k \sum_{k_1+\dots+k_n=k} \|Z_{k_1,\dots,k_n}\|$$

is convergent with probability 1 for all $c \in (0,b)$. Consequently,

(5) $$\limsup_k \sqrt[k]{\sum_{k_1,\dots,k_n=k} \|Z_{k_1,\dots,k_n}\|} \le \frac{1}{b} \qquad (P.1)$$

Moreover, by the elementary inequality

$$\max_{k_1+\dots+k_n=k} \|Z_{k_1,\dots,k_n}\| \le \sum_{k_1+\dots+k_n=k} \|Z_{k_1,\dots,k_n}\|$$

$$\le \binom{k+n-1}{n-1} \max_{k_1+\dots+k_n=k} \|Z_{k_1,\dots,k_n}\|$$

it follows that

$$\limsup_k \sqrt[k]{\max_{k_1+\dots+k_n=k} \|Z_{k_1,\dots,k_n}\|} \le \frac{1}{b} \qquad (P.1)$$

or, equivalently, for every $a > \frac{1}{b}$ we have

(6) $$P(\overline{\lim_k} \{ \max_{k_1+\dots+k_n=k} \|Z_{k_1,\dots,k_n}\| > a^k \}) = 0.$$

By Borel-Cantelli lemma the last condition is equivalent to the following:

(7) $$\sum_{k=0}^{\infty} P(\{ \max_{k_1+\dots+k_n=k} \|Z_{k_1,\dots,k_n}\| > a^k \}) < \infty$$

Putting $p_k = P(\{\|Z\| > a^k\})$, $q_k = 1 - p_k$, $r_k = \binom{k+n-1}{n-1}$ and taking into account the equality

$$P(\{\max_{k_1+\ldots+k_n=k} \|Z_{k_1,\ldots,k_n}\| > a^k\}) = 1 - q_k^{r_k}$$

we infer that (7) is equivalent to the following condition:

(8) $$\sum_{k=1}^{\infty} 1 - q_k^{r_k} < \infty .$$

Hence it follows that

(9) $$\lim_{k\to\infty} (1 - p_k)^{r_k} = 1$$

or, equivalently,

(10) $$\lim_{k\to\infty} r_k \log (1 - p_k) = 0 .$$

Since $\log 1/1-x \geq x$ if $0 \leq x < 1$ the condition (10) implies that

(11) $$\lim_{k\to\infty} r_k p_k = 0 .$$

Now, let $A_1,\ldots,A_m$ be independent events such that $p(A_i) = p$ for every $i=1,\ldots,m$. Then from inequalities

$$\sum_{i=1}^{m} P(A_i) - \sum_{i<j} P(A_i)P(A_j) \leq P(\bigcup_{i=1}^{m} A_i) \leq \sum_{i=1}^{m} P(A_i)$$

we get

(12) $$mp - \frac{m(m-1)}{2} p^2 \leq P(\bigcup_{i=1}^{m} A_i) \leq mp .$$

Consequently, by (12), we have

(13) $$r_k p_k - \frac{r_k(r_k - 1)}{2} p_k^2 \leq$$

$$\leq P(\{\max_{k_1+\ldots+k_n=k} \|Z_{k_1,\ldots,k_n}\| > a^k\}) \leq r_k p_k .$$

Further, by (11) it follows that $1 - \frac{r_k - 1}{2} p_k \geq \frac{1}{2}$ for sufficiently large k. Therefore for such k

$$r_k p_k - \frac{(r_k - 1)}{2} p_k^2 \geq \frac{1}{2} r_k p_k$$

which together with (13) imply that (7) holds if and only if

$$(14) \qquad \sum_{k=1}^{\infty} r_k p_k < \infty$$

Consider the sum $S_a(x) = \sum_{k=1}^{\infty} r_k X_{(a^k,\infty)}(x) \quad (x > 0)$, where X_A denotes the indicator of a set A. By a simple computition we have

$$(15) \qquad S_a(x) = \sum_{k=1}^{[\log_a x]} r_k = \sum_{k=1}^{[\log_a x]} \binom{k+n-1}{n-1} ,$$

where for a real number d by $[d]$ we denote its integer part. Furthermore, we have

$$(16) \qquad \frac{[\log_a x]^n}{(n-1)!} \leq S_a(x) \leq \frac{[\log_a x][\log_a x + n]^{n-1}}{(n-1)!}$$

Let G denote the distribution of $\|Z\|$. Then

$$\sum_{k=1}^{\infty} r_k p_k = \int_0^{\infty} \sum_{k=1}^{\infty} r_k X_{(a^k,\infty)}(x) G(dx) = \int_0^{\infty} S_a(x) G(dx) .$$

Hence it follows that (14) holds if and only if

$$(17) \qquad \int_1^{\infty} S_a(x) G(dx) < \infty ,$$

and by virtue of (16) the condition (17) is equivalent to the following

$$(18) \qquad \int_1^{\infty} \log^n x \, G(dx) < \infty .$$

But the condition (18) is the same as (3) hence we have proved that the convergence with probability 1 of the random series (2) implies (3).

Conversely, suppose that (3) is satisfied. Then, by the above argument, it follows that for every $a > 1$ the condition (6) is satisfied. Therefore the series (4) is convergent with probability 1 for all $c \in (0,1)$ and consequently, for any $c_1, \ldots, c_n \in (0,1)$, the power random series (2) is convergent with probability 1, which together with the above result show that the convergence with probability 1 of the random series (2) is equivalent to the condition (3).

Proceeding successively, we shall prove that the condition (1) is equivalent to (3). Let μ_{n+1} be a sulution of the convolution equations (1). Then for any $m_1,\dots,m_n = 1,2,\dots$ we have

$$\mu_k = T_{c_k^{m_k+1}}\mu_k * \mathop{*}_{r_k=0}^{m_k} T_{c_k^{r_k}}\mu_{k+1} \qquad (k=1,\dots,n). \tag{19}$$

Putting

$$\nu_{m_1,\dots,m_n} = T_{c_1^{m_1+1}}\mu_1 * T_{c_2^{m_2+1}} \mathop{*}_{r_1=0}^{m_1} T_{c_1^{r_1}}\mu_2 * \tag{20}$$

$$* \; T_{c_3^{m_3+1}} \mathop{*}_{r_1=0}^{m_1} \mathop{*}_{r_2=0}^{m_2} T_{c_1^{r_1}c_2^{r_2}}\mu_3 * \dots *$$

$$* \; T_{c_n^{m_n+1}} \mathop{*}_{r_1=0}^{m_1} \dots \mathop{*}_{r_{n-1}=0}^{m_{n-1}} T_{c_1^{r_1}\dots c_{n-1}^{r_{n-1}}}\mu_n$$

and taking into account the equations (19) we get the formula

$$\mu_1 = \nu_{m_1,\dots,m_n} * \mathop{*}_{r_1=0}^{m_1} \dots \mathop{*}_{r_n=0}^{m_n} T_{c_1^{r_1}\dots c_n^{r_n}}\mu_{n+1} \; . \tag{21}$$

$(m_1,\dots,m_n = 1,2,\dots)$. Let $\tilde{\mu}$ denote the symetrization of a distribution μ and let $Z'_{k_1,\dots,k_n}$ $(k_1,\dots,k_n = 0,1,2,\dots)$ be i.i.d. X-valued random variables with distribution μ_{n+1} such that $\{Z_{k_1,\dots,k_n}\}$ and $\{Z'_{k_1,\dots,k_n}\}$ are independent. By virtue of (21) we have

$$\tilde{\mu}_1 = \tilde{\nu}_{m_1,\dots,m_n} * \mathop{*}_{r_1=0}^{m_1} \dots \mathop{*}_{r_n=0}^{m_n} T_{c_1^{r_1}\dots c_n^{r_n}}\tilde{\mu}_{n+1} \tag{22}$$

which, by Ito-Nisio theorem ([1], Theorem 4.1) follows that the random series

$$\sum_{k_1,\dots,k_n=0}^{\infty} c_1^{k_1} \dots c_n^{k_n} (Z_{k_1,\dots,k_n} - Z_{k_1,\dots,k_n}) \tag{23}$$

is convergent with probability 1. From the first part of the proof it follows that

(24) $$E \log^n \max (1, \|Z_{k_1,\dots,k_n} - Z'_{k_1,\dots,k_n}\|) < \infty$$

Consequently, there exists a vector $x \in X$ such that

(25) $$E \log^n \max (1, \|Z_{k_1,\dots,k_n} + x\|) < \infty$$

Again by the first part of the proof it follows that the random series

(26) $$\sum_{k_1,\dots,k_n=0}^{\infty} c_1^{k_1} \dots c_n^{k_n}(Z_{k_1,\dots,k_n} + x)$$

is convergent with probability 1. But the last series is convergent with probability 1 if and only if the series (2) is convergent with Probability 1. Hence, by the first part of the proof, we have (3). Conversely, suppose that (3) holds. Then the series (2) is convergent with probability 1. Consequently, as it can be seen, (1) holds. The Theorem is thus fully proved.

<u>Remark</u>. For the particular case when $X = R^1$ and $n = 1$ the theorem was obtained by Zakusilo ([2], Theorem 1).

Let us denote

$$\phi_n(x) = \sum_{k=1}^{\infty} \frac{1}{2^k} \log^n \max (1, kx) \qquad (x \ge 0).$$

Then ϕ_n is a continuous nondecreasing function on R_+ satisfying the Δ_2-condition and vanishing only at the origin. Let $L_{\phi_n}(X) = L_{\phi_n}(\Omega,F,P,X)$ denote the Orlicz space of X-valued random variables Z such that

(27) $$E \phi_n(\|Z\|) < \infty .$$

Since the conditions (3) and (27) are equivalent we get the following corollary:

Corollary: The class of all $\mathcal{X}$-valued random variables Z with distributions μ_{n+1} satisfying the equations (1) coincides with the Orlicz space $L_{\phi_n}(\mathcal{X})$.

References

[1] K. Ito, M. Nisio, On the convergence of sums of independent Banach space valued random variables, Osaka Journal of Math. 5 (1968) pp.35-48.

[2] O.K. Zakusilo, On classes of limit distributions in some scheme of summing up (in Russian), Probability Theory and Mathematical Statistics, vol.12, Kiev 1975.

[3] K. Urbanik, Lévy's probability measures on Banach spaces, Studia Math. Tom LXIII, Fasc.3 (1978), pp.283-308.

Institute of Mathematics
Hanoi, Vietnam.
208 Đ Đội - cấn , Hànội.

BANACH SPACES RELATED TO α-STABLE MEASURES

NGUYEN ZUI TIEN and ALEKSANDER WERON*

ABSTRACT: A class V_α of Banach spaces is defined by the inequality (2) for α-stable measures, where $1 \leq \alpha \leq 2$. It is shown that if $\alpha < 2$ then there exists a Banach space of α-stable type which does not belong to V_α. A characterization of α-stable Radon measure in α-stable spaces from the V_α class is given for $1 < \alpha < 2$.

AMS(MOS) subject classification. Primary 60B05, 60E07; Secondary 46B20.
Key words and phrases: α-stable measures, cylindrical measure, Radon measure, α-stable type, V_α-class of Banach spaces.

*Partially written during the second author's stay at Southern Illinois University, Carbondale, IL 62901.

0. Let X be a real Banach space, X^* the dual space, $\mathcal{B}(X)$ the Borel σ-algebra and $A(X)$ the algebra of cylinders. A finitely additive set function $\nu: A(X) \to [0,1]$, $\nu(X) = 1$, which is σ-additive on each subalgebra consisting of all cylinders with a fixed collection of determinant functionals is called a cylindrical probability measure. Every functional $x^* \in X^*$ generates the measure ν_{x^*} in $\mathbb{R}$. The characteristic functional χ_ν of a cylindrical measure ν is defined by the equality

$$\chi_\nu(x^*) = \hat{\nu}_{x^*}(1), \qquad x^* \in X^*,$$

where on the right side is the ordinary Fourier transform. A σ-additive measure μ defined on $\mathcal{B}(X)$ is said to be tight if for each $\varepsilon > 0$ there exists a compact $K \subset X$ with the property $\mu(K) > 1 - \varepsilon$. A cylinder measure ν is a Radon measure if it admits the tight extension.

A cylinder measure ν is called α-stable, $1 \le \alpha \le 2$ if its characteristic functional has the form

$$(1) \qquad \chi_\nu(x^*) = \exp\{-\|Tx^*\|^\alpha\}, \qquad x^* \in X^*,$$

where T is a linear bounded operator from X^* into ℓ_α.

For the definitions of p-Rademacher type (cotype) and p-stable type (cotype) of Banach spaces and other facts related to Probability in Banach Spaces we refer to [10].

1. In finite dimensional spaces by employing the Bochner theorem the following inequality

$$|1 - f(x^*)| \le |1 - \chi_\mu(x^*)|, \quad x^* \in X^*$$

where functional $f(\cdot)$ is positive definite, $f(0) = 1$ and $\chi_\mu(\cdot)$ is the characteristic functional of a probability measure μ, implies that there exists a probability measure ν for which $f(\cdot)$ is its characteristic functional. Such result for Banach spaces was announced in [8] cf. also [9]. But unfortunately it turns out to be false in general. In this paper we concentrate on the case of α-stable measures $1 \le \alpha \le 2$. It is shown that for 2-stable (Gaussian) measures this fact holds in any Banach space, but for any $\alpha < 2$ there exists a Banach space for which it does not. This motivated us to introduce the following

DEFINITION. We say that a Banach space X belongs to the class V_α, $1 \le \alpha \le 2$, if for each α-stable Radon measure μ and for each α-stable cylindrical measure ν the inequality

$$(2) \qquad |1 - \chi_\nu(x^*)| \le |1 - \chi_\mu(x^*)|, \quad x^* \in X^*,$$

implies that ν is a Radon measure too.

It is easy to observe that Banach spaces of S-type (i.e., for which there exists the Sazonov topology, cf. [7]) belong to V_α for each α.

THEOREM 1. Each Banach space belongs to V_2.

Proof. Let μ be a 2-stable (Gaussian) Radon measure and ν be a cylindrical 2-stable (Gaussian) measure which satisfies the inequality (2). We assume without loss of generality that μ and ν are symmetric. The inequality (2) is in this case equivalent to the following one

$$\langle R_\nu x^*, x^*\rangle \leq \langle R_\mu x^*, x^*\rangle,$$

where R_ν and R_μ are covariance operators of ν and μ, respectively. But it is known, see for example [3], that if the above inequality holds then ν is a Radon measure too. □

Let X, Y be Banach spaces. B(X,Y) denotes the space of all linear and bounded operators from X into Y with the usual operator norm. The distance d(X,Y) between Banach spaces X, Y is defined as $\inf\|T\|\|T^{-1}\|$, where the infimum is taken over all invertible T in B(X,Y). A Banach space X is said to be embeddable in L^α (cf. [6]) if there is a measure m and a subspace E of $L^\alpha(m)$ with $d(X,E) \leq \varepsilon$.

THEOREM 2. Let X be a Banach space of α-stable type and let $X \in V_\alpha$ for $1 \leq \alpha < 2$. Then X is embeddable in L^α.

Proof. Let X be not embeddable in L^α. Then by Lindenstrauss-Pełczyński theorem ([6], Th. 7.3) there exist sequences $\{x_n\}$ and $\{y_n\}$ of elements of X such that

$$\sum_{n=1}^{\infty} \|x_n\|^\alpha < +\infty, \quad \sum_{n=1}^{\infty} \|y_n\|^\alpha = +\infty \tag{3}$$

and

$$\sum_{n=1}^{\infty} |\langle x_n, x^*\rangle|^\alpha \geq \sum_{n=1}^{\infty} |\langle y_n, x^*\rangle|^\alpha \qquad \forall x^* \in X^*. \tag{4}$$

Let $\mu = L\left(\sum_{n=1}^{\infty} \gamma_n^{(\alpha)} x_n\right)$, where $\{\gamma_n^{(\alpha)}\}$ is a standard α-stable sequence i.e., a sequence of independent identically distributed random variables with the characteristic functional $\exp\{-|t|^{\alpha}\}$. The series is convergent a.s. since X has α-stable type and $\sum_{n=1}^{\infty} \|x_n\|^{\alpha} < \infty$. Consequently μ is a Radon measure. Its characteristic functional has the form

$$\chi_{\mu}(x^*) = \exp\{-\sum_{n=1}^{\infty} |\langle x_n, x^*\rangle|^{\alpha}\}.$$

Let us put

$$(5) \qquad \chi_{\nu}(x^*) = \exp\{-\sum_{n=1}^{\infty} |\langle y_n, x^*\rangle|^{\alpha}\}.$$

Thus χ_{ν} is the characteristic functional of a cylindrical α-stable measure. By (4) we have that the characteristic functionals of μ and ν satisfy the inequality (2). Taking into account that $x \in V_{\alpha}$ we conclude that ν is a Radon measure.

On the other hand let us consider the series $\sum_{n=1}^{\infty} \gamma_k^{(\alpha)} y_n$, where $\{\gamma_k^{(\alpha)}\}$ and $\{y_n\}$ are as before. By the Levy's theorem for each $x^* \in X^*$, $\sum_{n=1}^{\infty} \langle \gamma_k^{(\alpha)} y_n, x^*\rangle$ is convergent a.s. in $\mathbb{R}$ if $\sum_{n=1}^{\infty} |\langle y_n, x^*\rangle|^{\alpha} < \infty$. Moreover, by a comparison with (5), we observe that χ_{ν} is the characteristic functional of $\sum_{n=1}^{\infty} \gamma_n^{(\alpha)} y_n$. Consequently by Ito-Nisio theorem (cf. [10]) the series $\sum_{n=1}^{\infty} \gamma_n^{(\alpha)} y_n$ is a.s. convergent in X and the Borel-Cantelli Lemma yields

$$(6) \qquad \sum_{n=1}^{\infty} P\{\|\gamma_n^{(\alpha)} y_n\| > 1 < +\infty.$$

Since it is known ([5], p. 544) that for each sufficiently

large n and all $\alpha < 2$

$$P\{\|\gamma_n^{(\alpha)} y_n\| > 1\} = P\{|\gamma_n^{(\alpha)}| \geq \|y_n\|^{-1}\} \sim \|y_n\|^{\alpha},$$

hence by (6) we have $\sum_{n=1}^{\infty} \|y_n\|^{\alpha} < +\infty$, which is in contradiction with the second part of (3). □

REMARK 1. Let us note that Theorem 1 is not true for the case $\alpha = 2$. Indeed in this case the assumptions reduce to the following: X has 2-stable type, which is equivalent to: X has 2-Rademacher type, see [10]. If we take for example $X = \ell_p$, where $p > 2$, then ℓ_p has 2-Rademacher type but it is not embeddable in L^2.

Now we shall show that for each $\alpha < 2$ there exists a Banach space X which does not belong to V_α.

EXAMPLE 1. Let $X = L^p(0,1)$, $p > 2$. Then X has 2-stable type and consequently has α-stable type for each $\alpha < 2$. If we suppose $X \in V_\alpha$ then by Theorem 1, $L^p(0,1)$ is embeddable in L^α for $\alpha < 2$ and has consequently 2-Rademacher cotype. But $\inf\{q: L^p(0,1)$ has q-Rademacher cotype$\} = p > 2$ (cf. [10]). Thus this contradiction establishes the fact that $X \notin V_\alpha$.

2. We consider the case when $1 < \alpha < 2$. Denote $\nu = N_\alpha(T)$ an α-stable cylindrical measure defined by (1). If ν is a Radon measure then it is known, [4], that T is a compact

operator and $T \in \Pi_0(X^*, \ell_\alpha)$. Let us recall that $\Pi_p(X,Y)$ denotes the class of all p-absolutely summing operators between the Banach spaces X and Y. Actually we are able to show

THEOREM 3. Let $X \in V_\alpha$ and X has α-stable type, $1 < \alpha < 2$. Then a cylindrical measure $\nu = N_\alpha(T)$ is a Radon measure iff $T^* \in \Pi_\alpha(\ell_\beta, X)$, where $\frac{1}{\alpha} + \frac{1}{\beta} = 1$.

Proof. To establish the "only if" part consider a sequence $\{a_n\}$ of elements from ℓ_β such that $\sum_{n=1}^{\infty} |\langle a_n, x^*\rangle|^\alpha < +\infty$ for every $x^* \in \ell_\alpha$. Let us define the mapping $A: \ell_\alpha \to \ell_\alpha$ by the formula $Ay^* = \langle a_n, y^*\rangle$.

It is easy to verify that $A \in B(\ell_\alpha, \ell_\alpha)$ and $A^* \in B(\ell_\beta, \ell_\beta)$ has the form $A^*b = \sum_{n=1}^{\infty} a_n b_n$, where $b = \{b_n\}$. Put $\lambda = N_\alpha(AT)$. We have

$$\|ATx^*\| \leq \|A\|\|Tx^*\| = \|\|A\|Tx^*\|.$$

Since $\nu = N_\alpha(T)$ is a Radon measure and $X \in V_\alpha$ then it follows that λ is a Radon measure on X and consequently the series

$$\sum_{n=1}^{\infty} \gamma_n^{(\alpha)} (AT)^* e_n^* = \sum_{n=1}^{\infty} \gamma_n^{(\alpha)} T^*A^*e_n^*$$

is convergent a.s., where $\{\gamma_n^{(\alpha)}\}$ is a standard α-stable sequence and $\{e_n^*\}$ is the natural basis in ℓ_β. Since each Banach space has a p-stable cotype for $p < 2$ we conclude that for $\alpha < 2$

$$\sum_{n=1}^{\infty} \|T^*a_n\|^\alpha = \sum_{n=1}^{\infty} \|T^*A^*e_n\|^\alpha < \infty$$

i.e., $T^* \in \Pi_\alpha(\ell_\beta, X)$.

To establish the "if" part we make use of the fact that $\sum_{n=1}^{\infty} |\langle x, e_n^*\rangle|^{\alpha} < \infty$ for $x \in \ell_{\alpha}$ and $e_n^* \in \ell_{\beta}$. Therefore the assumption $T^* \in \Pi_{\alpha}(\ell_{\beta}, X)$ implies $\sum_{n=1}^{\infty} \|T^*e_n^*\|^{\alpha} < \infty$. Since X has α-stable type thus $\sum_{n=1}^{\infty} \gamma_n^{(\alpha)} T^*e_n^*$ is a.s. convergent, where $\{\gamma_n^{(\alpha)}\}$ is as before. Hence $N_{\alpha}(T)$ is a Radon measure. □

REMARK 2. If $\alpha = 2$ the analogous result is known, (see [2], p. 198), but instead of assumption X has α-stable type is needed X has a 2-Rademacher cotype.

REMARK 3. The assumption $X \in V_{\alpha}$ was used in the "only if" part, when the assumption X has α-stable type in the "if" part. The following example shows that this last assumption is an essential one.

EXAMPLE 2. Let $X = \ell_{\alpha}$, $1 < \alpha < 2$ and $a = (a_n) \in \ell_{\alpha}$. Consider the diagonal mapping $T: X^* = \ell_{\beta} \to \ell_{\alpha}$ given by the formula $T\{t_n\} = \{a_n t_n\}$. In this case we have $T^* = T$, $T^*e_n^* = a_n e_n$ and $T^* \in \Pi_{\alpha}(\ell_{\beta}, \ell_{\alpha})$. Hence the series $\sum_{n=1}^{\infty} \gamma_n^{(\alpha)} T^*e_n^*$ converges a.s. in X iff $\sum_{n=1}^{\infty} |a_n \gamma_n^{(\alpha)}|^{\alpha} < \infty$ a.s. But it is well known ([1], p. 237) that the last condition holds iff

$$\sum_{n=1}^{\infty} |a_n|^{\alpha}\left(1 + \log \frac{1}{|a_n|}\right) < \infty. \tag{7}$$

Thus if we take $a \in \ell_{\alpha}$ for which (7) does not hold then $T^* \in \Pi_{\alpha}(\ell_{\beta}, X)$ but $\nu = N_{\alpha}(T)$ is not a Radon measure on ℓ_{α}.

REFERENCES

1. A. Badrikian, S. Chevet, Measures cylindriques. Espaces de Wiener et fonctions aléatoires Gaussiens. Lecture Notes in Math. 397 (1974), Springer-Verlag.
2. S. A. Chobanjan, V. I. Tarieladze, Gaussian characterizations of certain Banach spaces, J. Mult. Anal. 7 (1977), 183-203.
3. S. A. Chobanjan, A. Weron, The existence of the linear prediction for Banach space valued Gaussian process, J. Mult. Anal. (to appear).
4. Dang Hung Thang, Nguyen Zui Tien, On stable measures in Banach spaces, preprint.
5. W. Feller, An introduction to probability theory and its applications, Vol. II, 1966, John Wiley and Sons, New York.
6. J. Lindenstrauss, A. Pełczynski, Absolutely summing operators in L_p-spaces and their applications, Studia Math. 29 (1968), 275-326.
7. D. Mouchtari, La topologie du type Sazonov pour les Banach et les supports Hilbertiens. Annales L'Universite de Clermont, 61 (1976), 77-87.
8. A. Tortrat, Lois indefiniment divisibles dans un groupe topologique abelien metrisable X. Ces des espaces vectoriels, C.R. Adad. Sci. Paris, 261(1965), 4973-4975.
9. N. N. Vakhania, Probability distributions in linear spaces, Tbilisi, 1971 (in Russian).
10. W. A. Woyczynski, Geometry and martingales in Banach spaces, Part II, in Advances in Probability and Related Topics, Vol. 4, Ed. J. Kuelbs, Marcel Dekker, Inc., New York, 1978, 267-517.

INSTITUTE OF MATHEMATICS
WROCLAW TECHNICAL UNIVERSITY
50-370 WROCLAW, POLAND

Current address of the first author
DEPARTMENT OF MATHEMATICS
UNIVERSITY OF HANOI
HANOI , VIETNAM

ON SERIES REPRESENTATION OF SECOND ORDER RANDOM ELEMENTS AND STOCHASTIC PROCESSES

M.R. Żeberski

We study a connection between a series representations of second order random elements in Hilbert spaces (cf. [2]) and a series representation of second order measurable stochastic processes (cf. [1]).
It turns out that these results may be obtained as consequences of similar series representations of vector valued functions (see Proposition 1 and Proposition 2). The main difference is that Proposition 2 is a weaker result but it gives a concrete formula for coeficients, which is very important for applications.

1. Series representation of vector functions.

Let $x: T \to H$ be a vector function, where T is any interval of the real axis and H - complex Hilbert space. We suppose that x is strongly measurable i.e. x is a strong limit of a sequence of measurable simple functions.
In particular, if H is separable, then the strong and weak measurability are equivalent (x is called weakly measurable if $h^* \circ x$ is a measurable function for each $h^* \in H^*$).
There exists a measure μ on T equivalent to the Lebesque measure and such that

$$(1) \qquad \int_T (x(t)|\, x(t))\mu\,(dt) < \infty \quad .$$

We may choose μ as follows: let f be a real-valued positive integrable function on T and put

$$g(t) = \begin{cases} 1 & \text{for} \quad (x(t)|x(t)) \leq 1 \\ (x(t)|x(t))^{-1} & \text{for} \quad (x(t)\,|x(t)) > 1 \end{cases}$$

It is easy to verify that the formula $\frac{d\mu}{dLeb}(t) = f(t)g(t)$ defines the measure μ satisfying required conditions.

Consider the operator $S: L_2(\mu) \to L_2(\mu)$ given by

(2) $$(S\,f)(t) = \int_T (x(t)|x(s))\, f(s)\, \mu(ds).$$

Condition (1) implies that the operator S is a Hilbert-Schmidt operator and moreover it is clear that S is a positive self-adjoint operator.

Proposition 1. Let H be a Hilbert space. If $x(t),\ t \in T$ is a strongly measurable function, then there exist an orthonormal sequence $(\varphi_k(t))$ in $L_2(\mu)$ and an orthonormal sequence (x_k) of vectors in H such that

(3) $$x(t) = \sum_{k=1}^{\infty} \varphi_k(t)\, x_k ,$$

where the series is convergent in $L_2(\mu, H)$.

Proof. It is known that to any Hilbert-Schmidt operator we may relate a sequence of eigenvalues (λ_k) and a sequence of eigenfunctions $(\varphi_k(t))$. We have for the operator S:

(4) $$\lambda_k\, \varphi_k(t) = \int_T (x(t)|x(s))\ \ \varphi_k(s)\, \mu(ds)$$

and

(5) $$\int_T \varphi_k(t)\, \overline{\varphi_l(t)}\mu(dt) = \delta_{kl} \qquad k,l = 1,2,\ldots .$$

For the operator S defined by (2), the numbers λ_k $k = 1,2,\ldots$ are non - negative and satify

(6) $$\int_T (x(t)|x(t))\mu(dt) = \sum_{k=1}^{\infty} \lambda_k .$$

Put

$$x_k = \int_T x(t)\overline{\varphi_k(t)}\mu(dt) \qquad k = 1,2,\ldots .$$

These vectors are well - defined by the above Bochner integral because

$$(x_k| x_k) = \int_T \int_T (x(t)\overline{\varphi_k(t)}\,|\,x(s)\overline{\varphi_k(s)})\mu(dt)\mu(ds) \leq$$

$$\leq (\int_T |\varphi_k(t)|^2\mu(dt))^2\ (\int_T (x(t)|x(t))\mu(dt))^2 < \infty .$$

Moreover, by Fubini theorem and (4), (5) they are orthogonal :

$$(x_k|x_l) = \int_T \int_T (x(t)\overline{\varphi_k(t)}|x(s)\overline{\varphi_l(s)})\mu(dt)\mu(ds) =$$

$$= \int_T \int_T (x(t)|x(s))\varphi_l(s)\mu(ds)\overline{\varphi_k(t)}\mu(dt) =$$

$$= \int_T \lambda_l \, \varphi_l(t)\overline{\varphi_k(t)}\mu(dt) = \lambda_l \, \delta_{lk} \quad .$$

Hence, we have by (6)

$$\int_T \| x(t) - \sum_{k=1}^{n} \varphi_k(t)x_k \|^2 \mu \, (dt) =$$

$$= \int_T [(x(t)|x(t)) - (x(t)|\sum_{k=1}^{n} \varphi_k(t)x_k) - (\sum_{k=1}^{n}\varphi_k(t)x_k|x(t)) + + \sum_{k=1}^{n}|\varphi_k(t)|^2 \lambda_k]\mu \, (dt) =$$

$$= \int_T (x(t)|x(t))\mu(dt) - \sum_{k=1}^{n} \lambda_k \longrightarrow 0$$

when n tends to infinity.

Proposition 2. <u>If a measure μ is equivalent to the Lebesque measure and (1) holds, then</u>

(a) <u>every strongly measurable H - valued vector function has a representation</u>

$$x(t) = \sum_{k=1}^{\infty} a_k(t)x_k \quad , \tag{7}$$

<u>where (x_k) is an orthonormal sequence of elements in H , $(a_k(t))$ is a minimal sequence in the space $L_2(\mu)$ given by the formula (10) below and the series (7) is convergent in $L_2(\mu, H)$.</u>

(b) <u>A closed subspace $H° \subset H$ generated by x_k, $k = 1,2,...$ in (7) is independent of the choice of measure μ .</u>

Proof. Let μ be a measure on T equivalent to the Lebesque measure such that (1) holds and let $(f_k(t))$ be a countable orthonormal basis in $L_2(\mu)$ (which exists by the separabelity of $L_2(\mu)$).

From the Fubini theorem and Schwartz inequality it follows that the elements y_k defined by Bochner integral

$$(8) \qquad y_k = \int_T x(t)\, \overline{f_k(t)}\mu(dt) \qquad k = 1,2,\ldots$$

exist and belong to H. From among functions $(f_k(t))$ we choose successively only these for which vectors $y_1, y_2, \ldots, y_n$ are linearly independent for all $n = 1,2,\ldots$. Applying the Gramm-Schmidt procedure to (y_k) we obtain orthonormal vectors

$$(9) \qquad x_k = \sum_{j=1}^{k} c_{kj}\, y_j = \int_T x(t)\, \overline{g_k(t)}\mu(dt)$$

where

$$g_k(t) = \sum_{j=1}^{k} \overline{c}_{kj}\, f_j(t) \qquad k = 1,2,\ldots .$$

Put

$$(10) \qquad a_k(t) = (x(t)|x_k) = \int_T (x(t)|\, x(s))\; g_k(s)\mu(ds) .$$

Functions $a_k(t)$ belong to $L_2(\mu)$, it is a result of Schwartz, inequality and (1). Moreover, the construction of the vectors x_k implies that the functions $a_k(t)$ are minimal : $(a_k(\ ), g_n(\ )) = \delta_{k,n}$. Since in the proof of part (b) we will not use the convergence of the series given by (7) we may use here the arguments from the part (b).

Now the closed subspace generated by the vectors x_k in (9) is equal to the closed subspace generated by vectors of H in expansion (3). This, by the Parseval identity and Proposition 1, implies the convergence of the series (7) in $L_2(\mu, H)$.

(b) Since measure μ and orthonormal set of functions in $L_2(\mu)$ may be selected by many ways, $x(t)$, $t \in T$ has many different representations. We consider two different expansion of $x(t)$, $t \in T$ which we obtain in proof of part (a) with two bases $(f_k^i(t))$ in $L_2(\mu_i)$ $i = 1,2$, respectively. It suffices to show that the orthogonal complements of the subspaces H_1^o and H_2^o are equal. If $z \in H$ and $z \perp H_1^o$, then for all $k = 1,2,\ldots$ we have

$$0 = (z|y_k^{(1)}) = \int_T (z|x(t)\quad f_k^{(1)}(t)\, \mu_1(dt).$$

which implies $(z|x(t)) = 0$ in $L_2(\mu_1)$.

Since the measure μ_1 and μ_2 are equivalent (both are equivalent to the Lebesque measure), we have $(z|x(t)) = 0$ in $L_2(\mu_2)$.

Now we obtain

$$\int_T (z|x(t))\ f_k^{(2)}(t)\ \mu_2(dt) = (z|y_k^{(2)}) = 0 .$$

Consequently $z \perp H_2^o$ and $(H_1^o)^{\perp} \subset (H_2^o)^{\perp}$. In view of the symmetricity of the assumptions for subspaces H_1^o and H_2^o we conclude that $H_1^o = H_2^o$. Thus the proof is complete.

2. Applications.

2.1. Representation of H - valued random elements.

Let (Ω, Σ, P) be a probability space and H a complex Hilbert space. A strong measurable map $X : \Omega \to H$ (H - valued random element) such that $\|X(\omega)\| \in L_2(\Omega, \Sigma, P) = L_2(P)$ is called a strong second order random element.
Every random element X defines operator $S: L_2(P) \to L_2(P)$ of the form

$$(11) \qquad (S\xi)(\omega') = \int_\Omega (X(\omega')|\ X(\omega))\xi(\omega)\ P(d\omega).$$

The following two series representations of H-valued random elements, in the spirit of [2], are obtained as a consequence of the Proposition 1 and Proposition 2, respectively.

Corollary 1. Let H be a Hilbert space. If X is H-valued strong second order random element, then there exist an orthonormal sequence of random variables (ξ_k) belonging to $L_2(P)$ and an orthogonal sequence of vectors $(x_k) \in H$ such that

$$X(\omega) = \sum_{k=1}^{\infty} \xi_k(\omega)\ x_k ,$$

where the series convergens in $L_2(P,H)$.

Proof. It is known that the operator S given by (11) is a self-adjoint, positive Hilbert - Schmidt operator. Since Hilbert-Schmidt operators are compact, they have at most countably many eigenvectors. In addition it posseses only point spectrum. Hence the ortnogonal complement of the kernel of S ($\ker S = \{\xi \in L_2(P) : S\xi = 0\}$) is a separable subspace of $L_2(P)$. Now we repeat the arguments as in the proof of the Proposition 1 repleacing $L_2(P)$ by $L_2(P) \ominus \ker S$.

As a consequence of Proposition 2 in a similar way we get the following result.

Corollary 2. Each H-valued strong second order random element has a representation

$$X(\omega) = \sum^{\infty} \xi_k(\omega)\, y_k ,$$

where (y_k) is a minimal sequence of vectors in H, $(\xi_k(\omega))$ is an orthonormal sequence of random variables in $L_2(P)$ and the series is convergent in $L_2(P,H)$.

Proof. From the Proposition 2 it follows immediately, that

(12) $$X(\omega) = \sum_{k=1}^{\infty} \zeta_k(\omega)\, x_k,$$

where (x_k) is an orthonormal sequence of vectors in H, $(\zeta_k(\omega))$ is a minimal sequence of random variables in $L_2(P)$ and the series is convergent in $L_2(P,H)$. Moreover, the subspace $H^o \subset H$ generated by x_k in (12) is independent of the choice of the orthonormal basis in $L_2(P) \ominus \ker S$. Hence we have that H^o is separable. Now we repeat the arguments as in the proof of the Proposition 2 replacing $L_2(\mu)$ by H^o and formula (8) by the following one

$$\zeta(\omega) = (X(\omega)|x).$$

2.2. Series representation of second order stochastic processes.

Let $\{ x_t ;\ t \in T \}$ be a measurable second order process (i.e., measurable as a function of two variables ω and t) where T is any interval of the real axis. Autocorrelation function of $\{x_t ;\ t \in T \}$ is defined as $r(t,s) = E\, x_t \overline{x_s}\quad t,s \in T$.

Corollary 3 (S.Cambanis [1]). Each measurable second order stochastic process $\{ x_t ;\ t \in T\}$ has the form

(a) $$x_t = \sum_{k=1}^{\infty} a_k(t)\, \zeta_k + w_t \qquad t \in T$$

where

$a_k(t)$ $k = 1,2,\dots$ are functions in $L_2(\mu)$,

(ζ_k) is a sequence of orthonormal random variables in $L_2(P)$ and the series is convergent in $L_2(P)$.

Moreover,

$$\text{(b)} \quad r(t,s) = \sum_{k=1}^{\infty} a_k(t)\, \overline{a_k(s)} + r_w(t,s) \qquad t,s \in T$$

where the series is absolutly convergent and $r_w(t,s) = E\, w_t\, \overline{w}_s$.

(c) *If* $H(w)$ *and* $H(x)$ *are the closed subspaces generated by random variables of the process* $\{w_t \; ; \; t \in T\}$ and $\{x_t \; ; \; t \in T\}$ *respectively, then*

$$H(x) = H^o \oplus H(w) \; ,$$

where H^o *is the closed subspace generated by the sequence* (ζ_k) *in expansion (a).*

(d) $r_w(t,t) = 0$ *almost surely on* T *with respect to the Lebesque measure.*

Proof. Every second order stochastic process can be considered as a function on T into the Hilbert space $L_2(P)$ of random variables of second order. Since in this case $(x_t|x_s)_{L_2(P)} = E\, x_t\, \overline{x}_s$, we can take measure μ the same as in section 1, and by Proposition 2 we can write

$$(13) \qquad x_t = \sum_{k=1}^{\infty} a_k(t)\zeta_k \; ,$$

where the series is convergent in $L_2(\mu, L_2(P))$.
We denote by H^o the closed subspace generated by the sequence (ζ_k). Now, define w_t as the orthogonal projection of x_t on the closed subspace $H(x) \ominus H^o$. Let us note that w_t is not identically equal to 0 which is a consequence of the convergence of the series (13) in $L_2(\mu , L_2(P))$. Part (b) follows from the orthonormality of the random variables ζ_k and from the decomposition $H(x) = H^o \oplus H(w)$.

References

[1] S.Cambanis, Representation of stochastic processes of second order and its linear operators, J.Math.Anal.Appl.41 (1973) 603-620.

[2] A.Weron, On weak second order and Gaussian random elements, Lecture Notes in Math. vol.526 (1976), 263-272.

Institute of Mathematics
Wrocław Technical University
50-370 Wrocław Poland